Greedy Science

Greedy Science

*Creating Knowledge, Making Money,
and Being Famous in the 1980s*

Edited by Michael D. Gordin
and W. Patrick McCray

jh

Johns Hopkins University Press
Baltimore

Johns Hopkins University Press
2715 North Charles Street
Baltimore, Maryland 21218
www.press.jhu.edu

Library of Congress Cataloging-in-Publication Data

Names: Gordin, Michael D., editor. | McCray, W. Patrick,
   1967– editor.
Title: Greedy science : creating knowledge, making money,
   and being famous in the 1980s / edited by Michael D.
   Gordin, and W. Patrick McCray.
Description: Baltimore : Johns Hopkins University Press,
   2025. | Includes bibliographical references and index.
Identifiers: LCCN 2024021284 (print) | LCCN 2024021285
   (ebook) | ISBN 9781421450858 (hardcover) | ISBN
   9781421450865 (paperback) | ISBN 9781421450872 (epub)
Subjects: LCSH: Science—Research—Moral and ethical
   aspects. | Technology—Research—Moral and ethical
   aspects. | Science—Social aspects. | Research—Economic
   aspects. | Business—History—20th century. |
   Avarice—History—20th century.
Classification: LCC Q180.55.M67 G74 2025 (print) |
   LCC Q180.55.M67 (ebook) | DDC 178—dc23/eng/20240514
LC record available at https://lccn.loc.gov/2024021284
LC ebook record available at https://lccn.loc.gov/2024021285

Special discounts are available for bulk purchases of this book.
For more information, please contact Special Sales at
specialsales@jh.edu.

Contents

Acknowledgments　　　　　　　　　　　　　　　　　　　　*vii*

Introduction: Greed, Science, and Greedy Science　　　　1
Michael D. Gordin and W. Patrick McCray

PART I: To the Market

1. Taking the Marks to the Market: The Oil Industry
   and the Entrepreneurial Turn　　　　　　　　　　　　19
   Cyrus C. M. Mody

2. Taller Than a *T. rex*: Celebrity and Leftist Politics
   in the Public Career of Stephen Jay Gould　　　　　45
   Myrna Perez

3. VisiCalc, Personal Computing, and the Speculative
   Entrepreneur of 1980s America　　　　　　　　　　72
   Laine Nooney

PART II: Privatization

4. Thatcherism, Science, and Greed　　　　　　　　　　97
   Jon Agar

5. Kids, Commerce, and Communists: Access to Space
   in the 1980s　　　　　　　　　　　　　　　　　　123
   Margaret A. Weitekamp

6. Neoliberal Mutations　　　　　　　　　　　　　　147
   Angela N. H. Creager

7. "Drugs into Bodies": AIDS Activism and the
    Constitutional Limits of Biocapital   167
    Cathy Gere

    PART III: Regions

8. Far from Footloose: The Greedy Localization of
    Biotechnology in Cambridge   193
    Robin Wolfe Scheffler

9. Science as Speculation: State Capitalism, Real Estate,
    and Singapore's Jurong Town Corporation   217
    Hallam Stevens

10. Science, Texas Style: How the Lone Star State
    Embraced Science in a Big Way   246
    Peter Westwick

    PART IV: Speculations and Spectacles

11. "The Required Allocations Grew Considerably":
    Soviet Science, Military Imperatives, and the
    Ambivalent Response to Reagan's Star Wars   269
    Asif Siddiqi

12. Extinction, Insurance, or a New Weapons Industry:
    Asteroid Impacts and the Triumph of the Apocalyptic
    Lobbyist   297
    Matthew Stanley

13. Service with a Smile, Or, How Profit Made Japanese
    Robots Personal and Personable   322
    Yulia Frumer

    Afterword: From Groovy Science to Greedy Science   349
    David Farber

    List of Contributors   *359*
    Index   *361*

Acknowledgments

This volume had its origins in a series of three workshops, collectively entitled "Greedy $cience: Money and Knowledge in the Global 1980s," hosted by the Program in the History of Science at Princeton University on October 30, 2021, January 15, 2022, and April 2, 2022. Despite the hopes of the two organizers, the ongoing state of the COVID-19 pandemic and the various protective measures imposed by the university meant that the workshops were hosted entirely virtually as Zoom discussions that occupied several very intense hours on those Saturdays. Much to our great astonishment, we encountered dozens of participants at each meeting— including many repeat attendees—who stayed engaged in the breakout rooms, the live chat, and the back-and-forth of a transcontinental and transoceanic endeavor. Our first debt is to all those who joined us to make the experience so unexpectedly delightful.

None of that could have happened without the dedication of several individuals. We are especially grateful to Lee Horinko, who arranged every aspect of the workshops (including the still-live website, greedyscience .princeton.edu). It would literally have been impossible without her. During each session, a pair of graduate students, Julia Marino and Jack Klempay, took detailed notes on the conversations that later proved invaluable for the revisions of the presentations into drafts. Another student, Anin Luo, ably managed the Zoomscape when Lee was feeling under the weather for one of the sessions.

The editors also thank the contributors, who not only tolerated the upfront commitment to attend the three initial sessions on Zoom but also joined the two of us in several smaller Zoom workshops as we worked with both individuals and groups to refine the chapters and devise the final architecture of this book. After they turned in their manuscripts, and then again after they submitted extensive revisions to those manuscripts, Austen

Van Burns did a heroic job of standardizing the format and catching not a few errata.

Our final thanks go to Matt McAdam of Johns Hopkins University Press, who believed in the project and shepherded the manuscript through every stage.

All displayed a striking amount of selflessness to bring this story of greed to the reader.

Greedy Science

Introduction

Greed, Science, and Greedy Science

Michael D. Gordin and W. Patrick McCray

"I think," Ivan Boesky told the graduating class of Berkeley's business school in 1986, "greed is healthy. You can be greedy and still feel good about yourself."[1] Two years later, the stock trader was in prison. Nonetheless, the message endured, immortalized in a titanic monologue given by the amoral corporate raider Gordon Gekko in Oliver Stone's *Wall Street* (1987). For better or worse, *greed* stands as a potent keyword for the 1980s, both in the United States and elsewhere.[2]

The authors in this edited collection of chapters explore this particular attribute—greed—by looking how it informed, intersected, and interlaced with science and technology (and scientists and technologists) during the 1980s. During this decade, greed, although undeniably present in earlier eras, became an extensive, expansive, and at times explicit characteristic of science both in the United States and around the world.[3] The global scientific community was reshaped in a multitude of ways, large and small, by money, fame, and the pursuit of celebrity.

Ivan Boesky would likely not have considered this to be a critique, and although we come from a rather different position, neither do we. We see greed as a historically contingent characteristic that shines a penetrating light on many features of science and technology in the 1980s. Greed is a copious concept, encompassing a host of behaviors such as covetousness, acquisitiveness, rapaciousness, and conspicuous consumption. Furthermore, if we aim to treat greed seriously as both a descriptive term and a category of analysis, we must consider its various opposites. At particular times throughout the 1980s, many scientists also displayed more valorized qualities such as altruism, compassion, charity, and generosity. Our argument is that greed played a key role in structuring the territory, while its counterparts cast penumbras where countervailing movements could form and virtues could be displayed. *Greedy Science* seeks to survey and understand the full range of these changes.

Why Greedy Science?

Rightly or wrongly, historians and journalists have glossed the 1980s as one long consumption binge. From television shows and bestselling novels to the pulpits of megachurches, the lifestyles of the rich, famous, and infamous were exhibited, extolled, and excoriated. Meanwhile, capitalism rushed like a riptide into the Soviet Union and the People's Republic of China, unmooring and reshaping political and economic foundations. One way to characterize a popular view of the 1980s would be: the Cold War ended and greed won. It is a straightforward story, and like many such stories, it is compelling. There indeed was a lot of money circulating in the long 1980s, altering most things it touched from politics and culture to the geopolitical order. Why would science be any different?

In this volume, we explore to what degree this standard narrative holds (along with where it doesn't) and what that tells us about the 1980s. We do not try to adjudicate whether the many manifestations of greed were good or bad for scientists, for knowledge, or for the planet. Instead, we focus the reader's attention on a selection of distinct transformations that reshaped global science and explore how those changes revolved not just around money but around *greed*. The chapters that follow collectively argue that greed in its many varied forms—the desire for wealth, power, celebrity, commercialism, status—marked global science and technology in the 1980s. (Of course, some of that imprint has been carried into the decades since.[4])

By the late 1970s, a general consensus began to emerge across the political spectrum that attempts by governments—be they nations or cities—to address pressing problems had failed to deliver the promised results.[5] This appeared especially acute in technoscientific areas such as energy (the oil shocks of 1973 and 1979) and medicine (the questionable successes of Nixon's "War on Cancer"). Political actors from both the right and, interestingly, the left pushed for a sharp turn toward "free market" solutions, less government intervention, and deregulation. This new political order, sometimes labeled in specific national contexts as Reaganism or Thatcherism, coalesced and assumed a definite shape during the 1980s. In whatever guise, a common tenet of such economic thinking was the promotion of market-oriented solutions coupled with the idea that their success needed the strong support of both governments as well as nonstate actors like the World Bank.

In choosing the term "greedy," we are pointing to a phenomenon that is broader than other related terms, which could be understood as species within the greedy genus. Most prominent among these in today's scholarly literature is *neoliberalism*. The very term raises hackles for some and makes the eyes of others glaze over.[6] Given that certain scholars have castigated that term as a "linguistic omnivore" that subsumes other concepts—the most egregious being the use of "neoliberalism" as a lazy epithet for "political conservatism"—we have opted to work with a more limited definition that highlights its connections to while maintaining distinctions from the supervening category of *greed*.[7]

When it appears in this collection of chapters, neoliberalism does not emphasize the intellectual project linked to midcentury Chicago School figures such as Friedrich von Hayek or Milton Friedman.[8] Instead, our understanding situates and connects neoliberalism to specific political and economic activities. These include the creation of policies that prioritized the (ostensibly) frictionless movement of goods, people, labor, and capital across borders; the dismantling of rules and regulations thought to impede this circulation; and the promotion of the individual (and individual responsibility) over governments, seen perhaps most notably when Margaret Thatcher dismissed "society" as a fictional construct.[9] While we recognize the importance of philosophy in shaping the economic landscape, our authors adopt a less abstract position by focusing on how philosophical preferences were turned into policy and practice with concomitant implications for both scientific research as well as scientists' careers. In other words, while the abstract concept of neoliberalism is important to many chapters in this volume, it does not permeate all of them, which is an important aspect of our story. We believe that *greed* in all of its multivariate forms captures more of this diversity, reflecting features that are often concealed by the deployment of "neoliberalism" as a catchall explanation. At the same time, it's important to recognize that as neoliberalism—defined by a renewed focus on the individual, markets, deregulation, privatization, and the like—materialized around the globe in the 1980s, these features did not manifest themselves everywhere simultaneously or identically. To paraphrase that guru of globalization, Thomas Friedman, greed isn't flat.

Seen this way, neoliberalism represents, as historian Gary Gerstle has argued, the vigorous creation of a new political order with attendant ideologies, policies, and constituencies.[10] One example (or symptom) of this new regime was the emergence in the late 1970s of "leveraged buyouts," often financed by "junk bonds," a trend that accelerated until the party came to a crashing end with the savings and loan scandals of the late 1980s.[11] Greed, in its varied manifestations, was central to these political and economic projects. "Greed" and "greediness" encompass a whole ensemble of ideas, beliefs, attitudes, and activities that informed and guided political and economic decisions enacted on a scale ranging from nations, regions, and cities at one end to universities, research groups, and individual scientists at the other. The decisions that fostered the formation of the neoliberal political order did not, of course, spring unbidden like so many entrepreneurial Athenas from the heads of political Zeuses. An immense network of lawyers and lobbyists was essential in its realization. Likewise, restructuring the relationship between money, science, and scientists, whether in the United States, the Soviet Union, or Singapore, required a similarly varied array of actors traditionally seen as "outside" the realm of science.

The fit with science was not happenstance. Neoliberalism spoke directly to how and why knowledge was produced and organized. Science, of course, has always had some commercial element to it and has never been "pure."[12] What changed during the "greedy '80s"—and what this collection of chapters addresses—is a difference not in kind but in degree. During the high Cold War, a novel social contract was forged in the United States and other countries between the state and scientists. Knowledge production was funded by generous government support, much of it from the military; science became a potent symbol of national prestige; and scientists were on tap as a reserve labor force of knowledge workers in the event of a hot war.[13] By the end of the 1980s, the threads in this fabric were starting to fray; state support of universities contracted even as corporations cut back their in-house research and development in favor of contracts to university-based scientists. Our interest is the capacious concept of greed as both a catalyst and a response to these changes, of which neoliberalism was one variant.

In examining the various forms of knowledge production during the long 1980s that we are labeling as "greedy science," it is important to recognize that the goals of the scientific enterprise itself were shifting. In a controversial 2007 essay, historian Paul Forman argued that there was a "reversal in the primacy between science and technology" in the 1980s.[14] Forman contended that as the twentieth century neared its end, "technology" had become more prevalent and prominent for the public and policymakers than "science" (by which he generally meant the pursuit of "pure" knowledge done with little to no attention to its applications). While we have no intention of adjudicating Forman's densely argued thesis—or entering the definitional thicket of "science" versus "technology" or "pure" versus "applied"—we acknowledge that shifts in the character of production and purposes of scientific knowledge certainly can be seen in the 1980s. A related rhetorical conflation and confluence of "science" with "technology" ensued as scientists and their patrons embraced the market. One (of many) possible data points was the fact that, in 1984, the Reagan administration tapped Erich Bloch to head the National Science Foundation (NSF). Bloch came to the position without a PhD in science (he had earned a BS in electrical engineering) but rather from working at IBM as a technology manager. As the head of the NSF, Bloch emphasized industrial competitiveness, engineering, and applied science over the traditional areas of basic research the agency had long championed—indeed, that it had been created in 1950 to foster.[15] While recognizing the distinction between science and technology, this volume's authors understand that the boundaries and borders between the two were porous, often functioning as rhetorically flexible actors' categories.

Greed, with its links to many ancient cultures and religions (and tempered, of course, with a host of countervailing virtues), suggests much more than just a reshaping of political and financial structures and scientists' activities. It also forms a distinct moral arena where vices and virtues are at play. Historians of science have long deployed the concept of a moral economy to describe the often-unspoken rules, expectations, traditions, and behaviors that govern how resources—whether research materials and article preprints or less tangible goods such as telescope observing time—are shared within the scientific community. Scientists and their communities follow often implicit rules that define their mutual hopes and obligations; in turn, these "moral conventions" regulate how resources are shared and circulated.[16] During the epoch of greedy science, the moral economy of science and scientists transformed in ways that were just as profound as those happening in politics and finance. With scientists and science, greed is much more complicated than the heroes-and-villains morality tales of a Gordon Gekko. There was plenty of *that* kind of greed, to be sure, but avarice arrived with a sense of anxiety and awkwardness even as scientists became increasingly fixated on money, funding, success, and fame.

The refashioning of science's moral economy in the 1980s was also associated with a concern about the acquisition and disbursement of different forms of capital besides those usually associated with finance, state funding, and real estate. Social capital in the form of recognition and celebrity was not new to the scientific community, but in the 1980s, the volume on these qualities was significantly turned up. Scientists secured wealth and power due to their research and writing but also became famous (and wealthier) in the process. A researcher might be branded a "Showman of Science" (*Time* magazine, when placing Carl Sagan on its October 20, 1980, cover) or write a book that millions of readers bought (Stephen Hawking with his surprise 1988 bestseller *A Brief History of Time*). Likewise, one's demise in the pursuit of their scientific goals might bring posthumous celebrity and public martyrdom, as in the cases of primatologist Dian Fossey (d. 1985) and teacher-turned-astronaut Christa McAuliffe (d. 1986).

Why the 1980s?

The bounds of "the 1980s" in this volume are a little longer than ten years, but only just. From 1979, with the ascendancy of leaders like Thatcher and

Reagan as well as the breakdown of 1970s-era détente, to 1991 and the dissolution of the Soviet Union, the temporal bookmarks for this distinct era are well established. The chapters in this volume assume the task of using the intersections of science, technology, and money to add new dimensions for thinking about the 1980s. They also use this time period as a springboard to consider what would follow—after the end of the Cold War, the arrival of a new and very much neoliberal political and economic order, championed by leaders such as Tony Blair, Bill Clinton, and post-Soviet Russian Prime Minister Yegor Gaidar, that promoted free markets, globalization, and capitalism on a planetwide scale. Although the 1980s are increasingly receding into the past, the legacy of greedy science is still very much with us.

Historians, especially US historians, have not shied away from tackling the 1980s, although their general attention has been trained either on high politics (i.e., the Reagan Revolution or the end of the Cold War) or cultural warfare (political and cultural polarization, the New Right).[17] To some degree, these amount to two sides of the same coin. Historians of science and technology have also tested their mettle against some of these major developments with works often written during the period itself or shortly afterward. The Human Genome Project, information technologies, and the biotech industry have all received sustained attention, for example.[18] Yet two components have been absent. First, many of the extant accounts are somewhat one-sided, not fully combining the narratives of politics and culture directly into the science or vice versa. Second, we lack a global and transnational picture for the decade that was to end with "globalization" becoming a keyword, and one closely connected to the worlds of both science and technology. This book aims to address both missing pieces.

To illustrate both the ubiquity of science and technology at many of the salient points in the 1980s and the capaciousness of "greed" as a frame for thinking about them in the context of world developments, consider just a single year plucked from the sheaf: 1980. In the United States, at least, 1980 may be remembered as the year Ronald Reagan handily defeated incumbent Jimmy Carter and, a few months later, assumed the presidency. Together with Thatcherism—the signature governing philosophy of Margaret Thatcher, a professional chemist turned politician who was elected British prime minister in 1979—Reaganism became the era's dominant political and economic ideology. Both -isms championed market solutions to social problems while favoring deregulation, privatization, entrepreneurship, and individual responsibility. The legacy of both ideologies still reverberates in the twenty-first century.

In the scientific sphere, these political developments provided both a context and a catalyst for events engaged by several chapters in this volume. In June 1980, the US Supreme Court ruled in favor of Ananda Mohan Chakrabarty's quest to be granted a patent for a novel organism—a bacterium that could break down crude oil—which the microbiologist had created nine years earlier while working for General Electric. The Court ruled, by a vote of five to four, that human-engineered lifeforms could be patented. The decision set the stage for the biotechnology start-up Genentech to go public with an enormously successful public stock offering in October 1980.[19] Filling out the triumvirate of money-science events was Congress's passage of the Bayh-Dole Act in December 1980. Signed by President Carter during his last days in the Oval Office, the new law permitted nonprofit organizations (like universities) and small business firms to maintain control of inventions that their employees created, even though the research had been funded by public money. As the 1980s unfolded, Bayh-Dole proved to be one in a sequence of laws that reshaped the intellectual property landscape.[20]

Whether via causation or correlation, these events signaled new practices—a shift in both political and moral economies—as entrepreneurship, patenting, and commercialization of knowledge increasingly became part of scientists' working lives. Scientific knowledge became ever more closely associated with commerce and potential applications. Put simply: science, or at least certain areas of research, could make you rich, and perhaps even a celebrity. (In March 1981, *Time* magazine featured geneticist and Genentech cofounder Herbert Boyer on its cover.) Just as importantly: getting rich off your science was now understood to be a good thing, perhaps even a professional obligation. In the United States at least, if you didn't patent and commercialize your federally funded discovery, the government had "march-in rights" to force you to do so.

The various economies of greedy science came with a newfound sense of the scientist as celebrity and icon. The idea that scientists could become famous—even world-renowned—was itself not a new phenomenon.

The first decades of the twentieth century had witnessed figures like Marie Curie and, even more so, Albert Einstein becoming media personalities and stand-ins for entire scientific communities. But the expanding media landscape of the 1980s, with cable television and the rise of specialty magazines like *Omni* and *Discover* that made popular science their beat, boosted certain scientists into widely recognized media figures. A key milestone in this development was the broadcast of *Cosmos: A Personal Voyage* in 1980. The thirteen-part series was eventually seen by a planetary audience estimated in the hundreds of millions. It also transformed Carl Sagan from an obscure astronomer into a global celebrity whose visage and voice (if you are of a certain age, try saying "billions and billions" without falling into a Sagan-esque imitation) were equated with "scientist."

These snapshots of greedy science in 1980 were drawn from the history of the United States, a narrower focus that in some ways reflects the historical scholarship about the 1980s in general. One goal of this volume is to embrace a broader perspective of both the United States and the larger world. America may have given television viewers around the world a distorted, perhaps even dyspeptic, view of wealth via such cultural commodities as *Dallas* (first aired April 1978), *Dynasty* (January 1981), and *Lifestyles of the Rich and Famous* (March 1984), but the international scientific community was also deeply tied to changing fortunes and finances that came with the so-called Decade of Greed.[21]

By the end of the 1980s, the new economic order manifested in science and technology as well: not only in the growth of the biotech and computing sectors that are encompassed in the synecdoche of Silicon Valley but also in the creation of the H-1B visa status (established in the Immigration Act of 1990). This, combined with the "globalization" of previously state-controlled economic sectors in countries such as India, led to a massive and flexible migration of technical specialists around the world. Similar changes marked the major science projects at the end of the decade. They were mainly publicly funded, rendering them vulnerable to the vicissitudes of domestic politics, but were simultaneously transnational collaborations. Such ventures had existed before the decade of greedy science, of course, but the passage through this moment, as the chapters in this volume show, left imprints on their future.

Navigating Greedy Science

Perceptive readers may have noted a titular similarity to another book. In 2016, one of us (McCray) coedited a collection of chapters called *Groovy Science*.[22] It used examples from across the physical, biological, and social sciences to chart how members of various countercultures pursued and appropriated science and technology. Fundamentally, it challenged the prevailing view among historians that the 1970s were marked by a rejection of science and technology and instead revealed a more diverse picture of how Americans sought and found alternative forms of science. *Greedy Science* is less revisionist in orientation. We do not fundamentally disagree with historians' assertions that the 1980s were marked by distinctive shifts in the global political economy. Rather, we want to understand how science and scientists were part of that process. *Greedy Science* is both a sequel and a response to *Groovy Science*. It continues the historical trajectory laid out in the earlier collection but revises and revisits the intersection of money, culture, and science while placing it within a larger global frame.

This volume seeks to articulate a significant shift in the global political economy and cultural place of science and technology during the only-slightly-long 1980s. Although there are many ways to elaborate on the transformation, it would be very hard to do so without emphasizing the importance of money, branding, and capital. While the category of "greedy science" cannot of course capture absolutely everything about the production of knowledge in the 1980s, it ties together many strands that historians have previously treated separately. The chapters in this volume approach this characterization of greedy science from many directions yet converge on a few dominant themes.

The first of these, which in many ways set the agenda for the rest of the decade—and the rest of the volume—is the topic of **part I, "To the Market."** The three chapters in this section adopt a capacious and flexible understanding of markets, capturing meanings that go beyond economics. These contributors are interested in entrepreneurship, in branding, and in creating demand for a particular vision of the sciences. This process begins, in each of the three chapters, in the middle of the 1970s, but it reached a crescendo in the 1980s.

We begin with Cyrus C. M. Mody's "Taking the Marks to the Market: The Oil Industry and the Entrepreneurial Turn." Mody, following the money behind several iconic entrepreneurial initiatives of the 1980s, finds a common source: an oil industry seeking to diversify its investments after the boom and bust of the embargo era. While the oil industry did not single-handedly create the phenomenon of greedy science, it provided lubricant for a system in which budding entrepreneurs began to envision their scientific efforts as producing knowledge as well as profits.

Not all entrepreneurs, of course, floated in on a tide of Brent crude, and sometimes the objects of their branding and market invention were themselves. Perhaps no individual was so representative of self-promotion being identical with science promotion as American paleontologist Stephen Jay Gould. In "Taller Than a *T. rex*: Celebrity and Leftist Politics in the Public Career of Stephen Jay Gould," Myrna Perez illustrates how Gould's public writing began as an expression of a particular type of 1970s leftism. As the Reagan era commenced, however, Gould transformed himself into a depoliticized brand as he became a public intellectual and a celebrity.

Start-ups, sci-tech celebrities, and the dawn of the 1980s—these all converged in the personal computer, and Apple Computer made the machines of the moment. In Laine Nooney's "VisiCalc, Personal Computing, and the Speculative Entrepreneur of 1980s America," we see the rise and further rise of VisiCalc, the spreadsheet application that was all about juggling figures and making microcomputers like the Apple II indispensable to a new kind of accounting. From business marketing to consumer consumption, VisiCalc helped create the look and feel of the 1980s while fueling its financial engineering.

Part II, "Privatization," brings us to the heart of the political-economic strategies that are often labeled "neoliberalism." The phenomenon of privatizing the welfare state happened in many places at the same time, but it was especially pronounced in the United Kingdom under Prime Minister Margaret Thatcher. Jon Agar's chapter, "Thatcherism, Science, and Greed," traces the components of greedy science in its British variant, even as the Tories resisted the label of greed itself. An equivalent set of experiments across the Atlantic took place in seemingly the most state-oriented of all state-science programs: the National Aeronautics and Space Administration (NASA). In the 1980s, as Margaret A. Weitekamp documents in "Kids, Commerce, and Communists: Access to Space in the 1980s," NASA led the way in turning to corporations as well as direct public support to shore up the finances of the Space Shuttle program.

It was in biotechnology, however, that the great strides of privatizing science made their biggest imprint, moving away (at least in part) from the government-sponsored grants essential to fields like high-energy physics. The pharmaceutical industry—large established firms as well as legions of smaller start-ups—routinely exploited patents in the 1970s as mechanisms for securing tremendous profits and driving up their stock prices. In "Neoliberal Mutations," Angela N. H. Creager explores how neoliberal modes of reasoning, especially about individuals assuming responsibility for risk, informed the science being done in the laboratory. Meanwhile, as Cathy Gere's chapter " 'Drugs into Bodies': AIDS Activism and the Constitutional Limits of Biocapital" explains, the loosening of pharmaceutical regulations amid the carnage of the AIDS pandemic had strikingly different repercussions in the United States and in postapartheid South Africa.

One characteristic of greedy science was that, for all the commonality of themes, the manifestations in far-flung parts of the world were starkly unalike. Historical studies that take the nation-state as the basic unit of analysis can often miss the distinctive qualities of individual regions or even municipalities, which offered their own improvisations on the greedy melody. **Part III, "Regions,"** traces such stories—all heavily inflected by the 1980s' obsession with real estate—in seemingly incommensurable locales. By sidestepping Silicon Valley and the ubiquitous public valorization of it as the archetypal "innovation region" of the 1980s, the chapters in this section help provincialize the San Francisco Bay Area by showing how much of its ostensible uniqueness was in fact broadly shared.

There were certain places and spaces where greed was more effective than others. We begin in and around Cambridge, Massachusetts, with Robin Wolfe Scheffler's "Far from Footloose: The Greedy Localization of Biotechnology in Cambridge." The combination of an elaborate health care complex, a series of distinguished universities, and a flurry of entrepreneurialism produced a surprising phenomenon: companies that at first sought to diversify and extend their imprint to Europe ended up returning and then rooting themselves ever more firmly in the Boston area.

On the other hand, some regions attempted to deliberately engineer the kinds of synergies that Boston (and Silicon Valley) had inherited, using both authoritarian and pork-barrel populist politics. Hallam Stevens, in

"Science as Speculation: State Capitalism, Real Estate, and Singapore's Jurong Town Corporation," concentrates on that island nation's efforts to build science parks in the 1980s in order to catapult its economy through the messiness of industrialization and on to the knowledge-based realm of postindustrialization. Far away, and on an appropriately much larger canvas, Peter Westwick takes us to one of the most self-consciously regional regions in the United States in "Science, Texas Style: How the Lone Star State Embraced Science in a Big Way." As one might expect, we find hucksters and real estate swaps but also a concerted effort to connect the extant aerospace complex to a more 1980s' set of technological and scientific ventures. Yet even when these succeeded, they looked nothing like the military–industrial–academic complex that coalesced around places like Stanford or MIT.

Finally, the chapters in **part IV, "Speculations and Spectacles,"** look to how researchers attempted to deploy greedy science in order to create new futures for science and technology. As part I looked to the legacies of the 1970s, in this section, we see individuals groping and gazing toward the 1990s, finding themselves sometimes stymied by their inability to navigate the new landscape of technology and science that had grown around them.

In the first chapter in this section, we turn to the Soviet Union—the decade of greedy science ended up being its last decade—as Asif Siddiqi examines in "'The Required Allocations Grew Considerably': Soviet Science, Military Imperatives, and the Ambivalent Response to Reagan's Star Wars." The Soviet Union, at least until its very final years, might have seemed the one place whose science was not gripped by greediness. Nonetheless, greed as bureaucratic rent-seeking was quite present as different ministries blocked particular responses to Ronald Reagan's proposed missile defense system. While opposing technological solutions that already existed, bureaucrats insisted instead on researching and developing something new. In response to imagined nuclear destruction, an appetite for glory and innovation existed across the Iron Curtain.

Back in the United States, a small group of scientists were concerned with the chances of a different kind of apocalypse: an asteroid impact. Matthew Stanley, in "Extinction, Insurance, or a New Weapons Industry: Asteroid Impacts and the Triumph of the Apocalyptic Lobbyist," shows how their attempts to mobilize the old levers of the military–industrial complex—vestiges of pre–greedy science—failed, even though the risk calculus looked familiar. Reluctantly, they turned to different kinds of lobbying, struggling to navigate a changing environment of patronage much as the Soviet defense scientists had.

The last chapter in the volume turns to the emblematic economic nation of the 1980s: Japan. In "Service with a Smile, Or, How Profit Made Japanese Robots Personal and Personable," Yulia Frumer chronicles how the markets that had restructured the relationship between science and the consumer fared in Japan, where "personal" had a very different valence. Her story also shows some of the lingering tendrils of this era, as greediness survived in various sectors of science and technology even if the Japanese boom did not.

Finally, David Farber, with his "From Groovy Science to Greedy Science," provides an astute afterword to the entire volume. Farber notes that the image of the scientist as a market-driven actor can be found as early as 1925 in Sinclair Lewis's classic book *Arrowsmith*. He then pivots in a different direction by contrasting the community-oriented medical activism c. 1970 of the Black Panthers with a streetwise version of "greedy science" that emerged a decade later. Profit-driven "street chemists" created a deadly yet undeniably innovative new product—crack cocaine—that destroyed whole neighborhoods and traumatized people "who found themselves on the wrong side" of Reaganomics. Farber's chapter highlights how greedy science came in many forms and suggests that further possibilities for its use as an analytical frame remain.

Indeed, we believe that "greedy science" offers opportunities for historians that go beyond identifying particular examples where money, funding, speculation, and celebrity were elemental if not indeed paramount features. The chapters in this volume do more than describe instances of greedy science—they also ask how those situations came to be. They analyze how actors, ranging from individual scientists and political leaders to entire industries and nations, tightly interwove the production of new knowledge with new entrepreneurial activities, financial instruments, open markets, and a refashioned sense of what was permissible or even possible. Put another way: the authors in this volume all seek to find and understand historical questions to which greed was increasingly seen as the answer.

NOTES

1. Quoted in Bob Greene, "A $100 Million Idea: Use Greed for Good," *Chicago Tribune*, December 15, 1986. There are numerous other versions of this quote as well, but all present the same message and moral.

2. A number of books that survey the 1980s, largely from a US perspective, single out greed in various manifestations. Two examples are relevant sections of James T. Patterson, *Restless Giant: The United States from Watergate to Bush v. Gore* (New York: Oxford University Press, 2005), and Philip Jenkins, *Decade of Nightmares: The End of the Sixties and the Making of Eighties America* (New York: Oxford University Press, 2006).

3. Although not as capacious a concept as greed, a recent collection of essays examined the connections between science and capitalism: Lukas Rieppel, William Deringer, and Eugenia Lean, *Science and Capitalism: Osiris*, vol. 33 (Chicago: University of Chicago Press, 2018). Also relevant is Philip Mirowski, *Science-Mart: Privatizing American Science* (Cambridge, MA: Harvard University Press, 2011).

4. Profiting off science and technology was not a new development in 1980, nor were the soaring ambitions of some of the most fervent boosters. For a general survey of examples from the twentieth century in the United States, see Thomas P. Hughes, *American Genesis: A Century of Invention and Technological Enthusiasm, 1870–1970* (New York: Viking, 1989). It's worth noting that this book itself, which privileges entrepreneurs, inventors, and investors, was a product of the 1980s.

5. Marc Levinson, *An Extraordinary Time: The End of the Postwar Boom and the Return of the Ordinary Economy* (New York: Basic Books, 2016).

6. For one critique, see Philip Mirowski, "Hell Is Truth Seen Too Late," *boundary 2* 46, no. 1 (2019): 1–53, which was part of a special issue on "epistemology and neoliberalism."

7. See, for instance, Daniel Rodgers, "The Uses and Abuses of 'Neoliberalism,'" *Dissent* 65, no. 1 (Winter 2018): 78–87. For a lengthier discussion of the term, including insights into historians' general reluctance to use it, see Kim Phillips-Fein, "The History of Neoliberalism," in *Shaped by the State: Toward a New Political History of the Twentieth Century*, ed. Brent Cebul, Lily Geismer, and Mason B. Williams (Chicago: University of Chicago Press, 2019), 347–362.

8. Daniel Stedman Jones, *Masters of the Universe: Hayek, Friedman, and the Birth of Neoliberal Politics* (Princeton, NJ: Princeton University Press, 2012). For an introduction to conservative thought in general, see Jerry Z. Muller, ed., *Conservatism: An Anthology of Social and Political Thought from David Hume to the Present* (Princeton, NJ: Princeton University Press, 1997); Corey Robin, *The Reactionary Mind: Conservatism from Edmund Burke to Donald Trump*, 2nd ed. (New York: Oxford University Press, 2017); and, more recently, Andrew Bacevich, ed., *American Conservatism: Reclaiming an Intellectual Tradition* (New York: Library of America, 2020).

9. For work on neoliberalism by historians, see, for example, Philip Mirowski and Dieter Plehwe, eds., *The Road from Mont Pèlerin: The Making of Neoliberal Thought* (Cambridge, MA: Harvard University Press, 2015); Angus Burgin, *The Great Persuasion: Reinventing Free Markets since the Depression* (Cambridge, MA: Harvard University Press, 2012); and Quinn Slobodian, *Globalists: The End of Empire and the Birth of Neoliberalism* (Cambridge, MA: Harvard University Press, 2018). If one expands to social scientists' engagement with neoliberalism, the bibliography grows quite large. Some relatively recent and representative examples include Jamie Peck, *Constructions of Neoliberal Reason* (New York: Oxford University Press, 2005); Wendy Brown, *Undoing the Demos: Neoliberalism's Stealth Revolution* (New York: Zone Books, 2015); and Monica Prasad, *The Politics of Free Markets: The Rise of Neoliberal Economic Policies in Britain, France, Germany, and the United States* (Chicago: University of Chicago Press,

2006), as well as the anthology edited by Damien Cahill, Melinda Cooper, Martijn Konings, and David Primrose, *The SAGE Handbook of Neoliberalism* (London: SAGE, 2018).

10. Gary Gerstle, *The Rise and Fall of the Neoliberal Order: America and the World in the Free Market Era* (New York: Oxford University Press, 2022).

11. Brendan Ballou, *Plunder: Private Equity's Plan to Pillage America* (New York: Public-Affairs, 2013).

12. Steven Shapin, *Never Pure: Historical Studies of Science as if It Was Produced by People with Bodies, Situated in Time, Space, Culture, and Society, and Struggling for Credibility and Authority* (Baltimore: Johns Hopkins University Press, 2010), and Shapin, *The Scientific Life: A Moral History of a Late Modern Vocation* (Chicago: University of Chicago Press, 2008).

13. See, for example, Daniel J. Kevles, *The Physicists: The History of a Scientific Community in Modern America*, rev. ed. (Cambridge, MA: Harvard University Press, 1995).

14. Paul Forman, "The Primacy of Science in Modernity, of Technology in Postmodernity, and of Ideology in the History of Technology," *History and Technology* 23, no. 1/2 (2007): 1–152.

15. Irwin Goodwin, "Erich Bloch: On Changing Times and Angry Scientists at NSF," *Physics Today* 41, no. 8 (1988): 47–52. One perspective on this focus on patents, intellectual property, and profits is in Elizabeth Popp Berman, *Creating the Market University: How Academic Science Became an Economic Engine* (Princeton, NJ: Princeton University Press, 2012).

16. Robert Kohler, *Lords of the Fly: Drosophila Genetics and Experimental Life* (Chicago: University of Chicago Press, 1994), 3; Lorraine Daston, "The Moral Economy of Science," *Osiris* 10 (1995): 3–24; and W. Patrick McCray, "Large Telescopes and the Moral Economy of Recent Astronomy," *Social Studies of Science* 30, no. 5 (2000): 685–711.

17. For example, Laura Kalman, *Right Star Rising: A New Politics, 1974–1980* (New York: W. W. Norton, 2010); Rick Perlstein, *The Invisible Bridge: The Fall of Nixon and the Rise of Reagan* (New York: Simon and Schuster, 2015); and Kevin Kruse and Julian E. Zelizer, *Fault Lines: A History of the United States Since 1974* (New York: W. W. Norton, 2019).

18. For a sampling, see Daniel J. Kevles and Leroy Hood, *Code of Codes: Social and Social Issues in the Human Genome Project* (Cambridge, MA: Harvard University Press, 1992); Paul Rabinow, *Making PCR: A Story of Biotechnology* (Chicago: University of Chicago Press, 1996); and Michael Riordan and Lillian Hoddeson, *Crystal Fire: The Invention of the Transistor and the Birth of the Information Age* (New York: W. W. Norton, 1997).

19. Daniel J. Kevles, "Ananda Chakrabarty Wins a Patent: Biotechnology, Law, and Society, 1972–1980," *Historical Studies in the Physical and Biological Sciences* 25 (1994): 111–135; Sally Smith Hughes, *Genentech: The Beginnings of Biotech* (Chicago: University of Chicago Press, 2011); Nicolas Rasmussen, *Gene Jockeys: Life Science and the Rise of Biotech Enterprise* (Baltimore: Johns Hopkins University Press, 2014).

20. Sheila Slaughter and Gary Rhoades, *Academic Capitalism and the New Economy* (Baltimore: Johns Hopkins University Press, 2004).

21. Richard McKenzie, "Decade of Greed?" *National Review*, June 10, 2004.

22. David Kaiser and W. Patrick McCray, eds., *Groovy Science: Knowledge, Innovation, and American Counterculture* (Chicago: University of Chicago Press, 2016).

Part I / To the Market

CHAPTER ONE

Taking the Marks to the Market

The Oil Industry and the Entrepreneurial Turn

Cyrus C. M. Mody

This volume explores a turn, around 1980, toward scientific practices driven by profit and "greed." Here, I examine one aspect of that turn: the increasing frequency of patenting by universities, founding of start-up companies by academic researchers, and collaboration between firms and universities. As these activities overlapped, I refer to them collectively as "academic entrepreneurship" and their growing popularity as the "entrepreneurial turn."

In US universities, that turn began in the late 1960s, prompted by various factors relating to the Vietnam War–era milieu. Various studies have explained why US academia initiated the entrepreneurial turn and how academic entrepreneurship gained legitimacy in the 1970s.[1] We know less, though, about why *industry* wanted to work with universities—and *which* industries specifically, apart from the isolated cases of semiconductor manufacturers and pharmaceutical makers.[2] This chapter examines an industry that studies of academic entrepreneurship have neglected: the oil industry. As the next section shows, oil ties were ubiquitous in US research long before the entrepreneurial turn. The following section argues, however, that

events from the late 1960s to the mid-1980s encouraged oil firms to intervene more directly in the academic landscape. I then offer case studies
of oil industry support for academic entrepreneurship in three era-
defining technologies: alternative energy, biotechnology, and artificial
intelligence (AI).

Although the sums invested were tiny for multinational oil firms, they
were enormous for nascent high-tech industries. The biotech industry
might not exist today were it not for oil firms' early intervention, and the
solar energy and artificial intelligence industries might have reached their
present size earlier if oil firms had not withdrawn their patronage in the
late 1980s. Oil firms *did* exit solar and AI largely because they received few
short-term benefits from supporting academic entrepreneurs, despite those
entrepreneurs' extravagant promises. Thus, contrary to popular and scholarly conceptions of an all-powerful oil industry, oil firms were more used
by entrepreneurial scientists than using them.[3]

This chapter takes a wide but shallow approach by presenting snippets
of many examples containing few details. The reason stems from lack of
sources. Oil firms offer little access to their private archives, and the few
public oil archives have been scrubbed of unflattering documents. Moreover, many oil-funded academic ventures failed, leaving little trace. Still,
by reading many thinly sketched examples in aggregate, we can better
understand why industry (in general) and oil firms (specifically) promoted the entrepreneurial turn.

Oil, Oil Everywhere, . . .

Oil firms didn't foster the entrepreneurial turn alone, though; they worked
in concert with universities, the state, journalists, intellectuals, investors,
and so on. Oil, however, so suffuses the research system that the distinction between oil and non-oil actors blurs. Former oil executives occupied
influential positions in government agencies such as the National Aeronautics and Space Administration (NASA) and the Central Intelligence
Agency (CIA) that funded academic research, and many research universities have buildings and professorships named after oil barons.[4] Were
NASA or Rockefeller University "oil actors"? Mostly no, but partially yes.

The historic and ongoing oil ties of two kinds of organizations did, however, directly stimulate new forms of academic entrepreneurship in the
1970s. First, philanthropic foundations founded in the early twentieth

century to launder oil barons' reputations were ubiquitous in US and global science.[5] True, by the 1970s, the Rockefeller, Pew, Macy, Mellon, and other longstanding oil baron philanthropies were not very oily. Still, those foundations' grants, especially to universities, promoted a modernity that relied heavily on oil, as seen for example in the Rockefeller Foundation's efforts to establish mechanized, irrigation- and petrochemical-intensive agriculture in the Global South.

The 1970s, though, saw new "conservative foundations willing to question public sector activity" and the emergence of philanthropy inspired by corporate social responsibility (CSR).[6] Those two trends unfolded in oil philanthropies such as the ARCO Foundation, Exxon Education Foundation, and Robert O. Anderson Foundation.[7] The "New [Oil] Philanthropy" promoted neoliberal-ish ideas about innovation, interdisciplinarity, improving universities' efficiency, and translating academic research to society.[8] Such philanthropies also funded academic economists and business historians to communicate free market values to the nation's students.[9]

The second kind of organization with similar historic *and* ongoing oil ties was the venture capital fund. Various studies have shown that venture capital (VC) stimulated the entrepreneurial turn; historically, oil stimulated venture capital.[10] One of the first VC firms was Rockefeller Brothers (later renamed Venrock), which funded academic start-ups, including Mosaic and Centocor. Other early VCs, such as Georges Doriot, also benefited from connections to oil.[11] In fact, the legal model underlying modern venture capital—the limited partnership—is adapted from a template wildcatters used to pool the risks of oil exploration.[12]

True, as with older oil philanthropies like the Rockefeller and Macy Foundations, the oil roots of the older venture capital funds would have been invisible by the 1970s, when venture capitalists started investing in academic entrepreneurs. Still, those roots indicate that oil money was a background condition for the entrepreneurial turn. More concretely, starting in the 1960s, many oil firms formed their own venture capital units, some of which invested in the fields that I will examine.

. . . Nor Any Drop to Drill

The entrepreneurial turn began in the late 1960s, coincident (and coevolving) with chaotic upheaval in the global oil industry. The preceding two decades saw "free world" oil production grow faster than demand, thanks to exploitation of Middle Eastern fields. In the late 1960s, however, Organization of the Petroleum Exporting Countries (OPEC) nations began demanding greater control over (and profits from) oil produced in their territories.[13] Production in most Western countries was near maximum capacity, and thus oil firms could not counter OPEC's demands by increasing domestic production. To regain that ability, Western oil firms began moving into the North Sea and North Slope of Alaska, further offshore in the Gulf of Mexico, and into non-oil energy technologies.

Those new energy sources took time to mature, however, and could not prevent the price of oil rising gradually after 1970 following many years of decline. Then, when Arab OPEC members cut production during the 1973 Arab–Israeli war, the (inflation-adjusted) price of oil more than tripled in a few weeks, stabilized, and then doubled again in 1979–1980. For oil firms, the two "oil shocks" meant healthy profits; for everyone else, they meant economic stagnation, rationing, and "panic at the pump."[14] Hence, there was widespread suspicion that oil firms were manipulating the oil shocks. Both crises also followed years of growing public concern over oil and petrochemicals in the air and waterways.[15] Enterprising politicians channeled that public ire: Senator James Abourezk mounted investigations of the industry's supposedly monopolistic and unscrupulous activities, and his colleagues Robert Packwood and Edward Kennedy tried to force oil firms to divest their non-oil subsidiaries.[16]

Oil companies responded variously. Mobil confronted calls for divestiture with advertising widely seen as transgressing norms for political speech.[17] Exxon tried "educating" citizens about economic "realities."[18] Gulf Oil's proxies intervened in diplomatic negotiations to assert US sovereignty over offshore deposits.[19] ARCO, Standard Oil of Ohio, and others assembled the resources and political will to build the Trans-Alaska Pipeline System.[20]

All these strategies drew on academic expertise, from economics to geophysics to ecology.[21] Other industry strategies for overcoming physical and political constraints were even more obviously technical: digitized seismic surveys and advanced rigs to find and produce new oil, sophisticated spectroscopy to detect pollutants, and so on.[22] Universities hosted research on such technologies and, especially after 1973, oil firms had the cash to help academic scientists commercialize that research.

Energetic Entrepreneurship

Admittedly, some technologies that fit this strategy were not conducive to academic entrepreneurship. For instance, oil actors largely agreed that nuclear power offered medium-term relief for energy scarcity, yet by 1970, the nuclear industry presented high barriers to entry for academic entrepreneurs. The oil industry *had* supported academic nuclear research in the 1940s, for example, at MIT and the University of Chicago.[23] Yet as nuclear energy matured, oil companies increased their foothold by buying or building uranium mines, power plants, and waste-processing facilities, rather than by assisting entrepreneurial professors.[24]

A telling exception was nuclear *fusion*. Several oil majors that formed nuclear fission ventures in the late 1960s—such as Exxon's Jersey Nuclear or Gulf and Shell's jointly owned subsidiary General Atomics—dabbled in fusion too.[25] Because fusion was expensive and uncertain, however, oil firms spread the risk via early university–industry–government research *consortia*. For instance, in 1972, Exxon and General Electric backed the Laser Fusion Feasibility Project at the University of Rochester's Laboratory for Laser Energetics; six other firms and government agencies, including Standard Oil of Ohio, joined them in 1976.[26]

There was even at least one oil-sponsored academic fusion start-up, KMS Fusion, founded in 1971 by former University of Michigan physicist Kip Siegel. The US government was KMS Fusion's main funder, with Burmah Oil coming second. Texas Gas Transmission Corporation (a natural gas pipeline firm) also funded KMS Fusion's research into using fusion energy to create methane from water and limestone.[27]

Note that oil firms could tap an existing industry for proven expertise in fission but not fusion, since no one in fusion had a successful track record. Thus, firms had difficulty telling whether fusion experts' promises were overblown. In emerging fields, overpromising and "froth" are necessary for standing out enough to attract investment, and in those fields, there are few heuristics for identifying overpromisers.[28] It was in exactly such fields that oil firms most readily invested in academic entrepreneurs.

The Prodigal Sun

One technology particularly prone to froth and overpromising in the 1970s was solar energy. While solar heating technologies are ancient, the solar *industry* is usually dated to 1954 and the invention of the modern photovoltaic cell.[29] At first, companies primarily developed solar technology for space applications; firms only began seriously pursuing terrestrial photovoltaic energy in the late 1960s. Oil firms were among the first to do so, both through in-house units and investments in solar start-ups. Most early solar entrepreneurs came from the aerospace and microelectronics industries rather than universities and are thus outside this chapter's scope. Two examples, however, are worth noting for their ties to later academic entrepreneurship.

The first comes from Texas Instruments (TI), a leading semiconductor manufacturer that was originally an oilfield services firm (with a large oilfield services unit until the late 1980s). Following the 1973 embargo, two former TI employees—Jack Kilby and Jay Lathrop—invented a solar energy system for single-family homes.[30] Kilby was at that point an independent inventor, while Lathrop taught at Clemson University. A little later, Rice University professor Arthur "Skip" Porter joined them. Then, in 1975, TI licensed their patents and formed an in-house solar team. In the early 1980s, TI almost commercialized its Solar Energy System, but the declining price of oil and disastrous losses across the company meant it was abandoned in 1983. Notably, as the program collapsed, TI tried to interest several oil firms, and Saudi royals, in rescuing it.[31]

In 1980, meanwhile, Porter assumed directorship of A&M's Texas Engineering Experiment Station (TEES), transforming it from a traditional engineering extension agency into a modern technology transfer office (and growing its budget from $3 million/year to $30 million).[32] Then, in 1985, he became founding president of the Houston Area Research Center (HARC), a contract research/academic incubator backed by four Texas universities plus natural gas and real estate magnate George Mitchell, the "father of fracking."[33] Kilby was HARC's first chief technology officer. As Elizabeth Popp Berman has shown, the transformation and expansion of organizations like TEES, and the founding of consortia such as HARC, were important steps in the entrepreneurial turn.[34] Peter Westwick's chapter in this volume (chapter 10) is a rare study that connects oil and gas firms and industry leaders such as Mitchell to the incubators, consortia, and tech transfer offices of the 1980s. Yet passing references elsewhere confirm the trend. Oil firms were important early clients of the famous Stanford Office of Technology Licensing, for instance.[35] An early tech incubator, the Uni-

versity of Pittsburgh's Applied Research Center, was housed in Gulf's former research center, which Chevron donated after buying Gulf in 1984.[36]
And Martin Kenney's *Biotechnology* lists a 1979 Exxon–MIT combustion
research agreement as a precursor for similar arrangements in the life
sciences.[37]

Another prominent microelectronics-turned-solar entrepreneur was
Stanford Ovshinsky of Energy Conversion Devices, Inc. (ECD).[38] Ovshinsky initially persuaded Atlantic Richfield (ARCO) to fund ECD's solar project; as Kilby noted, though, by 1981, "ARCO has told [Ovshinsky] that
they will cut off his funds. . . . He will probably find another sponsor. Anyone who has been able to raise $75 million on pure promises should not
be underestimated, but it looks like a good chance to sell short."[39] Ovshinsky did, indeed, find new sponsors: Standard Oil of Ohio and a Japanese
oil refinery, Toa Nenryo Kogyo, in which both Exxon and Mobil owned
25 percent interests.[40]

Ovshinsky thus fits my secondary thesis that entrepreneurs' wild over-
promises were magnets for oil money. Even entrepreneurs operating in
good faith, such as Bill Yerkes of Solar Technology International (another
ARCO investment), struggled to collaborate with oil firms: "ARCO didn't
really understand why they were buying a PV company, nor what the business was exactly, nor where it might go, nor where the costs were."[41] Less
scrupulous entrepreneurs just grabbed the money and ran.

Ovshinsky also inspired an early academic solar entrepreneur, Karl-
Wolfgang Böer of the University of Delaware.[42] Around 1967, Ovshinsky
asked Böer to consult for ECD. Initially put off by Ovshinsky's theatrical-
ity, Böer became an admirer and imitator, founding Solar Energy Systems
(SES) in 1972. Post-embargo, Shell invested in SES, then bought the company. Shell also supported the on-campus center Böer founded, the Institute for Energy Conversion (IEC), indirectly with money funneled via SES
and directly with donations of surplus equipment.[43] Similarly, in the early
1980s, the IEC worked with Chevron, yielding patents that university officials considered important to their intellectual property portfolio.[44]

The IEC–SES–Shell relationship ended acrimoniously, perhaps due to
fluid norms governing academic entrepreneurship. In 1975, Böer lost the
IEC directorship thanks to the National Science Foundation's concerns
about conflict of interest.[45] In 1978 or 1979, his successor at IEC, Allen Barnett, reportedly cut a deal with two US oil firms and two international

companies to pay $1 million each to finance "experimental production of solar cells." Around the same time, the university offered SES and Shell right of refusal on IEC's solar research—in Barnett's view, tying up IEC's patents and voiding the deal he negotiated. In the aftermath, "Barnett was ousted and four of his closest aides either fired or resigned." Then, in 1981, Shell merged SES into a joint venture with Motorola and, in 1983, moved the operation to that company's base in Arizona—leaving the University of Delaware's foray into solar entrepreneurship in disarray.[46]

Biotech

Let's turn now to biotechnology. There is a vast literature on the early biotech start-ups, the norms and practices they established, the hopes of economic and organizational transformation they elicited, the copycats they inspired, and so on.[47] Routinely, contributions to that literature briefly note the individual oil companies hovering around the biotech revolution. But those studies do not ask *why* oil firms invested in biotechnology, nor do they observe that *multiple* oil firms supported multiple biotech firms—these were not one-off links. A partial exception is Robert Bud's *The Uses of Life: A History of Biotechnology*, but while Bud is excellent on the oil industry's pre-1973 interest in biology, he is less interested in oil's role in the post-1973 era.[48]

Oil's minimal presence in conventional biotech narratives is mystifying, because oil firms sponsored many important early academic biotech start-ups. Chevron and Amoco owned (as of 1978) over 50 percent of the "first" such venture, Cetus.[49] In 1980, Amoco almost put $30 million into Biogen, too, but pulled out because of its Cetus play (see chapter 8 for more on Biogen).[50] Again in 1980, and again Amoco, along with Elf Aquitaine, expressed interest in Hybritech.[51] In the end, they didn't invest directly, but Elf's pharmaceutical arm, Sanofi, did.[52] Elf also joined the six-company consortium financing a Stanford spin-off, Engenics, and the complementary on-campus institute, the Center for Biotechnology Research, in 1981.[53] And Elf bought "a sizeable share of Transgene, the French equivalent of Genentech."[54]

Genentech itself received money from Lubrizol (a specialty chemicals and oilfield services firm) and Fluor (a construction firm specializing in refineries and petrochemical plants) for microbe-enhanced oil recovery

(MEOR) research.[55] Amgen took off thanks to funding from Tosco (The Oil Shale Co.) and Harrison Capital, Texaco's venture capital arm.[56] SDS Biotech, the venture capital unit of Diamond Shamrock, a refinery and gas station chain, invested in Immunex in 1982.[57] Immunex's chair, Steve Duzan, was formerly with Atlantic Richfield, and Immunex's name was "a play on Esso's recent rebranding as Exxon."[58] And those are just oil firms' investments (or near misses) in academic biotech start-ups. There were also joint ventures, co-development of products, and patent licenses: Centocor and Dupont (owner of Conoco) in 1985,[59] Amoco and Integrated Genetics (later acquired by Genzyme) in 1986,[60] Genex and Nippon Oil in 1987,[61] Chiron and Phillips Petroleum in 1989, and so on.[62]

Woody Powell and Karl Sandholtz identify eleven companies as establishing the academic biotech start-up business model; nine or ten (depending on criteria) made substantial deals involving oil companies in the 1980s.[63] Oil money went into less successful ventures too, including some associated with notable people or organizations. In 1981, for instance, the Salk Institute created the Salk Institute Biotechnology Corporation (later Salk Institute Biotechnology/Industrial Affiliates, or SIBIA); that year, Phillips Petroleum bought a 37 percent share for $10 million.[64] In 1983, SIBIA also formed a joint venture with Diamond Shamrock to develop animal vaccines.[65] Similarly, Cold Spring Harbor Laboratory founded a company, Cellbiology, in 1980. The next year, Cellbiology was commissioned to research MEOR for Southern Pacific Petroleum of Australia.[66] A collaboration with a specialty chemicals company, Petroferm, almost ensued; after Cellbiology folded, Petroferm commercialized similar technology from the University of Tel Aviv instead.[67] Finally, in 1977, Occidental Petroleum bought Zoëcon, a spin-off from Carl Djerassi's more famous start-up, Syntex; debt restructuring forced Zoëcon's sale to Sandoz in 1983.[68]

Moreover, biotechnology's most important lawsuit, *Diamond v. Chakrabarty*, concerned a microbe engineered to eat oil. That patent originated in Ananda Chakrabarty's work on single-cell protein (SCP)—microbes designed to turn oil into animal feed—while a postdoc at the University of Illinois in the late 1960s.[69] Many Western oil firms had SCP programs and were a likely audience for Chakrabarty's work.[70] When Chakrabarty moved to General Electric in 1971, he realized the same microbe could consume oil slicks and petrochemical pollutants.[71] Then in 1982, Chakrabarty's lab at Illinois (to which he had returned) was funded by Petrogen to develop bugs for MEOR.[72] That is, oil firms wanted Chakrabarty's bugs for at least *three* applications.

Of those three, MEOR has endured the longest. Cetus and Genentech worked on it in the mid-1980s, as did Bethesda Research Labs, Pfizer, Synergen (20 percent owned by Getty Oil), and others.[73] For Amgen, MEOR was key to bringing in $3.5 million of crucial early investment from Tosco. Yet as George Rathmann, Amgen's CEO, admitted later, he never took oil recovery seriously: "It's embarrassing to think that you ranked fragmenting oils with bacteria down a well as equivalently interesting a project as EPO [erythropoietin]. . . . [F]rom the beginning, EPO was the one that we had a lot of hopes for."[74] There was much hype about MEOR and related use of engineered microbes in mining around 1980, even from establishment institutions such as the National Academy of Sciences, Engineering Foundation, and NSF.[75] Yet MEOR wasn't economically feasible then; *promises* of MEOR's potential were very useful, however, in raising cash from oil companies that start-ups like Amgen diverted into products with higher likelihood of near-term profit.

Extravagant promises abounded at the oil–biotech nexus. For instance, the president of Occidental Petroleum justified the purchase of Zoëcon and Iowa Beef Processors by telling a reporter, "We think food will be in the 1990s what energy has been to the 1970s and 1980s."[76] Martin Apple, a former University of California, San Francisco professor and CEO of a start-up controlled by ARCO, told the same reporter, "We are going to make pork chops grow on trees."[77] Apple later claimed he wanted the quote retracted, but the chair of his board told him not to, as "it's great publicity."[78]

Apple's start-up also developed ice-nucleating microbes for agricultural cloudseeding—another much overhyped technology.[79] Advanced Genetic Sciences (AGS), though, promised even more visionary applications of ice-nucleating bugs. AGS was founded in 1979 by a venture capitalist, Daniel Adams, but the company had a close relationship with UC Berkeley, and "the director, John Bedbrook (who came from academia), hired a bunch of academics."[80] In 1987, AGS merged with a Ghent University start-up, Plant Genetic Systems; the CEO of the combined firm, Jos Bouckaert, led Belgian academia's entrepreneurial turn, having founded Catholic University Leuven R&D in 1972, "the first tech transfer office in our country [Belgium], the second in Europe."[81] In 1985, AGS marketed a product called Snomax that used a special protein to nucleate water droplets to form snow. Its development was funded by Standard Oil of Ohio (one of the firms pushing to drill in Alaska), supposedly for "the construction of ice islands for use as drilling platforms in the Arctic."[82] No artificial ice islands entered service, but in 1986, the Environmental Protection Agency *did* charge AGS with falsifying scientific data on Snomax![83]

Not all oil industry investments in biotech were far-fetched, but the people and organizations who turned oil money into progress in biotech were those who moved *away* from oil. The clearest example is Monsanto, which owned an oil refinery from 1955 to 1975. After 1973, Monsanto's feedstocks became disastrously expensive, so it moved toward markets where it would depend less on oil, especially biotech-based herbicides and genetically modified organism (GMO) crops.[84] Oil producers such as Shell, ARCO, Occidental, and Elf Aquitaine also bought seed companies and invested in biotech, but those companies stayed in oil and mostly exited biotech in the 1980s or 1990s, whereas Monsanto (and its main competitor, Dupont, which owned Conoco from 1981 to 1999) stayed in GMOs and exited oil.[85]

Artificial Intelligence, Artificial Expectations

Finally, artificial intelligence. Historians have long studied AI's first wave of the 1950s and 1960s, ending with the first "AI Winter" of the 1970s.[86] Recently, studies of AI's 1980s second wave have started to appear.[87] Yet the literature does not note that oil lubricated the hinge between first- and second-wave AI and was partially responsible for the second AI winter circa 1990.[88] Yet many prominent second-wave AI researchers allied with oil firms, and some influential figures in third-wave AI also got their start thanks to oil patronage.

The main cause of the first AI winter was state patrons' (especially the US military's) disappointment with elite academic AI programs' overpromising. As Harry Law notes, the "winter" framing is arguable, since "many breakthroughs occurred in the middle of periods that have been described as AI winters."[89] Yet the term "winter" was used contemporaneously, especially by actors who turned to oil patronage because of it.[90] For example, *The Times* reported in 1987 that the United Kingdom's most prominent academic AI researcher, Donald Michie, "sat out what he calls the long winter of artificial intelligence" following the devastating 1973 Lighthill report.[91] During that time, Michie and his wife (also a prominent AI researcher) Jean Hayes-Michie founded Intelligent Terminals Limited (ITL), which developed expert systems for "luminescent fingerprinting on offshore oil rigs," "to design large vessels for the offshore separation of oil and gas," and "a system [commissioned by Enterprise Oil] which would 'read' geological maps and determine the optimum spots for test drilling."[92] BP and Shell invested in ITL.[93]

As those applications indicate, the expansion of North Sea drilling benefited British AI start-ups, such as Oilfield Expert Systems (which had a relationship with an MIT spin-off, Symbolics).[94] Admittedly, the Michies were unusual in being British academic entrepreneurs, but even if there wasn't an *academic* entrepreneurial turn in British AI, there was a Thatcherite *neoliberal* turn (see chapter 4 in this volume). Both national oil firms that underwent privatization under Thatcher, BP and Britoil, established AI initiatives—BP's through the parent company *and* its subsidiary, Sci-Con.[95] BP's neoliberal-scented Venture Research Unit, running from 1980 to 1990, also funded the noted computer scientist Edsger Dijkstra—not an AI fan but influential in the field—during which time Dijkstra also became Schlumberger Centennial Chair at Texas A&M.[96]

British *government* research organizations also commercialized expert systems for oil exploration. One AI start-up, Exploration Consultants, was founded by staff from the Atomic Engineering Research Establishment and marketed its Eclipse program for analyzing rock formations containing oil reservoirs.[97] Another government lab, BHRA Fluid Engineering, had by 1987 "for the past six years . . . been transferring its fluid engineering knowledge to a branch of artificial intelligence known as expert systems with assistance from the trade and industry department and major organizations like BP, Brown and Root [an oilfield services firm], the CEGB [Central Electricity Generating Board] and ICI [Imperial Chemical Industries—a major beneficiary of North Sea gas]."[98] And Shell worked with the UK government's Alvey Program as a potential lead user for experimental speech recognition and natural language processing systems.[99] Oil, Thatcher, and science formed an iron triangle: North Sea oil sustained Thatcherism, Thatcher promoted scientific entrepreneurship, and that entrepreneurship preferentially benefited oil firms.

Studies that mention the US oil–AI nexus, meanwhile, mostly ascribe it to large firms' irrational mania for diversification in the 1970s.[100] That is, indeed, one way to understand *some* oil firms' AI investments. Take, for instance, one of AI's biggest names, Ed Feigenbaum, who started a company, Intelligenetics, in 1980 to commercialize the molecular biology expert systems programs, DENDRAL and MYCIN, which his Stanford group developed.[101] Amoco took a majority stake in Intelligenetics in 1986 and bought it entirely in 1990, perhaps to complement its biotech investments.[102] Given poor returns from both AI and biotech, in 1994, Amoco sold Intelligenetics to a drug development company, Oxford Molecular.[103]

An even stranger example is Exxon's two voice recognition software companies, Periphonics and Dialog Systems. Periphonics was started in 1969 by a Brookhaven National Lab physicist and Dialog Systems in 1971 by an MIT engineer.[104] In the early 1970s, both received funding from Exxon's venture capital arm, Exxon Enterprises, to develop automated phone banking.[105] Then, in the late 1970s, Exxon Enterprises bought twenty-odd computing and information technology companies and integrated their products in an Exxon Office Information Systems suite, including a voice recognition package.[106] Dialog (renamed Verbex Voice Systems) expanded as a result, hiring top talent such as James and Janet Baker, former PhD students in Carnegie Mellon's renowned AI program. By 1982, Office Information Systems was collapsing, so the Bakers formed their own company, Dragon Systems.[107] Once the second AI winter ended in the 1990s, Dragon obtained funding from DARPA and Analog Devices (an electronics company partly owned by Amoco) and became one of the preeminent voice recognition firms; today, the Bakers' code is almost certainly in Siri and her peers.[108] For Exxon, though, the foray into computing became a byword for disastrous corporate diversification.

Yet the oil–AI nexus can't be dismissed entirely as irrational diversification mania. For instance, even before Amoco invested in Intelligenetics, Schlumberger was working with Feigenbaum to adapt MYCIN to geological expert systems. Feigenbaum's other start-up, Teknowledge, later built expert systems to help Elf Aquitaine with its rigs.[109] LISP Machines (an MIT start-up) was by 1985 "one of the two major players in the artificial intelligence computer market . . . [and] working with Texaco Inc., on a computer system to help operators make critical decisions during an oil refinery emergency."[110] A LISP spin-off, Gensym, also marketed process control systems, for instance, to the Japanese Petroleum Energy Center in 1988.[111] Likewise, Roger Schank, director of the Yale Artificial Intelligence Project, founded Cognitive Systems in 1979, which by 1982 was "designing a system for oil companies that will retrieve information on oil wells using plain English commands."[112]

The most influential oil–AI nexus resulted from Schlumberger's poaching Stanford and SRI International's top AI researchers (SRI was formerly Stanford's contract research unit; after Stanford divested it in 1970, the two organizations still cooperated, especially in AI). In 1980, Schlumberger hired one of SRI's AI leaders, Peter Hart, to direct its Fairchild Laboratory for Artificial Research (FLAIR) and bring along his Stanford and SRI colleagues.[113] Hart was possibly chosen because of PROSPECTOR, a program he wrote at SRI with John Harbaugh, a Stanford petroleum engineer.[114]

The FLAIR group produced expert systems such as Dipmeter Advisor, which "attempted to emulate human expert performance in interpretation of the dipmeter well-logging tool," and SlurryMinder, "a design tool to aid field engineers" in sealing oil and gas wells.[115] Yet some FLAIR employees later believed their "key learnings back in the mid-1980's [sic] were not related to AI *per se*" but instead to graphics, database techniques for workstations, and refinement of neural networks.[116] Unsurprisingly, therefore, Schlumberger abandoned FLAIR in 1989. By then, most of FLAIR's Stanford and SRI veterans had left to found companies such as "industry giants Cisco and Netflix."[117] One "entire [Schlumberger] team wound up at Pixar, where they won two technical Academy Awards . . . [and were] responsible for Pixar's now iconic bouncing Luxo Lamp animation."[118]

So in the early 1980s, Schlumberger reasonably expected its AI hires to contribute to oilfield exploration, based on those hires' earlier academic work. Yet those hires' contributions to Schlumberger were dwarfed by their achievements *after* they left the oil industry. Not that Schlumberger was exceptional in its AI optimism. Chevron, Phillips, Elf Aquitaine, Exxon, MW Kellogg, Bechtel, Stone & Webster, Texaco, and many other oil (or oil-related) firms made forays into second-wave AI.[119] Those investments resulted in some useful tools, for example, in computer-assisted design, but in the short term, they yielded little "artificial intelligence" that oil firms could use.

The Spoils of Greed

Note the modifier "short term." Until the mid-1980s, oil firms had the will and capacity for long-term investment. There was consensus among CEOs of the oil majors, for instance, that nuclear fusion and solar power would not be commercially feasible until the twenty-first century, yet they invested in those technologies anyway.[120] Over the 1980s, however, oil firms mostly exited fusion, solar, and the other technologies I have examined. Long-range investment dwindled, replaced by a focus on short-term profit. The reason? The declining price of oil after 1981. In the early 1980s, that decline was manageable and oil firms' withdrawal from the academic landscape orderly. In 1986, however, the Saudis raised production and the price per barrel plunged; the ensuing retreat from long-term investment was an omnishambles.

With respect to solar power, for instance, companies like Texas Instruments and Shell always planned on the price of solar energy improving relative to that of oil; as the latter crashed, commercialization of solar no longer seemed like a near-term possibility. Most corporate solar programs were sharply cut back or abandoned. The same logic applied to the oil shale and "tight oil" programs that earlier underwrote investments in biotech companies; conventional oil now became too cheap to justify pursuing exotic fossil fuels.[121]

Indeed, in the 1980s, it was cheaper to buy another company's reserves of *conventional* fossil fuels than to explore for new drillable formations.[122] The oil industry was thus swept by hostile takeover attempts. Giant companies like Exxon and Mobil mostly escaped the frenzy, but everyone else witnessed bids, counterbids, poison pills, white knights, and lawsuits over failed bids. Some companies, such as Cities Service, were subjected to multiple takeover attempts.[123] Some battles—for example, for Getty or Conoco—became contests among multiple bidders at once.[124]

Thus, oil was central to the "greed is good" mantra. The inspirations for Gordon Gekko—T. Boone Pickens, Carl Icahn, Michael Milken, Ivan Boesky—derived much of their money and infamy from oil and gas deals.[125] Moreover, the era's contentious financial innovations, such as the poison pill, originated in the oil industry's takeover wave.[126] The industry depicted in *Dallas* and *Dynasty* was more riddled with gossip, scandal, and backstabbing than its fictional counterpart.

Research, especially on speculative, long-range technologies, fell victim to this chaos. To block takeover bids, firms had to persuade shareholders that current management would bring faster, more certain returns than Pickens or Icahn would; dumping ventures (e.g., in solar) that were projected to lose money for at least twenty years was one way to display sensitivity to investors. Firms also diverted cash from R&D and non-oil operations to defend against takeovers.[127] Falling profits meant all oil firms (even companies not facing takeovers) cut their payrolls; as early as 1982, even Exxon Research and Engineering let go 8 percent of its personnel.[128] The Department of Energy worried that US oil firms might lose 100 percent of their research capacity, and Michel Halbouty—a prominent wildcatter and Ronald Reagan's "energy guru"—warned that takeovers threatened the industry's innovative capacity.[129]

I have argued elsewhere that oil firms' withdrawal from alternative energy encouraged their drift toward climate denialism.[130] Even into the early 1980s, there was consensus among the oil majors that oil usage would decline relative to nuclear fission in the 1980s, with solar and unconventional nuclear (fusion and fast breeder reactors) gaining share sometime after 2000.[131] Oil firms could therefore claim that once climate change posed a major problem, they would have the solution. In abandoning that solution, they opted to instead claim that climate change would never be a problem and that customers could greedily consume without consequence.

Thus, the two 1970s oil shocks and the 1980s countershock encouraged greed in general and greedy science in particular. Was that a bad thing? Not necessarily. I admire many academic entrepreneurs, and good science can be motivated by profit. But if greed guided *some* academic entrepreneurs toward robust, well-regarded discoveries, it steered others toward scientific misbehavior. Specifically, greed motivated *some* academic entrepreneurs to make exaggerated promises in order to attract investment from oil companies. More perniciously, from the late 1980s onward, greed encouraged the "merchants of doubt" to market their services as climate denialists.[132] Both perversions of academic entrepreneurship haunt us today.

In some sense, the oil–greedy science nexus is unsurprising. Meg Jacobs, Matthew Huber, and Caleb Wellum have shown that the 1970s oil shocks legitimated neoliberalism.[133] Philip Mirowski argues that neoliberalism encouraged the academic entrepreneurial turn.[134] So, transitively, the oil shocks *must* have facilitated the entrepreneurial turn. But it wasn't just the general postshock environment that encouraged academic entrepreneurship; oil firms *directly* sponsored many early academic entrepreneurs. Possibly an entrepreneurial turn would have happened even without that intervention. But oil firms did intervene, and tracing their sponsorship of early academic entrepreneurs illuminates greedy science in its complexity and variety, the hopes pinned to it, promises made for it, and fortunes and misfortunes resulting from it.

NOTES

Research for this chapter was supported by ERC Synergy grant no. 951393 NanoBubbles and by Nederlands Organisatie voor Wetenschappelijk Onderzoek (Dutch Research Council) grant VI.C.191.067 Managing Scarcity and Sustainability. The author thanks both funders and his collaborators in both projects.

1. Elizabeth Popp Berman, *Creating the Market University: How Academic Science Became an Economic Engine* (Princeton, NJ: Princeton University Press, 2012); Paula Stephan, *How Economics Shapes Science* (Cambridge, MA: Harvard University Press, 2012); David C. Mowery, Richard R. Nelson, Bhaven Sampat, and Arvids A. Ziedonis, *Ivory Tower and Industrial Innovation: University-Industry Technology Transfer before and after the Bayh-Dole Act in the United States* (Stanford, CA: Stanford Business Books, 2004); Eric J. Vettel, *Biotech: The Countercultural Origins of an Industry* (Philadelphia: University of Pennsylvania Press, 2006); Elizabeth Popp Berman, "Why Did Universities Start Patenting? Institution-Building and the Road to the Bayh-Dole Act," *Social Studies of Science* 38, no. 6 (2008): 835–871; Albert N. Link and Martijn van Hasselt, "On the Transfer of Technology from Universities: The Impact of the Bayh-Dole Act of 1980 on the Institutionalization of University Research," *European Economic Review* 119 (2019): 472–481; Matthew R. Keller and Fred Block, "Explaining the Transformation in the US Innovation System: The Impact of a Small Government Program," *Socio-Economic Review* 11, no. 4 (2013): 629–656.

2. Cyrus C. M. Mody, *The Long Arm of Moore's Law: Microelectronics and American Science* (Cambridge, MA: MIT Press, 2017); Maureen McKelvey, *Evolutionary Innovations: The Business of Biotechnology* (Oxford: Oxford University Press, 2000); Louis Galambos and Jeffrey L. Sturchio, "Pharmaceutical Firms and the Transition to Biotechnology: A Study in Strategic Innovation," *Business History Review* 72, no. 2 (1998): 250–278.

3. For depictions of oil omnipotent, see Daniel Yergin, *The Prize: The Epic Quest for Oil, Money, and Power* (New York: Free Press, 1991); Timothy Mitchell, *Carbon Democracy: Political Power in the Age of Oil* (London: Verso, 2011). My take on oil firms' constrained agency is indebted to Francesco Petrini, "Eight Squeezed Sisters: The Oil Majors and the Coming of the 1973 Oil Crisis," in *Oil Shock: The 1973 Crisis and Its Economic Legacy*, ed. Elisabetta Bini, Giuliano Garavini, and Federico Romero (London: I. B. Tauris, 2016), 89–116.

4. James Webb (NASA's administrator from 1961 to 1968; a Kerr-McGee employee before NASA and a company director afterward) and George H. W. Bush (CIA director in 1976, later US president; cofounder of Zapata Petroleum).

5. Robert E. Kohler, *Partners in Science: Foundations and Natural Scientists, 1900–1945* (Chicago: University of Chicago Press, 1991); Tim B. Mueller, "The Rockefeller Foundation, the Social Sciences, and the Humanities in the Cold War," *Journal of Cold War Studies* 15, no. 3 (2013): 108–135; Deborah Fitzgerald, "Exporting American Agriculture: The Rockefeller Foundation in Mexico, 1943–1953," *Social Studies of Science* 16, no. 3 (1986): 457–483; Tara H. Abraham, "The Macy Conferences on Cybernetics: Reinstating the Mind," in *Oxford Research Encyclopedia of Psychology* (Oxford University Press, 2020); Alexei Kojevnikov, *The Copenhagen Network: The Birth of Quantum Mechanics from a Postdoctoral Perspective* (Cham, Switzerland: Springer, 2020).

6. James Allen Smith, "The Evolving American Foundation," in *Philanthropy and the Nonprofit Sector*, ed. C. T. Clotfelter and T. Ehrlich (Bloomington: Indiana University Press, 1999), 34–51; on the extractive industries' role in the rise of CSR, see Jessica M. Smith, *Extracting Accountability: Engineers and Corporate Social Responsibility* (Cambridge, MA: MIT Press, 2021), chap. 3.

7. The preeminent contemporary expert on philanthropic foundations was Waldemar Nielsen, author of *The Big Foundations* (New York: Columbia University Press, 1972) and *Golden Donors: A New Anatomy of the Great Foundations* (New Brunswick, NJ: Transaction Publishers, 1985). When Nielsen wrote *The Big Foundations*, he was simultaneously consulting for Robert O. Anderson (founder of the Atlantic Richfield oil company, ARCO) and his personal foundation, ARCO itself, the ARCO Foundation, and the Aspen Institute, which Anderson controlled and to which the ARCO Foundation was a major donor. See memo from Nielsen to Robert O. Anderson and T. F. Bradshaw, November 23, 1971, re: A First Effort to Define My Functions and Priorities, and report from Nielsen to Mr. Anderson and Mr. Bradshaw, July 25, 1974, re: ARCO in Indonesia, both in Folder 9, Box 12, Waldemar A. Nielsen Papers, 1930–2004, Ruth Lilly Special Collections and Archives, University Library, Indiana University–Purdue University Indianapolis.

8. See, for instance, excerpts from the Exxon Education Foundation's 1975 and 1977 annual reports in Folder "Exxon Education Foundation," Box 273, Office of the President Records, Martin Meyerson Administration 1970–1980 (UPA 4 Meyerson), University Archives and Records Center, University of Pennsylvania, or the description of an Aspen Institute conference on "Creativity, Invention, and Technology," funded by the ARCO Foundation in a memo from Sven B. Lundstedt to the conference working group, July 25, 1980, in Folder "Aspen Inst Innovation Conf Wye, MD 1980," Box 3, Papers of Harvey Brooks (HUGFP 128), Harvard University Archives.

9. For example, Enno Hobbing, "Economics Can Be Interesting," *The Lamp* (Spring, 1969): 16–19.

10. Case studies illustrating the importance of VC money include Mark Peter Jones, "Biotech's Perfect Climate: The Hybritech Story," PhD dissertation, University of California, San Diego, 2005, and Sally Smith Hughes, *Genentech: The Beginnings of Biotech* (Chicago: University of Chicago Press, 2011).

11. Martin Kenney, "How Venture Capital Became a Component of the US National System of Innovation," *Industrial & Corporate Change* 20, no. 6 (2011): 1677–1723; Spencer E. Ante, *Creative Capital: Georges Doriot and the Birth of Venture Capital* (Cambridge, MA: Harvard Business Press, 2008).

12. Leslie Berlin, "Robert Noyce and Fairchild Semiconductor, 1957–1968," *Business History Review* 75, no. 1 (2001): 63–101; David H. Hsu and Martin Kenney, "Organizing Venture Capital: The Rise and Demise of American Research & Development Corporation, 1946–1973," *Industrial and Corporate Change* 14, no. 4 (2005): 579–616.

13. Petrini, "Eight Squeezed Sisters." Also, Giuliano Garavini, *The Rise and Fall of OPEC in the Twentieth Century* (Oxford: Oxford University Press, 2019); Tyler Priest, "The Dilemmas of Oil Empire," *Journal of American History* 99, no. 1 (2012): 236–251; Robert D. Lifset, "A New Understanding of the American Energy Crisis of the 1970s," *Historical Social Research* 39, no. 4 (2014): 22–42.

14. Rüdiger Graf, *Oil and Sovereignty: Petro-Knowledge and Energy Policy in the United States and Western Europe in the 1970s* (New York: Berghahn, 2014).

15. Adam Rome, " 'Give Earth a Chance': The Environmental Movement and the Sixties," *Journal of American History* 90, no. 2 (2003): 525–554; Teresa Sabol Spezio, *Slick Policy: Envi-*

ronmental and Science Policy in the Aftermath of the Santa Barbara Oil Spill (Pittsburgh: University of Pittsburgh Press, 2018).

16. Meg Jacobs, *Panic at the Pump: The Energy Crisis and the Transformation of American Politics in the 1970s* (New York: Farrar, Straus and Giroux, 2016).

17. Rawleigh Warner Jr. (chairman of Mobil), "Energy Resources—and the Public" (speech to annual convention of the Edison Electric Institute), June 3, 1974, in Folder "Correspondence—Mobil Oil—April–May–June 1974," Box 44, George C. McGhee papers (WHC-M-2168), Western History Collection, University of Oklahoma.

18. The words in quotes nod (verbatim) to *The Other Dimensions of Business: A Report on Exxon's Participation in Areas of Public Interest*, undated but circa late 1977 or early 1978, in Folder "Exxon Education Foundation Reports the Other Dimensions of Business," Box 2.207-G252, ExxonMobil Collection, Briscoe Center for American History.

19. John N. Garrett letter to Hollis Hedberg, September 6, 1978, and Hollis Hedberg letter to the United Nations Mission of India, August 25, 1978, both in Folder "Law of the Sea General Correspondence," Box 47, Hollis Hedberg Papers (RH MS 546), Kenneth Spencer Research Library, University of Kansas.

20. Peter A. Coates, *The Trans-Alaska Pipeline Controversy: Technology, Conservation, and the Frontier* (Bethlehem, PA: Lehigh University Press, 1991).

21. Coates, *The Trans-Alaska Pipeline Controversy*; Simone Schleper, "Caribou Crossings: The Trans-Alaska Pipeline System, Conservation, and Stakeholdership in the Anthropocene," *British Journal for the History of Science* 55, no. 2 (2022): 127–143.

22. Tyler Priest, "Seismic Innovations: The Digital Revolution in the Search for Oil and Gas," in *Energy in the Americas: Reflections on Energy and History*, ed. Amelia M. Kiddle (Calgary: University of Calgary Press, 2021), 179–210; Tyler Priest, *The Offshore Imperative: Shell Oil's Search for Petroleum in Postwar America* (College Station: Texas A&M University Press, 2007); Hugh S. Gorman, *Redefining Efficiency: Pollution Concerns, Regulatory Mechanisms, and Technological Change in the U.S. Petroleum Industry* (Akron, OH: University of Akron Press, 2001).

23. On Chicago, see Joseph D. Martin, "The Simple and Courageous Course," *Isis* 111, no. 4 (2020): 697–716. On MIT, see the correspondence (circa 1946) from various administrators soliciting (and receiving) support from various oil companies such as Atlantic Refining Company and Cities Service for the Institute's nuclear research in Folders 8 through 13, Box 161, MIT Office of the President, records of Karl Taylor Compton and James Rhyne Killian 1930–1959 (AC-0004), Distinctive Collections, MIT Libraries.

24. Wesley M. Cohen, "Firm Heterogeneity, Investment, and Industry Expansion: A Theoretical Framework and the Case of the Uranium Industry," PhD dissertation, Yale University, 1981.

25. Michiel Bron, "The Sun on Earth: How the Netherlands Dealt with the Promise of Nuclear Fusion, 1951–1979," master's thesis, Utrecht University, 2020.

26. "Powerful Laser Fusion Project Launched," *Chemical & Engineering News* 54, no. 11 (1976): 7; Laboratory for Laser Energetics, "The Laser Fusion Feasibility Project," https://www.lle.rochester.edu/media/about/timeline/3_Timeline_1972_73.pdf, accessed June 1, 2024.

27. Keeve M. Siegel, "The Search for Fusion," *The Detroit Engineer*, November 1974. A vice president of another oil and gas firm, Gibraltar Exploration Limited, was also on the board of KMS Fusion. See the firm's annual report for 1981, in Folder 2, Box 1, Series I, subseries I, Keeve Milton Siegel papers, 1953–1983 (89273 Aa2), Bentley Historical Library, University of Michigan.

28. Jonathan Coopersmith, "Fraud and Froth: Free-Riding the 3D Printing Wave," in *3D Printing: Legal, Philosophical and Economic Dimensions*, ed. Bibi van den Berg, Simone van der Hof, and Eleni Kosta (Berlin: Springer, 2016), 137–152.

29. Frank N. Laird, *Solar Energy, Technology Policy, and Institutional Values* (Cambridge: Cambridge University Press, 2004).

30. For a study of the TISES project, with full sourcing, see Cyrus C. M. Mody, *The Squares: US Physical and Engineering Scientists in the Long 1970s* (Cambridge, MA: MIT Press, 2022), chap. 5.

31. Although they approached firms in other industries, it is clear that they thought oil firms were the most likely investors.

32. W. Arthur Porter, *The Knowledge Seekers: Creating Centers for the Performing Sciences* (Austin: IC² Institute, University of Texas, 1998).

33. Jurgen Schmandt, *George P. Mitchell and the Idea of Sustainability* (College Station: Texas A&M University Press, 2010); Loren C. Steffy, *George P. Mitchell: Fracking, Sustainability, and an Unorthodox Quest to Save the Planet* (College Station: Texas A&M University Press, 2019).

34. Berman, *Creating the Market University*.

35. Personal communication, June 29, 2020.

36. "Chevron Turns over Gulf Research Center," *Associated Press*, March 17, 1986.

37. Martin Kenney, *Biotechnology: The University-Industry Complex* (New Haven, CT: Yale University Press, 1986), 55.

38. Lillian Hoddeson and Peter Garrett, *The Man Who Saw Tomorrow: The Life and Inventions of Stanford R. Ovshinsky* (Cambridge, MA: MIT Press, 2018).

39. Jack Kilby, report to Fred Bucy, June 5, 1981, Folder 10, Box 100, Kilby papers, Library of Congress. Kilby's papers include press clippings about Ovshinsky on which Kilby or his colleagues scrawled comments such as "I don't know how he gets such publicity."

40. Robert Metz, "New Products, New Hopes," *New York Times*, June 1, 1978.

41. John Berger, *Charging Ahead: The Business of Renewable Energy and What It Means for America* (Berkeley: University of California Press, 1997), 81–82.

42. Information in this paragraph and the next partly taken from Karl W. Böer and Esther Riehl, *The Life of the Solar Pioneer Karl Wolfgang Böer* (Kelly, NC: Toplink, 2017). See also Gerhard Mener, "Zwischen Labor und Markt: Geschichte der Sonnenenergienutzung in Deutschland und den USA, 1860–1986," PhD dissertation, Deutsches Museum, 2001, 379–387.

43. Letter from Thomas Baron (president, Shell Development Company) to E. A. Trabant (president, University of Delaware), May 30, 1974, II.C.3, Folder "Equipment (Surplus)," Box 27, and letter from Theodore Sherbow (Sherbow, Shea and Doyle) to Robert Clement (Shell Development Company), October 12, 1973 (and accompanying contract), VI.A.3, Folder 101, Box 58; both Karl-Wolfgang Böer papers (MSS 0483), Special Collections, University of Delaware Library (hereafter Böer papers).

44. Notes taken by K.E.L. (full name unknown) titled "Areas with commercial potential for which the university has a patent position," October 6, 1987, in I.A.1-42, Folder 42, Box 1, Böer papers; Ralph Flood, "Big Oil Reaches for the Sun," *New Scientist* 92, no. 1279 (1981): 426–428.

45. This paragraph (including quotations) is based on an admittedly tendentious article championing Barnett: Kent Stoppard, "How the U of D Sold the Sun," *Delaware Today*, January 1980, 21–23. Böer's version is less inflammatory, yet his frustrations with being forced out of the IEC directorship, and with Shell's business decisions, are still evident. Karl-Wolfgang Böer, draft transcript of interview with Gerhard Mener, August 28, 1996, in VI.A.3, Folder 48, Box 2, Karl-Wolfgang Böer papers supplement (MSS 0632), Special Collections, University of Delaware Library.

46. W. A. Zama (Solavolt) interoffice memorandum and holding statement, June 20, 1983, in VI.A.3, Folder 116, Box 58, Böer papers.

47. Vettel, *Biotech*; Nicolas Rasmussen, *Gene Jockeys: Life Science and the Rise of Biotech Enterprise* (Baltimore: Johns Hopkins University Press, 2014); Walter W. Powell and Jason Owen-Smith, "Knowledge Networks as Channels and Conduits: The Effects of Spillovers in the Boston Biotechnology Community," *Organization Science* 15 (2001): 5–21; Paul Rabinow, *Making PCR: A Story of Biotechnology* (Chicago: University of Chicago Press, 1996); Fiona Murray, "The Role of Academic Inventors in Entrepreneurial Firms: Sharing the Laboratory Life," *Research Policy* 33 (2004): 643–659; Kenney, *Biotechnology*; McKelvey, *Evolutionary Innovations*; Jones, "Biotech's Perfect Climate"; Hughes, *Genentech*; Doogab Yi, *The Recombinant University: Genetic Engineering and the Emergence of Stanford Biotechnology* (Chicago: University of Chicago Press, 2015); Jeannette Colyvas, "From Divergent Meanings to Common Practices: Institutionalization Processes and the Commercialization of University Research," PhD dissertation, Stanford University, 2007.

48. Robert Bud, *The Uses of Life: A History of Biotechnology* (Cambridge: Cambridge University Press, 1994).

49. "Biotechnology's Drive for New Products," *Chemical Week*, February 24, 1982.

50. Letter from Gordon C. McKeague (Corporate Development, Standard Oil of Ohio) to Moshé Alafi (Biogen), February 15, 1980, in Series 3, Folder "Biogen 1980," Box 5, Walter Gilbert Collection (WAG), Cold Spring Harbor Library and Archives. My thanks to Robin Scheffler for this document.

51. Jones, "Biotech's Perfect Climate."

52. "Drug Export Law Will Aid Biofirms' Early Cash Flow, Improve Foreign Contracts," *McGraw-Hill's Biotechnology Newswatch*, December 1, 1986.

53. Ann Crittenden, "Universities' Accord Called Research Aid," *New York Times*, September 12, 1981.

54. "French Oil Giant Is Building $16.7-million R&D Facility," *McGraw-Hill's Biotechnology Newswatch*, March 1, 1982.

55. "How to Be a Winner with Venture Capital," *Chemical Week*, December 14, 1983; "Fluor Buys Shares in Genentech, Inc.," *New York Times*, January 16, 1981.

56. "Oil Refiner Tosco Buys into Bioengineering," *Chemical Week*, February 11, 1981. Harrison Capital's investment in Amgen is harder to trace except in CVs of Harrison's principals; for example, Catalina Ventures, "Randall R. Lunn," http://catalinaventures.com/?page _id=30, accessed November 16, 2023. There is some contemporary evidence that Harrison's owner, Texaco, collaborated with Amgen: "Amgen," *PR Newswire*, October 24, 1984.

57. "Immunex Enters Agreement with SDS Biotech to Develop Shipping Fever Therapeutic," *Business Wire*, October 26, 1983.

58. Walter W. Powell and Kurt Sandholtz, "*Chance, Nécessité et Naïveté*: Ingredients to Create a New Organizational Form," in *The Emergence of Organizations and Markets*, ed. John F. Padgett and Walter W. Powell (Princeton, NJ: Princeton University Press, 2012), 379–433.

59. "Centocor Unveils r-DNA AIDS Assay; Faces FDA," *McGraw-Hill's Biotechnology Newswatch*, August 5, 1985.

60. "Amoco Invests in Diagnostics Joint Venture," *Crain's Chicago Business*, August 4, 1986.

61. "r-DNA Vitamin B12 Process Acquired," *McGraw-Hill's Biotechnology Newswatch*, May 4, 1987.

62. "Chiron Licenses Phillips' Pichia Expression System," *McGraw-Hill's Biotechnology Newswatch*, September 4, 1989.

63. That is, Cetus, Genentech, Genex, Hybritech, Centocor, Amgen, Chiron, Genzyme, and Immunex. Biogen could perhaps be included because it received investment from

Monsanto, which owned a refinery from 1955 from 1975, but sold it because of the 1973 crisis. As noted above, Biogen also came close to a deal with Amoco. Powell and Sandholtz, *"Chance, Nécessité et Naïveté."*

64. "A New Joint-Venture Duo: Phillips, Salk Institute," *Chemical Week*, June 17, 1981.

65. "Diamond Shamrock, SIBIA Form Joint Vaccine Firm," *McGraw-Hill's Biotechnology Newswatch*, April 18, 1983.

66. Angus McIntyre, memo to Joe Sambrook, October 8, 1981, in Cold Spring Harbor Laboratory Office of Technology Transfer Collection, Cellbiology series, Folder "Oil Shale Project," Box 2, Cold Spring Harbor Laboratory Archives. Document courtesy of Robin Scheffler.

67. "Oil-Eating Germ," *The Bulletin*, June 1, 1982, 107.

68. Carl Djerassi, oral history interview with Jeffrey L. Sturchio and Arnold Thackray, July 31, 1985.

69. Daniel J. Kevles, "Ananda Chakrabarty Wins a Patent: Biotechnology, Law, and Society, 1972–1980," *Historical Studies in the Physical and Biological Sciences* 25, no. 1 (1994): 111–135.

70. Bud, *The Uses of Life*; Martha M. Trescott (with Dev Bickston and Pat Cheek), "Proteins from Petroleum and Industrial Wastes: Their Manufacture, Nutrition, and Economics: A Bibliography of World Literature (1969–1973)" (June–September 1973), semi-published report with call number Z5524.P83 P7 1973 ESL at Linda Hall Library.

71. Stephen Strauss, "Patented Superbug Bombs as Oil Eater, Left to Die," *The Globe and Mail*, May 13, 1983.

72. "Petrogen to Test, Patent Chakrabarty's Newest Oil-Recovery Organism," *McGraw-Hill's Biotechnology Newswatch*, February 21, 1983.

73. "MEOR=MORE Oil Recovery Some Day, Especially If Impending Field Trials Pan Out," *McGraw-Hill's Biotechnology Newswatch*, September 20, 1982; "Oilmen Enlisting Microbiologists to Aid Tertiary Oil Recovery," *McGraw-Hill's Biotechnology Newswatch*, June 7, 1982; "Pfizer to Scale up 'Emulsan' Output, Marketing, Research for Petroferm Patent Holder," *McGraw-Hill's Biotechnology Newswatch*, September 19, 1983.

74. George B. Rathmann, oral history interview with Sally Smith Hughes, Regional Oral History Office, Bancroft Library, UC Berkeley, 2004.

75. Summary of Ad Hoc Committee Meeting on New Separation Processes for Genetic Engineering and Chemical and Energy Industries, May 20, 1982, in Folder Ad Hoc Committee on New Separation Processes, Box 118, H. Guyford Stever Papers, Gerald R. Ford Presidential Library. Document Courtesy of Matt Wisnioski. Also, *Proceedings of 1976 Engineering Foundation Conference the Role of Microorganisms in the Recovery of Oil* (Washington, DC: National Science Foundation, 1975).

76. Ann Crittenden, "The Gene Machine Hits the Farm," *New York Times*, June 28, 1981.

77. Crittenden, "The Gene Machine Hits the Farm."

78. Daniel Charles, *Lords of the Harvest: Biotech, Big Money, and the Future of Food* (Cambridge, MA: Perseus, 2001).

79. "Biologically Triggered Ice Nucleation," *Chemical Week*, August 11, 1982, 33; James Rodger Fleming, *Fixing the Sky: The Checkered History of Weather and Climate Control* (New York: Columbia University Press, 2012).

80. Catarina Vicente, "An Interview with Caroline Dean," *Development* 142, no. 16 (August 2015): 2725–2726.

81. Jos Bouckaert, *Merkstenen: Op weg naar de ondernemende onderzoeksuniversiteit* (Leuven, Belgium: Leuven University Press, 2021). My translation.

82. "ADV-GENETIC-SCIENCE-2; Announces First Product Sales of Snomax™ and EPA Approval for Field Testing of Frostban™," *Business Wire*, November 21, 1985.

83. Keith Schneider, "Genetic Company Loses Permit for Field Testing," *New York Times*, March 25, 1986, A20.

84. Bartow J. Elmore, *Seed Money: Monsanto's Past and Our Food Future* (New York: W. W. Norton, 2021).

85. Richard M. Harley, "Feast or Famine," *Christian Science Monitor*, August 20, 1981; Louis Galambos and Jeffrey L. Sturchio, "Pharmaceutical Firms and the Transition to Biotechnology: A Study in Strategic Innovation," *Business History Review* 72, no. 2 (1998): 250–278.

86. For example, Esther-Mirjam Sent, "Sargent versus Simon: Bounded Rationality Unbound," *Cambridge Journal of Economics* 21 (1997): 323–338; Hunter Crowther-Heyck, *Herbert A. Simon: The Bounds of Reason in Modern America* (Baltimore: Johns Hopkins University Press, 2005); Ronald Kline, "Cybernetics, Automata Studies, and the Dartmouth Conference on Artificial Intelligence," *IEEE Annals of the History of Computing* 33, no. 4 (2011): 5–16; Jonathan Penn, "Inventing Intelligence: On the History of Complex Information Processing and Artificial Intelligence in the United States in the Mid-Twentieth Century," PhD dissertation, University of Cambridge, 2020.

87. For example, Matthew L. Jones, "Querying the Archive: Data Mining from Apriori to Pagerank," in *Science in the Archives*, ed. Lorraine Daston (Chicago: University of Chicago Press, 2017), 311–328, and "How We Became Instrumentalists (Again): Data Positivism since World War II," *Historical Studies in the Natural Sciences* 48, no. 5 (2018): 673–684; Stephanie Dick, "Artificial Intelligence," *Harvard Data Science Review* 1, no. 1 (2019), https://doi.org/10.1162/99608f92.92fe150c; Xiaochang Li, "Divination Engines: A Media History of Text Prediction," PhD dissertation, New York University, 2017; James Steinhoff, "Industrializing Intelligence: A Political Economic History of the AI Industry," in *Automation and Autonomy: Labour, Capital, and Machines in the Artificial Intelligence Industry*, ed. James Steinhoff (Cham, Switzerland: Springer, 2021), 99–131.

88. On the first AI winter, see Jon Agar, "What Is Science For? The Lighthill Report on Artificial Intelligence Reinterpreted," *British Journal for the History of Science* 53, no. 3 (2020): 289–310; Nils J. Nilsson, *The Quest for Artificial Intelligence: A History of Ideas and Achievements* (Cambridge: Cambridge University Press, 2010); Melanie Mitchell, "Why AI Is Harder than We Think," arXiv:2104.12871v2, April 28, 2021.

89. Harry Law, "An Introduction to AI History," *Learning from Examples*, August 7, 2023, https://www.learningfromexamples.com/p/an-introduction-to-ai-history.

90. For example, Alan Kane, "Year When Computers Became 'Experts,'" *Financial Times*, January 4, 1982; Michael Tucker, "AI Woos MIS Anew," *Computerworld*, June 1, 1988, 27.

91. Agar, "What Is Science For?," examines the Lighthill report in the context of Michie's conflicts with its author regarding the real-world applications of AI. Quote is from Ann Kent, "Computer Horizons: One Man's Wonderland/Profile of Donald Michie," *The Times* (London), September 22, 1987.

92. Colin Donald, "Connected: The Scientist Whose Life Inspires a Major New Film, an Obscure Research Unit in Scotland, Flying the Space Shuttle, and the Invention of the Internet," *Sunday Herald*, November 16, 2014; Alan Cane, "Computing: Expert System Brought in for Space Shuttle Test Firings," *Financial Times*, September 11, 1986; Intelligent Terminals FAQ.

93. Intelligent Terminals corporate overview and FAQ, Turing: The Papers of Alan Mathison Turing, GBR/0272/AMT/A/43 (Information leaflets relating to Intelligent Terminals, 1985).

94. Peter Marsh, "He Means to Help Reverse Britain's Drain of Computer Talent," *Christian Science Monitor*, May 15, 1984, 24.

95. Mark Meredith, "N. Sea Oilmen Face New Technology Dilemma," *Financial Times*, July 11, 1984, 14; Maggie Mclening, "Computer Horizons: The Human Factor Systems Cannot Ever Replace," *The Times* (London), May 6, 1986; Donald Michie, "Computer Guardian: A New Neutral Fever to Use Our Noggins—A Rush of Investment Blood over Neural Nets," *The Guardian*, March 16, 1989.

96. BP Venture Research Unit, *Venture Research* (London: BP International, 1990).

97. Meredith, "N. Sea Oilmen."

98. "BHRA Launches New Knowledge Handling System," *FT Energy Newsletters—North Sea Letter*, April 8, 1987.

99. Brian Oakley, *Alvey: Britain's Strategic Computing Initiative* (Cambridge, MA: MIT Press, 1989), 152–153.

100. See Michael Ollinger, "The Limits of Growth of the Multidivisional Form: A Case Study of the U.S. Oil Industry from 1930–90," *Strategic Management Journal* 15, no. 7 (1994): 503–520 for a summary of the irrational diversification argument specifically with respect to oil firms.

101. David C. Brock, "Learning from Artificial Intelligence's Previous Awakenings: The History of Expert Systems," *AI Magazine*, Fall 2018, 3–15, presents participants' reminiscences of the Feigenbaum group—though only one mentions the oil industry (and that mention makes it seem as though Feigenbaum's team received no oil support).

102. "INTELLICORP / AMOCO; Announce Joint Venture," *Business Wire*, March 18, 1986; "IntelliCorp Sells Remaining Interest in Joint Venture to Amoco," *Business Wire*, November 9, 1990.

103. "Oxford Molecular Crosses Atlantic to Pick up Intelligenetics for GBP5.2m," *Computergram International*, August 26, 1994.

104. Jeff Matthews, "Winfield Entrepreneur Passes on Knowledge," *Town Talk*, https://eu.thetowntalk.com/story/news/2016/06/23/winnfield-entrepreneur-passes-knowledge/86243808/, accessed October 1, 2021, and "CMU Jones Center Director Brings Wealth of Experience to New Role," *Pittsburgh Business Times*, August 25, 2000, https://www.bizjournals.com/pittsburgh/stories/2000/08/21/daily23.html, accessed October 1, 2021; Jim Cook, "James E. Cook, 'Jim,'" https://cha4mot.com/jim/cook-en.html, accessed October 1, 2021.

105. Jeffrey Kutler, "Technology, System Sharing Improve Phone Banking Outlook," *The American Banker*, December 7, 1979.

106. Michael Byrne, "When Exxon Wanted to Be the Next Apple," *Vice*, April 26, 2015, https://www.vice.com/en_us/article/5394x5/when-exxon-wanted-to-be-a-personal-computing-revolutionary, accessed October 1, 2021; Anthony J. Parisi, "Exxon Offers Laser Devices," *New York Times*, January 24, 1979, D1.

107. Dragon Naturally Speaking, "History of Speech and Voice Recognition and Transcription Software," retrieved from the Internet Archive, https://web.archive.org/web/20150813223326/http://dragon-medical-transcription.com/history_speech_recognition.html, accessed October 1, 2021.

108. Babs Carryer, "Innovation in Speech Recognition," http://newventurist.com/2013/07/innovation-in-speech-recognition/, accessed October 1, 2021.

109. Reid Smith, Eric Schoen, and Jay Tenenbaum, "Early AI Applications at Schlumberger," *IEEE Annals of the History of Computing*, January–March 2022, 88–102; Andrew Pollack, "Selling Artificial Intelligence," *New York Times*, September 13, 1982, D1; Steven Marcus, "Computer Systems Applying Expertise," *New York Times*, August 29, 1983, D1.

110. Kenneth N. Gilpin and Todd S. Purdum, "Digital Officer to Fill Chief's Post at LISP," *New York Times*, May 27, 1985, 34.

111. "C. Itoh, Gensym Announce Japanese Distribution Agreement," *Business Wire*, February 24, 1988.

112. Andrew Pollack, "Selling Artificial Intelligence," *New York Times*, September 13, 1982, D1.

113. Smith, Schoen, and Tenenbaum, "Early AI Applications."

114. Nilsson, *The Quest for Artificial Intelligence*.

115. Smith, Schoen, and Tenenbaum, "Early AI Applications"; Subrata Dasgupta, *The Second Age of Computer Science: From Algol Genes to Neural Nets* (Oxford: Oxford University Press, 2018).

116. Smith, Schoen, and Tenenbaum, "Early AI Applications."

117. Smith, Schoen, and Tenenbaum, "Early AI Applications."

118. Smith, Schoen, and Tenenbaum, "Early AI Applications."

119. J. Scot Finnie, "Sudden Shower Enriches MIS Turf," *Computerworld*, October 17, 1988, 77; Steven J. Marcus, "Firms Begin to Industrialize Artificial Intelligence Systems," *The Globe and Mail*, September 30, 1983; Kenneth Brooks, "Process Simulators Give Improved Training, Plus," *Chemical Week*, October 22, 1986, 24; Kenneth Brooks, "AI Tackles Real-Time Process Control," *Chemical Week*, September 10, 1986, 38; "Wary but Interested, Construction Stretches to Make the AI Connection," *Engineering News-Record*, March 28, 1985, 20; "LISP Machine; Gains $7.6 Million in Equity Financing; Add $2.5 Million Lease Line," *Business Wire*, December 5, 1984.

120. For example, Rawleigh Warner [chair of Mobil], "U.S. Energy: To Have and to Have Not?" speech to the World Affairs Council of Philadelphia, November 6, 1980, found in a binder with call number HD9569.M62.Z99, or C.C. Garvin Jr. [chair of Exxon], "The Energy Outlook: Near and Long-Term," speech to University Club, New York, November 14, 1980, in binder with call number HD9569.E99.Z99, both at Baker Library, Harvard Business School.

121. Douglas Martin, "Exxon Abandons Shale Oil Project," *New York Times*, May 3, 1982.

122. Thomas C. Hayes, "Why Socal Sought Oil on Wall Street," *New York Times*, March 7, 1984.

123. Richard S. Ruback, "The Cities Service Takeover: A Case Study," *Journal of Finance* 38, no. 2 (1983): 319–330.

124. Richard S. Ruback, "The Conoco Takeover and Stockholder Returns," *Sloan Management Review* 23, no. 2 (Winter 1982): 13–33; Steve Coll, *The Taking of Getty Oil: Pennzoil, Texaco, and the Takeover Battle That Made History* (New York: Scribner, 1987).

125. Gary Putka and Brenton Schlender, "Mesa's High-Stakes Takeover Fight Reflects Chairman's Style," *Wall Street Journal*, June 7, 1982; Coll, *The Taking of Getty Oil*; Steven Brill, "The Roaring Eighties," *The American Lawyer*, May 1985, 10–19; Connie Bruck, *The Predator's Ball: The Inside Story of Drexel Burnham and the Rise of the Junk Bond Raiders* (New York: Simon & Schuster, 1988).

126. Jeff Madrick, *Taking America: How We Got from the First Hostile Takeover to Megamergers, Corporate Raiding, and Scandal* (Toronto: Bantam, 1987).

127. Charles F. McCoy, "Phillips Petroleum Is Seeking to Sell Coal Reserves, Geothermal Operations," *Wall Street Journal*, May 3, 1985.

128. Letter from Edward E. David (president of ER&E) to Dick (last name unknown), September 1, 1982, in Folder "August–October 1982," Box 20, Edward E. David Jr. papers (MC-0582), Distinctive Collections, MIT Libraries.

129. John Metzler (director, Energy Research Advisory Board, Department of Energy), vugraphs for presentation "The Impending United States Energy Crisis," November 14, 1986, in Folder 8, Box 36, Ruth Patrick Papers, Academy of Natural Sciences; Michel T. Halbouty,

"Mergers and Hostile Takeovers: Effects on R&D Programs in the Petroleum Industry," *Issues in Science and Technology*, Winter 1986, 15–16.

130. Cyrus C. M. Mody, "Surveying the Landscape: The Oil Industry and Alternative Energy in the 1970s," in *Electrical Conquest: New Approaches to the History of Electrification*, ed. Erik Conway and W. Bernard Carlson (Cham, Switzerland: Springer, 2024), 51–79).

131. A representative forecast is contained in *The National Energy Outlook 1980–1990* (Shell, 1980) in a binder with call number HD9502.U52.Z99, Baker Library, Harvard Business School.

132. Naomi Oreskes and Erik M. Conway, *Merchants of Doubt: How a Handful of Scientists Obscured the Truth on Issues from Tobacco Smoke to Global Warming* (New York: Bloomsbury, 2010).

133. Jacobs, *Panic at the Pump*; Matthew Huber, *Lifeblood: Oil, Freedom, and the Forces of Capital* (Minneapolis: University of Minnesota Press, 2013); Caleb Wellum, *Energizing Neoliberalism: The 1970s Energy Crisis and the Making of Modern America* (Baltimore: Johns Hopkins University Press, 2023).

134. Philip Mirowski, *Science-Mart: Privatizing American Science* (Cambridge, MA: Harvard University Press, 2011).

CHAPTER TWO

Taller Than a *T. rex*

Celebrity and Leftist Politics in the Public Career of Stephen Jay Gould

Myrna Perez

When Stephen Jay Gould was five, his father, Leonard Gould, took him to the American Museum of Natural History (AMNH) on Central Park West in Manhattan. As he later described the experience, Gould was awestruck by the skeleton of the *Tyrannosaurus rex* and seized by a fascination with paleontology and evolution that lasted for the rest of his life.[1] His father not only encouraged Gould at home but also wrote several times to the staff at the AMNH, asking advice about the high school and university preparation necessary for young Steve to pursue paleontology. One letter written in 1957 from Leonard to Allison Palmer, a US Geological Survey paleontologist at the AMNH, referenced the family story of Steve and the *T. rex*: "As I may have told you in my original letter, from the age of 5, when I took Steve for a look at the dinosaurs in the N.Y. Museum of Natural History, that has been one subject about which an otherwise annoyingly-indecisive family has had little or no doubt."[2] The Gould family even went fossil collecting in California's Mojave Desert during the summer of 1957 to satisfy the teenage Steve's dinosaur mania.[3]

That moment both inspired Gould's career and became an important part of his transformation from a junior professor of paleontology at Harvard in the late 1960s into an international scientific celebrity in the glittery world of 1980s popular science media. Along the way, the story of the *T. rex* grew into a deliberate piece of self-fashioning for Gould to emphasize his fascination with the history of life to his audiences. As with all larger-than-life figures, Gould needed an origin story. The story disappeared and reappeared throughout Gould's life; it changed shape and adapted to the circumstances of his career.[4] He became taller than a *T. rex* through a set of relationships that he formed through his column "This View of Life," which ran in *Natural History* magazine from 1974 to 2000. The column, the books that anthologized his essays, and the attendant press provided Gould with a regular source of public engagement and gave him a forum for airing his ideas to an audience beyond his discipline. It set the foundation for Gould's public intellectual voice and made him a worthy subject of a *NOVA* special. By studying the material circumstances of Gould's growing celebrity, we can learn something about the transformations to the persona of the public scientist from the activist 1970s into the greedy 1980s.

When Gould began writing his *Natural History* column, his expressed motivation was not fame and fortune but to have a venue in which to engage with "social and political questions bearing upon scientific issues."[5] His first efforts weighed in on topics at the intersection of biological research and social contentions. In the early 1970s, for instance, he dedicated columns to the relationship between race and IQ, the biological study of human aggression and racial diversity, and leftist criticisms of E. O. Wilson's 1975 book *Sociobiology: The New Synthesis*.[6] The column was one kind of response to the generational challenge faced by Gould and other leftist academics of his time—it allowed them to reimagine their intellectual work as part of broader cultural efforts to reform American society. Gould came to Harvard fresh from his student activist days in the civil rights and anti-nuclear protest movements. Writing a public science column was a deliberate attempt to bring that political framework to his work as an evolutionary scientist. How did that column—touted by the radical science organization, the Sociobiology Study Group, as one of its publications—become the basis for Gould's scientific celebrity and land him a feature in *People* magazine by 1984? And what does this tell us more broadly about the transformations of the academic left after the protest years during the greedy decade of the 1980s?

Between 1975 and the end of the greedy 1980s, Gould's public writing moved from venues and subjects that existed alongside leftist work in the academy to products that were barely recognizable as political efforts. What accounts for this change? There are many possible answers, of course—ranging from personal motivations to the new political landscape of the era. For the purposes of this volume, however, I am going to concentrate on one specific aspect: the creation of a Gould "brand," which was developed through television productions and further marketed and transformed as Gould's career moved from the world of public science writing to full-blown celebrity.

Gould's transformation into a celebrity happened in a series of (somewhat unexpected) opportunities, and certainly aspects of it were mostly out of his control. For instance, a significant driver of his notoriety in the early 1980s was the dramatic surge of public interest in evolution, which was motivated by the new creationism of the Christian Right. But although Gould was alternatively gracious and irascible when it came to the public interest in his life and his work, it is undeniable that he sought public venues to express what he contended was a unique view of the evolutionary process. Over the course of his career, Gould came to argue that evolution was characterized more by historical contingency than by the engineering design of selection. And he furthermore asserted that understanding evolution in this way was a matter of moral urgency because it allowed us to set aside our collective faith in evolutionary progress, making it possible to also dismantle our conviction in racial hierarchies.

Public science writing was Gould's answer to the New Left dilemma of how to reimagine the American academy as an institution oriented toward social justice. And at the outset of his writing for *Natural History*, his solution to this cause existed in relative harmony alongside other leftist actions to create a science for the people—including protests, public workshops, teach-ins, Black Panther health care networks, and the activism of the Combahee River Collective. Even at the height of the sociobiology controversy, Gould was not the most politically active member of the radical science movement—but his writing was one attempt to give voice to its political causes. But this writing also laid the foundation for the creation of a "Gould brand" that was cultivated, packaged, and circulated to audiences through Gould's material experiences with editors, journalists, and marketing departments. By the end of the 1980s, Gould's public science writing had earned him the status of a celebrity—whether it had also carried forward the political convictions of a generation of leftist scientists was less certain.

Science Writing as Leftist Politics

Gould grew up the grandson of Hungarian immigrants in the Queens borough of New York City. This background, along with his father's work as a Marxist social organizer and service on the local board of the National Association for the Advancement of Colored People (NAACP), were a foundation for a leftist approach to social change in Gould's early years.[7] He took this political perspective with him to Antioch College, where he was a member of student activist organizations, including Students for a Democratic Society (SDS), the Student Peace Union, and the Congress for Racial Equality (CORE).[8] It was through these organizations that Gould participated in integration sit-ins and protests against nuclear armament.

Not long after his final sit-in, he began graduate school at Columbia University as a student of the paleontologist Normal Newell. By the fall of 1967, he had filed his dissertation and joined the faculty of Harvard's geology department.[9] Like many other universities across the country, the Harvard campus was filled with political protests and social unrest. A Harvard-Radcliffe chapter of SDS had formed in 1964, largely to protest the escalating draft calls emanating from the Gulf of Tonkin Resolution.[10] The Harvard SDS organized its first antiwar march that spring, prompting President Nathan Pusey to comment on "a new and rather disturbing seriousness of tone" in the student agitations against Vietnam War.[11] Harvard SDS organizers formulated an antidraft program and encouraged students to file as conscientious objectors to the war. Over the next few years, the rhetoric of the organization became increasingly militarized, prompting a physical confrontation between Harvard undergrads and a visiting government dignitary in the fall of 1966.[12] By 1968, a series of national events—including the assassinations of Martin Luther King Jr. and Robert Kennedy—caused the situation to boil over in student communities across the country and around the world. At Harvard, students marched against the university administration, demanding the end of "blatantly racist policies" and the eradication of the university's ROTC program.[13] Into the early 1970s, this atmosphere of radical politics infused the campus culture. And it was in this setting that Gould began writing his column for *Natural History* magazine.

When Gould was approached by Alan Ternes in the fall of 1973, he was already familiar with *Natural History* from his time working at the magazine's institutional home at the AMNH. As editor of the magazine as well as director of the museum, Ternes brought his experience as a journalist for publications such as *Stars and Stripes* and the *Times Herald Record* to his work cultivating popular writing with working scientists.[14] After a conversation in the summer of 1973, Gould wrote to Ternes, "I hope that you haven't read my month of silence as a sign of doubt or indifference to the prospect of writing regularly for *Natural History*. Quite the contrary: the prospect intrigues me more and more and I have every intention of pursuing it."[15] He suggested that the column would be "firmly based in evolutionary theory and its implications but trying to synthesize under that rubric my divergent interests in the history and philosophy of science, social and political questions bearing upon scientific issues, and the phenomena of life's history on a grand scale." According to Gould, this combination would "lead to a potpourri of articles," including "straight and quaint pieces of old-fashioned natural history" and "controversial diatribes on the politics of pop ethology."[16] From its earliest moments, Gould saw the column as space to combine his primary intellectual motivations: revolutionizing evolutionary theory, engaging with the history of science, and expressing his dissatisfaction with what he thought were empirically vacuous scientific theories that lent support to racism.[17]

A month after Gould's proposal, Ternes sent Gould a hurried reply apologizing for the intervening period of silence. His letter revealed the flurry of work that was involved in the routines of working with magazine deadlines. "I have been, I am sorry to say, too busy to really sit down and analyze the mix of pieces you suggested," was Ternes's only substantial response to the list Gould sent him. But he needed some "personalized columns," having had difficulty getting anything out of another potential contributor, so he felt that Gould may as well make a definite move and "try a couple." Ternes added pragmatically, "When can I expect the first one?"[18]

Ternes's eagerness for "personalized columns" was largely due to the mandate of *Natural History* itself. Originally founded in 1900, it had originally been conceived as a venue to promote the expeditions and research of the AMNH, but in the latter half of the twentieth century, the publication transformed. By the time Gould started writing his column, the magazine had become one of several popular science magazines (along with *Scientific American, Discover, Smithsonian,* and *National Geographic*) that commanded a significant audience.[19] *Natural History* found its niche among the competition by focusing on the scientist's gaze.

Since the magazine wanted the perspective of scientists, the editorial staff needed to find working scientists who were willing and able to write regular columns. Ternes's willingness to have Gould try a few columns was borne out of the need for fresh copy and not necessarily an assumption that Gould would become a long-standing columnist.

With the go-ahead, Gould began working on three essays for the magazine. Less than a month later, he sent Ternes a draft of all three, in a letter bubbling over with suggestions for future columns: "I can hardly believe it myself, but here (and on time) are the three columns I promised a month ago."[20] He sent all three to Ternes at once "to give you some idea of how I might manage a loosely coherent series."[21] His proposals for future columns included "a critique of 'biological determinism'" in the work of the ethologist Konrad Lorenz and Robert Ardrey, as well as in the writings of Richard J. Herrnstein and Arthur Jensen. Along with Ardrey, Lorenz argued that aggression was a fundamental driver of humanity's differentiation from earlier anthropoid species.[22] Gould was eager for a chance to add his voice to the "criticism (and anger) directed towards these views" and believed that he would bring a fresh critical perspective by countering these theories "in a unitary way."[23]

In the early years of the column, Gould's writing was edited by Ternes to meet the perceived needs of the magazine. Interacting with Ternes helped Gould to adapt to the pace of the publication but also the experience of having his work thoroughly edited.[24] In their correspondence, Ternes wrote authoritatively to Gould—Gould was neither a celebrated writer nor a scientific celebrity in 1974. He was simply an enthusiastic paleontologist, whom Ternes hoped would have enough interest and aptitude to become a regular columnist. In the years to come, Ternes would continue to be an active and forthright editor of Gould's work. In particular, he watched that Gould did not stray into an overly academic tone. For instance, an essay from 1975 earned Gould these comments from Ternes: "This was, I think one of the hardest editing jobs you've presented to me.

It has, as you'll see, been worked over heavily. I would rather not do such editing, but something had to be done, in my opinion."[25]

When asked by a writer's magazine in 1991, "Whom do you see as your audience for your essays?" Gould replied, "All I can do is infer from people who write letters, which is an obviously biased percentage of those who read them. I like to think of my essays as equally available and useful to three groups of people: to intelligent laypeople . . . to students, and to professionals."[26] And Gould's impression of his readers did come largely through the mail he received in response to the columns each month, either sent directly to him or forwarded by the editorial staff. Over the course of the column, Gould received thousands of letters from a variety of people. Many were fellow biologists and other natural scientists or academics; others were simply interested laypersons. In the decade after the start of the column, it was letters to and from readers along with correspondence with professional biologists that equally dominated Gould's daily letter writing. This continual interaction with a reading public was often touted by Gould as evidence for the persuasiveness and importance of his theoretical and social ideas.

As Edwin Barber would later relate, he was reading old issues of *Natural History* magazine in the New York public library when he ran across Gould's monthly column. He contacted Gould, demanding to know why Gould didn't have a book in print. Barber was an editor for Norton, which was the largest independently owned publishing company in the country. The publishing house specialized in several areas (including fiction, nonfiction, poetry, college textbooks, cookbooks, art books, and professional books) but prided itself on releasing titles meant "not for a single season but for the years" in both trade and textbook offerings.[27] Eventually, Gould published ten books with Norton, the first five anthologies of his column, two popular science books, and two collaborations with photographer Rosamond Purcell. Importantly, the book anthologies created an experience of Gould's ideas that was tidier and more manageable than the columns. Rather than encountering a column monthly, a reader could pick up an anthology and read through Gould's writing in one coherent experience. But when Gould's first book anthology, *Ever Since Darwin*, was published in 1977, he was relatively unknown. The book was promoted as a pleasure read in places such as the *New York Times Book Review* and *Notable Books* but not as a book by a celebrity author.[28] *Ever Since Darwin* did not make Gould an overnight literary sensation, but it did enter him into the editorial world of book publishing and publicity. The first anthology opened an important set of doors that set the precedent for his nonfiction writing for the next decade.

Both Barber and Gould were concerned that the book should appeal to a large audience, from the practical desire that it be a commercial success. Gould felt that there was a good chance that it could reach a variety of readers. As he wrote to Barber, he based this idea partly on his existing readership, "I doubt that you would lose on the book (N.H. readership is nearing half a million, and I would be surprised if you didn't sell a couple of thousand right there)," and partly on the market for college texts that he was anecdotally aware of: "I am also certain (based on the expressed desire of many colleagues who teach courses in evolution) that it would have a respectable sale . . . as a supplementary text . . . for courses in evolutionary biology."[29] But both Barber and Gould wanted the book to be appealing and accessible to a general lay audience—readers who were difficult to characterize and perhaps even more difficult to reach.

After months of planning and editing, both Barber and Gould were pleased that reviews of the book were highly favorable. After the book's release, Barber captured the reviewers' tone in a note to Gould: "I hesitate, really, to send along this review from the *Chicago Tribune*, for fear it will turn your head. They do seem to have liked the book. Actually, go ahead and let your head turn all it will."[30] And the early book reviews were venues for a variety of audiences to assess and categorize Gould along with his writing. Many of the reviewers focused on the potential contradiction between Gould's identity as a professional biologist and a writer. Jim Buckley, a reviewer for the conservative magazine *American Spectator*, proclaimed, "It is Mr. Gould's special gift to be able to illuminate often complex scientific ideas with a special blend of erudition, wit and clarity of expression."[31] His facility as a writer was discussed as an effortless, natural-born gift that marked him out from other scientists. The stories about Gould's life and background helped to turn Gould into a character for readers—one who was relatable and interesting. But the descriptions of his professional life, particularly his technical work, cast Gould as more than a science writer. In reviews such as these, Gould promoted his identity as a serious professional scientist and as a defender of his audience's interest against the potential biases and hubris of the scientific establishment.

Gould's identity, however, was neither essential nor immediately obvious. Reviews of his books, beginning with these initial pieces, were spaces in which his identity was formed and negotiated. In their attempt to express why Gould's writing was good and interesting, the reviews used language about Gould that was repeated and codified. It is not the case that Gould formulated an identity where he worked in the halls of Harvard's Museum of Comparative Zoology that was then projected out to the American public. The line between public writer and private scientist was forged in the language of these reviews, assessments, and even interviews of Gould. Although they differed, depending on their region and readership, most of the reviews settled on a central message about Gould. Here was a professional scientist—concerned with the esoteric, even dull—who had the "natural" gift to write like an angel. Another strategy for explaining Gould's place in the world of science writers was to couple reviews of *Ever Since Darwin* in pieces on other already well-known scientific popularizers. The *Chicago Sun Times* combined their review of *Ever Since Darwin* with a review of *Origins* by Richard Leakey.[32] This was a signaling device to potential readers. It was a way of giving context for Gould and what sort of reader might be interested in his work.

These reviews of the first anthology laid a foundation for Gould's later fame, but they did not make him an overnight sensation. Gould was not a household name, but he was becoming a prominent academic with greater than average public recognition. The hardcover edition of the book sold about 9,000 copies. As Barber put it, "We go along nicely with sales."[33] But *Ever Since Darwin* was nothing like an immediate sensation. It did modestly well and was featured in literary publications such as the *New York Times Book Review*.[34] It was the beginning of Gould's introduction into a particular kind of literary sphere—one that reached larger audiences than *Natural History*. The book reviews provided an opportunity for Gould to feature at least once in almost every daily and weekly periodical in the country, everything from *Nature* to the *New York Times* to the *American Socialist*. Since the reviews appeared in not only major mainstream dailies but also numerous regional newspapers and special interest magazines, the reviews served as a grassroots publicity campaign to the various potential audiences for Gould's work.[35] The spread of the reviews was not very unusual for a nonfiction book for a general audience, but it would have been almost unthinkable for an academic monograph. By the publication in

1980 of his second anthology, titled *A Panda's Thumb*, Gould began to develop a niche for himself—a scientific popularizer who was unique for his insight into evolutionary theory and his continued original scientific research.

A Growing Celebrity

In the summer of 1984, Gould was still recovering from his bout with cancer. He had been diagnosed with peritoneal mesothelioma in July 1982, an experience that resulted in one of the most heartfelt articles of his career.[36] His illness had temporarily placed a hold on his usual activities, although he faithfully published his *Natural History* columns through the seven months of intensive treatment. When he was approached by the television series *NOVA* that summer to feature in a special, there was much to consider for both his usual career responsibilities and his still weak health.

Gould was well acquainted with *NOVA*, having served on its board since 1975.[37] Part of the vision of *NOVA* was to convey to the general public in the United States the best that American science had to offer. Thus, it drew upon the expertise of academics in the Cambridge area, such as Gould. The popular science television show, created in 1974 by Michael Ambrosino, was intended to be broadly educational for the American public. Ambrosino's ambition was to fill what he saw as an "appalling gap" in the offerings on the Public Broadcast Station.[38] In his proposal to the vice president for programs at WGBH, he stated that the show "would attempt to explain and relate science to a public that must be aware of its impact."[39] *NOVA* was a part of a concerted effort in the mid-1970s to harness government funding, academic expertise, and public good will to translate current science and technology findings to the American public. *NOVA* went on to be the longest-running documentary series in American history (with over twenty-five years on the air) and the recipient of numerous awards, including Emmys and Peabodys.[40]

The film required many hours of filming, and the whole production occupied eight months during 1983. The director and producer was Linda Harrar, who had been with *NOVA* since 1977. At the center of the documentary was Gould's childhood story of the encounter with the *T. rex*. While standing above the *T. rex* in the hall of the American Museum of Natural History, Gould opened the program by directly addressing the television audience. He began with his own origin story: "Most naturalists get their start in the country. But for a handful of city street kids they fall in love with the natural world by standing near specimens like these. It's not rare for kids to see dinosaurs and want to study them. What's rare is to stick with it."[41] Woven together in this comment was Gould's assertion of his identity as just a "city street kid" who nonetheless had a unique capacity for science and a potential for greatness. Harrar also arranged for Gould to meet his childhood hero, Joe DiMaggio, on screen (and therefore was able to capture Gould playing catch with his young son Ethan on film). The segment of the film showed Gould as "a regular guy," really just a grown-up boy from Queens who idolized only Darwin above his baseball hero.

The film concentrated on the multifaceted nature of Gould's identity, both to explain his own understanding of what it was but also to capture his audience's imagination. His childhood influences were examined: his father, Joe DiMaggio, and Charles Darwin. His credentials as a working naturalist were touted. "As a working scientist, Gould adds to the data of his field by intensive study of the Bahamian land snail."[42] As Gould explained to the audience, calm and straight-faced, "It takes years and years to get to know a creature well, and I love Cerion with all my heart and intellect."[43] He evoked the spirit of the naturalist, inviting the reader into a sympathy with the natural world, with a particular cognitive orientation toward scientific discovery: "The pure joy of discovering something new and understanding a creature as well as anyone in the world has ever understood it."[44] Gould as the scientist-expert had a love of peering into the complexities of the natural world that somewhat escaped the purview of his audience, but he shared it with them.

The film helped to cultivate Gould as a brand—one that combined his views on various evolutionary topics but also created him as a figure who could be recognized, packaged, and circulated. The program performed typically well for a *NOVA* documentary, although not spectacularly.[45] Gould's readership in *Natural History* was half a million, whereas the numbers for the *NOVA* special ran in the several million and across various markets (including Los Angeles, Chicago, San Francisco, and New York).[46] Gould was by no means an instant household name. But his exposure through the film was much greater than before. Also, the film bundled together the ideas found in several of Gould's published works. To get the full breadth of what the film covered in the space of an hour, a person in the general public would have had to read countless professional articles, a decade's worth of his column, and his books *Ontogeny and Phylogeny* and *The Mismeasure of Man*. The film put Gould into a package that could be remembered and enjoyed, introducing even more of the American public to the paleontologist from Harvard.

Two years after the *NOVA* special aired, Gould was featured in the June 2 issue of *People* magazine. Although he didn't make the cover (which was reserved for "*Top Gun* Heartthrob Tom Cruise" and "Great Trash for Summer"), the magazine devoted a full spread to Gould, complete with black-and-white photographs that captured several intimate family moments. He had truly arrived as a public icon—an evolutionary expert, a Harvard professor, and a celebrity. The centerpiece of the article, "Stephen Jay Gould: Driven by a Hunger to Learn and to Write," was Gould's courage in facing cancer but also his scientific critics. "When he arrived with his freshly minted Ph.D. from Columbia, the rumpled, kinetic Gould was an exceptionally promising paleontologist; in the years since, he has become a popular symbol of erudition and scholarship."[47] The piece recast Gould's professional career into a heroic arc—Gould attained evolutionary enlightenment by doing battle with the forces of racism, creationism, and even cancer. "He has done battle with creationists, testified before congressional committees concerning nuclear winter and lectured in South Africa on the history of racism."[48] And as always, there was the story of the *T. rex*: "Steve chose his career at the age of 5, when his father, a court stenographer, took him to the Hall of Dinosaurs in the American Museum of Natural History. The towering skeleton of Tyrannosaurus rex triggered his decision. 'I had no idea there were such things—I was awestruck,' says Gould. Although he didn't know the word, he left the museum determined to become a paleontologist." This story was no longer a simple device to lead readers into Gould's world. No matter how true, the anecdote had become part of Gould's public personality, a crucial piece in his narrative of self-discovery and scientific enlightenment.

In the decade before a significant transition in 1995, Gould gained increasing public success through his popular writings, magazine features, and television appearances. During this period, Gould managed, largely on his own and with the assistance of his faculty secretary Agnes Pilot, several different careers.[49] He was a professional paleontologist who published in technical journals, a popular writer who wrote books on various topics and for literary magazines, and a media personality who appeared on television. But in 1995, Gould switched to a new publishing house, Crown Harmony, which would publish the remainder of the anthologies of his column. The transition to Harmony signaled a general shift in how Gould's time was spent, how much he earned for his appearances, and the general tone and character of his correspondence and public activities. This was a shift that affected only the last seven years of Gould's career—but this was the period in which Gould became a celebrity of a different caliber than he had been in the 1980s, even when he did make the cover of *Newsweek* and appeared in *People*. By the middle of the 1990s, Gould was rubbing elbows with an increasingly famous crowd—with prestige far beyond what most other academic biologists were acquainted with.

Throughout his life, Gould kept most of his correspondence, variously organized, and habitually saved copies of his outgoing letters as well as those he received.[50] He began to make a habit of keeping letters from "notable" correspondents in a set of folders. At first, this folder primarily contained other noteworthy biologists—figures such as E. O. Wilson, Richard Dawkins, Sewall Wright, George Gaylord Simpson, Ernst Mayr, and Ledyard Stebbins. But as his career progressed, this folder collected letters from a wider range of persons: literary figures and artists, actresses, politicians, judges, and professional sports players. In the decade and a half before his passing, Gould exchanged letters with a variety of famous authors and other celebrities, including Jean Auel, Julia Child, Noam Chomsky, Stephen King, Gary Larson, Glen Close, Alan Alda, President Bill Clinton, First Lady Hillary Clinton, Vice President Al Gore, Henry Kissinger, and Justice Stephen Breyer. Some of these correspondences were more sustained than others and represented either friendships or collaborations. Cartoonist Larson, for instance, wrote Gould to express his "profound thanks for the extremely flattering (and entertaining) foreword to 'The Far Side Gallery III.'"[51] Stephen King and Gould had a long-standing, affectionate correspondence. King sent extracts of stories to Gould for Gould's children, reported on the doings of his own, and sent Gould funny postcards. In 1992, King sent Gould an advance copy of Michael Crichton's *Jurassic Park* with the note, "You may have a copy of this by now . . . I think you'll get a kick out of it. A few too many cute 'Spielberg moments' to suit me, but Crichton appears to have his facts fairly straight and he's *playful*."[52]

King no doubt thought that Gould would be amused by the prominence that paleontology played in the fictional thriller.

Perhaps the most significant sign of Gould's entrée into rarefied circles during the 1990s was the attention he received from the current president, Bill Clinton, and his wife, Hillary. Gould was triumphant when Clinton defeated George H. W. Bush in the 1992 election—writing in various letters that he was "thrilled to vote for someone he actually liked for once."[53] Gould commented often on his high regard for Clinton, and he occasionally received personal notes from the president, including one in the winter of 1994 in which Clinton professed, "Thanks so much for the signed copies of your book . . . Hillary and I were delighted to hear from you again."[54] Gould was invited by the Clintons to the presidential inauguration and was photographed alongside them in the late 1990s.[55]

In the late 1970s and early 1980s, Gould's *Natural History* column had entered him into a different world outside of professional science academia—he was newly connected to a landscape of publishers, filmmakers, broadcasters, editors, and producers. By the 1990s, he had arrived in a world of true celebrity. In the final years of Gould's career, the first world to which he had belonged, the scientific circle, began to occupy less of his time and energy. This is an empirical observation, not a subjective evaluation—but it is clear from a survey of Gould's correspondence in the late 1990s that he could no longer claim, as he had in a letter from 1979, "that he wrote to several biologists around the country just a matter of course over the past week."[56] Gould maintained his place in biological circles. He continued to teach at Harvard at least once a week until his final illness, and then the *Structure of Evolutionary Theory* captured much of his attention in the last decade of his work. But Gould became increasingly part of a literary set—writing more and more reviews for the *New York Review of Books*, as well as appearing as a talking head on *Charlie Rose* and on *CNN Talkback Live*. He was an intellectual luminary who perhaps only happened to be an evolutionary biologist.

Celebrity versus Social Justice

Gould began writing for *Natural History* in 1974, his first Norton anthology *Ever Since Darwin* was published in 1977, and *The Mismeasure of Man* was released (also by Norton) in 1981. At the outset of this public work, Gould had expressed his motivation to have a space in which to address broad questions of evolutionary theory as well as the relationship between science and politics. And by all accounts, in these years, he was eminently successful in this endeavor. It was in this period that the strategy of public science writing as a venue for justice seemed most harmonious with other leftist work in the academy. Although Gould was not the most active member of the radical science circles that worked in Boston or New York, he had a conversant relationship with Science for the People and with the Sociobiology Study Group. His publications, particularly his columns on sociobiology, race and IQ, and, of course, *The Mismeasure of Man*, were broadly understood to be publications that expressed the leftist arguments that science had been part of a history of racial and sexual oppression in the West.

We have already seen, however, that Gould's engagement with public science writing was materially shaped by his relationships with editors and by marketing departments. Over the course of the next two decades, the subjects of Gould's public science writing would drift away from the urgent political topics of the civil rights era. Although he continued to express a conviction that the structure of evolutionary models shaped our understanding of society and human nature, these points were expressed less boldly than in his earlier works. Consider, for instance, another stand-alone nonfiction work he published with Norton in 1989, *Wonderful Life: The Burgess Shale and the Nature of History*. This book, like *The Mismeasure of Man*, took aim at the question of race and science. But it did it much more obliquely, by arguing that it was the logical structure of a pervasive account of evolutionary history that was responsible for the tendency toward racism in Western thought. Gould's argument in the book was for the vital importance of historical contingency—the view that history was not merely the unfolding of a set of predetermined steps in a teleology designed by the engineering prowess of natural selection. Viewing evolutionary history as inevitably ordered by the work of natural selection, Gould argued, led us to believe in evolutionary progress. And faith in evolutionary progress, he further argued, would always lead to a belief in racial hierarchies.

Because of this, it is of course possible to read *Wonderful Life* as a continuation of Gould's early work at the intersections of race, science, and social justice. But the book was not marketed in this way, nor is it generally now understood to be part of the canon of works about the politics of race and the history of science. And as Gould's career went on, his treatment of the subject became increasingly oblique. A 1996 book that also addresses the question of progress and racism, *Full House: The Spread of Excellence from Plato to Darwin*, is hardly understood to be about race by most readers and publicists.

The relationship between scholarship, social justice, and celebrity is a pressing issue as we consider the ongoing reimagining of universities, particularly in the political context of the United States. Historians of science, such as Andrew Jewett, have argued for the vital role of universities in adjudicating the relationship between science and democracy in the United States since the Civil War.[57] As we begin to more fully assess the relationship between global New Left movements and the place of science in the American academy, understanding how and why leftist approaches to science institutionalized in universities is vital. Following scholars such as Jenna Tonn, my own work has emphasized the unique role that feminists had in transforming the radical science movement from activism into scholarship during the 1970s and 1980s.[58] Feminists in the natural and social sciences, as well as in the humanities, translated the work of collectives like the Genes and Gender group into feminist epistemologies of science and society. Figures such as Ruth Hubbard and Marian Lowe conducted workshops, held teach-ins, and also published collected volumes and special issues in new feminist journals such as *Signs: Journal of Women in Culture and Society* on the ways that science must be reimagined and understood through a gendered analysis of its history and epistemology.[59]

In other words, there were other models of justice and science available in the 1970s besides the mode of public science writing that Gould pursued. Feminists, of course, also followed the route of individual popular science publications. But feminist efforts were more squarely directed at collective action and communicated through the more general effort to institutionalize feminist studies in the academy. This meant the establishment of undergraduate and graduate programs and the creation of faculty lines devoted to feminist and feminist science studies. These efforts held the possibility of long-standing change in university institutions and structures that could be communicated across and be built by generations. And feminist science studies also benefited from the ongoing influence of the broader field's debates over the relationship between activism and scholarship, between community and research, between public and university. That is not to say that all this was ideal. The relative dearth of communication between the radical science movement and the emergence of critical race theory and Black studies in the early 1980s, for instance, is still arguably echoed in the disciplinary structure of feminist science studies and the history of science. My point, though, is that there were other possibilities for Gould than public science writing—other means by which he could have attempted to bridge the gap between professional scientific research and the political work of social justice.

Most of the assessment of Gould's science and legacy has focused on the tensions between his identity as a practicing scientist and his blossoming roles as a public intellectual and scientific celebrity.[60] This type of analysis is important for our understanding of the changing persona and public role of "the scientist" during the late Cold War, particularly in the United States. But in this chapter, I have wanted to understand a different dynamic—that is, the tension between Gould's celebrity and his expressed conviction that the academic work could be repurposed for social justice. Like many other academics of his generation, Gould came to work as a professor after engaging in New Left protest movements during his undergraduate years. It was only a few short years after anti-nuclear protests and civil rights sit-ins that he began writing for *Natural History* in 1974. But by the end of Ronald Reagan's presidency, Gould's public science writing had created a brand that celebrated the genius of scientific innovation and the glories of evolutionary thinking. His leftist critiques of the history of scientific racism and sexism had been subsumed underneath the glossier covers of his public brand.

I suggest there are two main factors for this transformation, beyond any speculation about Gould's individual ambition or psychological motivations. The first, as we have seen, was the structure of the publishing industry and media markets during the 1980s that produced a "Gould brand" across his columns, trade publications, and television appearances. The second was the broader political context of the 1980s that drove public interest in evolutionary science—especially the rising prominence of creationism in the context of the growing political power of the Christian Right.

Creationism didn't pop out of thin air in 1980. After all, religious responses to Darwinian evolution stretch back before the first American edition of *On the Origin of Species* was published in 1860. Historians of

American creationism have worked carefully to document the deep and complex roots of the movement, to differentiate the diverse theological perspectives that gave rise to creation-science and locate it in the changing landscape of American Protestantism during the twentieth century.[61] Moreover, this scholarship has been vital in emphasizing that evolution and creationism are not timeless entities at constant war and that conflict between them is the result of highly specific cultural contexts. But it was in this moment that creation-science collided with the world of evolutionary biology and expanding markets for popular science.

The igniting event came in 1981, when the case *McLean v. Arkansas* came before the Arkansas State Supreme Court. The media coverage of the trial propelled young-earth creationism to national attention in a fashion unseen since the infamous Scopes Monkey Trial of 1925. *McLean v. Arkansas* was a response to a creation-science bill passed by the Arkansas state legislature that mandated "balanced treatment for creation-science and evolution-science" in Arkansas public schools. After the American Civil Liberties Union (ACLU) brought a suit against the state, Gould was one of a number of expert witnesses at the trial who helped the ACLU make the case that creation-science was not science but religion. The plaintiffs carried the day, but the trial sparked an avalanche of attention. The increase in public discussion of creationism outside of conservative religious circles was remarkably rapid. This was evident in the dramatic rise in the use of the term "creationism" in the American book market in the years directly after the *McLean* trial. And unlike the 1920s, this antievolution movement was propelled into a colorful world of 1980s television programs and glossy magazine covers. Creationism featured in popular science outlets such as *NOVA*, *Discover*, and *Scientific American*. It splashed across the pages of *Newsweek*, *People* magazine, and news dailies. Although the creation-science movement had been building for decades, this was the moment when the entire country began to pay attention.

Crucially, creationism was more than an attack on evolutionary science; it was part of the political aims of the new Christian Right. One of the key leaders of the Christian Right, Tim LaHaye, was also a founder of the Institute for Creation Research, the primary institutional home of creation-science. LaHaye believed that evolutionary science was the root cause of secular humanism—and he blamed this philosophy for all manner of evils in contemporary American society. LaHaye was a significant influence on another Christian Right leader, the Southern Baptist pastor Jerry Falwell. Alongside their political advocacy group, the Moral Majority, Falwell and LaHaye were instrumental in bringing the conservative evangelical platform into the Republican Party in 1980. Through their efforts, combatting evolution was folded into the quest to bring morality back to American society.

The rise of the Christian Right and the consequent transformation of the Republican Party is the most significant realignment within US domestic politics of the past half-century. It is a framework that continues to shape party politics over a series of issues into the twenty-first century—including abortion, climate change, immigration, race, and higher education, as well as gay and trans rights. As the introduction to this volume discusses, however, the 1980s was an era of global economic deregulation and a time when many across the ideological spectrum increasingly doubted that political entities could enact meaningful social change. One way of understanding the story of the 1980s is to see it as the end of collective action and the rise of "neoliberalism"—as Gordin and McCray describe it, "the creation of policies that prioritized the (ostensibly) frictionless movement of goods, people, labor, and capital across borders, the dismantling of rules and regulations seen to impede this circulation, and the promotion of the individual (and individual responsibility) over governments."[62] But on many points, evangelical politics in fact sought government interventions (e.g., on matters of reproductive health, abortion, and access to contraceptives) and pushed forms of collective identity (e.g., the identification with Christian nationalism). How the greedy 1980s intersects with the rise of the Christian Right is a complex matter, one that historians of US history and American empire continue to debate.

Although a full account of the relationship between neoliberalism and Christian nationalism is an immense subject, beyond the scope of this chapter, I suggest there are fruitful avenues that histories of greedy science can offer to the subject. Several authors in this volume have highlighted the need to consider the relationship between markets and political formations with care. In chapter 4, John Agar suggests that we recognize the role that Thatcherite science policy had in promoting forms of "curiosity-driven research," which explicitly promoted science as an entrepreneurial enterprise that was meant to be divorced from collective politics.[63] And in chapter 10, Peter Westwick observes in his study of

Texas during this decade that conservative social politics and money worked together to support the expansion of scientific research in the state.[64] These authors direct our attention to the triangulation between scientific research, the market, and the state in ways that help us to consider modes of collective politics with fresh eyes. As Cathy Gere reflects in chapter 7, "'Drugs into Bodies,'" leftist AIDS activists aligned themselves with the private sector against the US government to get access to drug treatments—a historical configuration that disrupts a neat organization of the political right with the market and the political left with the state.[65] Through the lens of greedy science, we are able to see the work of AIDS activists in the 1980s as the inheritor of the radical health collectives of the 1970s—whether of the Black Panther Party, women's health movements, or the Young Lords Party—movements that protested credentialed science as part of political representation. The antiregulatory instincts of AIDS activists, I would suggest, have roots in the radical protests of these earlier leftist health movements. And in fact, what the AIDS activism episodes emphasize is that the American left— with its historical ties to global anticolonialism—was often antagonistic to the liberal state.

What was remarkable about the 1980s was the leftist move toward an embrace of capitalism in the face of state oppression. How the left came to view deregulation as an avenue toward community survival is to recognize that by the 1980s, science was prominently intertwined with American governance—both nationally and across the American empire. Through the earlier part of the century, the military–industrial–academic complex had indelibly transformed funding as well as research priorities and structures within American science. Therefore, we can imagine how engaging with science—as an abstract category or as a series of institutions— represented a powerful means for various individuals and communities to represent themselves to the American state. When creationists attacked the teaching of Darwinian evolution in public school, or when leftist health collectives protested credentialed medical authorities, they did so with dramatically opposing political aims and metaphysical worldviews. Nevertheless, the commonality of the political tactic—that is, the protest of mainstream science—should tell us something about the strength of science's identification with national authority. Recognizing the cultural power that could be gained by protesting, holding, or shaping scientific authority can help us to make sense of new forms of collective identity and new modes of global sociality in this period. Criticizing Darwinian science was more than a feature of evangelical theology—it was a mode of shaping a political identity that held together a nostalgia for America's Anglo Protestant past and an eschatological faith in its white Christian future. In other words, even if Thatcher declared at the time that society was not a "thing," engagement with science across the greedy 1980s was a powerful mode of creating and sustaining political communities.

Conclusion

After 1980, Gould's public writing about evolution was inextricably bound up in the larger political contests that were articulated in the "evolution versus creationism" culture war. And this overshadowed the earlier political efforts of the radical science movement to critique biological theories of sex and race difference. The altercations between Darwin and creationism, as well as between science and religion, were far more compelling, intertwined in the nation's cultural and political debates. Although evolutionary accounts of sex, race, and gender did not abate, these years cast the politics of evolution into a war of cosmologies.

Critically, in the face of these religiously inflected criticisms, liberal definitions of science were best equipped to defend the authority of the scientific community. The transcendence of enlightenment science was a much better rebuttal to creationism than the leftist fears of the political abuses and historical oppressions of science. This fact drove the celebrity of many of Gould's contemporaries. In the years after the *McLean* trial, ethologist and scientific celebrity Richard Dawkins began a public campaign to argue for an evolutionary model in which the ability for natural selection to fashion adaptations was proof against intentional design and thus evidence against a supernatural designer. His 1986 book *The Blind Watchmaker: Why the Evidence of Evolution Reveals a Universe without Design* contended that natural selection was the ultimate proof against a theistic worldview. In the face of the political threat of creationism, the public defense of evolution was distilled into the capacity of natural selection to explain the adaptive complexity of life without recourse to the action of God.

But even for Gould—who had begun his public science writing with the convictions of leftist politics—these years softened the radical edge of his public critiques of science. Although it is possible, as we have seen, to read across Gould's technical publications and works such as *Wonderful Life* and *Full House* to see his conviction that evolutionary frameworks could be responsible for racism and other social oppressions, after the publication of *The Mismeasure of Man*, this was no longer the centerpiece of his public brand. In the face of the cultural power of right-wing creationism, Gould's public writing morphed into a more straightforward celebration of the power of science and the triumphs of scientific genius, particularly of his greatest hero, Charles Darwin.

Gould began writing for *Natural History* in 1974 to caution the American public against trusting in the pronouncement of scientific theories of race and sex difference. He used his column, his status as a Harvard scientist, and his rich interest in the history of science to communicate a nuanced view of scientific practice as shaped by context and the social views of the scientist. By the end of the Reagan administration, he wrote glowingly about the powerful truths of science and the inviolability of the scientific process. The intervening culture war with the religious right had tempered Gould's willingness to publicly foreground the history of racism and sexism in science. Reckoning with this can help us to reflect on our own time, as we consider how—and whether—to press the critique of scientific racism and sexism in the face of religiously inflected criticisms of scientific authority. In an era when the public doubt of science is entangled so strongly with Christian ethnonationalism and right-wing populism, there is much at stake as we consider the role of historians and scholars of science in public conversations over the authority of science. Ought we to highlight histories of scientific violence? Or should we defend the authoritative truth of scientific epistemology in the face of religious attacks? Gould's biography can give us at least one lens for a reflective (and perhaps empathetic) way forward.

Does Gould's rise to celebrity represent a "greedy science"? By framing his public career through the lens of greed, it is tempting to consider this question primarily as an issue of individual character. What Gould achieved through his public writing was in many ways the opposite of the expressed purpose of his original *Natural History* column. That is, public science writing was meant to solve the problem of transforming the academy into an institution oriented to justice rather than an ivory tower of elite experts. Gould's interest and willingness to engage with the public audiences of his columns held the seeds of democratic possibilities. Particularly in the early years, when his writing was featured alongside the efforts of radical science collectives, it was conceivable that popular science might manifest in a form of democratic science.

By looking closely at the archive of the material production of Gould's popular science, however, what we can see is not the opening of an elite science "to the people" but instead the use of Gould's credentialed expertise as a Harvard evolutionary scientist to create a brand of knowledge that was packaged, marketed, and sold to audiences. Gould's science was not greedy because it required the amassing of capital to produce knowledge about the natural world. Rather, it was greedy because it sought the power of cultural influence—and it gained that influence by the material creation of wealth and social status. Gould wanted to change the American public's understanding of evolution because he believed that a view of evolutionary history that emphasized contingency was a powerful idea in the service of racial justice. But this form of popular science—which relied on audiences trusting and admiring a scientific public intellectual—created a necessary form of fame. And by the end of his career, Gould's brand as a celebrated evolutionary thinker occupied more space than his voice as a champion of racial justice.

At the outset of Gould's relationship with *Natural History*, it was not obvious that it was a path that could lead as much to fame as to political change. And that celebrity overtook leftist politics in his public writing didn't happen overnight; it was built over decades through a series of (sometimes small) contingent steps and transformations. And the drivers of his celebrity were built on material and social circumstances that extended far beyond Gould's individual motivation or actions. Which is why I offer this analysis of greed and justice in Gould's career not as a critique or a final pronouncement on his legacy. Rather, it is a lesson to me and perhaps to anyone else who longs for justice but is sorely tempted by dreams of glory.

NOTES

1. Stephen Jay Gould, *The Panda's Thumb: More Reflections in Natural History* (New York: W. W. Norton, 1980), 267.

2. Leonard Gould to Allison Palmer [after October 15, 1957], Box 123, Folder 2, Stephen Jay Gould Papers, M1437, Department of Special Collections, Stanford University Libraries, Stanford, California.

3. Alison Palmer to Leonard Gould, July 18, 1957, Box 123, Folder 2, Stephen Jay Gould Papers, M1437, Department of Special Collections, Stanford University Libraries, Stanford, California.

4. An example of a different version of the story describes, "The eminent paleontologist Stephen Jay Gould conceded . . . that his passion for dinosaurs dated from childhood, when his father took him to the Walt Disney movie 'Fantasia'"; see David Lambert, "Book Notice: Review of *Dinosaurs*," *New York Times*, April 3, 1979.

5. Stephen Jay Gould to Alan Ternes, June 5, 1973, Box 230, Folder 3, SJG Papers, Stanford University, California.

6. Stephen Jay Gould, "The Race Problem," *Natural History* 83, no. 3 (April 1974): 8–14; Stephen Jay Gould, "The Nonscience of Human Nature," *Natural History Magazine* 83, no. 4 (May 1974): 21–24; Stephen Jay Gould, "Racist Arguments and I.Q.," *Natural History* 83, no. 5 (June 1974): 24–29. All reprinted in Stephen Jay Gould, *Ever Since Darwin: Reflections in Natural History* (New York: W. W. Norton, 1977).

7. Stephen Jay Gould, "In Praise of Charles Darwin," *Discover* 3, no. 2 (1982): 20; Leonard Gould, "FBI File," February 12, 1968, Box 663, Folder 6, Stephen Jay Gould Papers, M1437, Department of Special Collections, Stanford University Libraries, Stanford, California.

8. Literature kept by Gould on activities in CORE and SDS, Box 122, Folders 10–12, Stephen Jay Gould Papers, M1437, Department of Special Collections, Stanford University Libraries, Stanford, California; material from Cuba sit-in, Box 122, Folders 12–14, Stephen Jay Gould Papers, M1437, Department of Special Collections, Stanford University Libraries, Stanford, California.

9. Stephen Jay Gould, "Pleistocene and Recent History of the Subgenus Poecilozonites (Poecilozonites Gastropoda: Pulmonata) in Bermuda: An Evolutionary Microcosm" (Columbia University, 1967).

10. John T. Bethell, *Harvard Observed: An Illustrated History of the University in the Twentieth Century* (Cambridge, MA: Harvard University Press, 1998), 220.

11. Quoted in Bethell, *Harvard Observed*, 220.

12. Bethell, *Harvard Observed*, 225.

13. Bethell, *Harvard Observed*, 226–228.

14. In Memoriam, Alan Ternes, *Potash Hill: Magazine of Marlboro College*, Fall 2015, 48.

15. Stephen Jay Gould to Alan Ternes, June 25, 1973, Box 230, Folder 3, Stephen Jay Gould Papers, M1437, Department of Special Collections, Stanford University Libraries, Stanford, California.

16. Stephen Jay Gould to Alan Ternes, June 25, 1973, Box 230, Folder 3, Stephen Jay Gould Papers, M1437, Department of Special Collections, Stanford University Libraries, Stanford, California.

17. Stephen Jay Gould to Alan Ternes, June 25, 1973, Box 230, Folder 3, Stephen Jay Gould Papers, M1437, Department of Special Collections, Stanford University Libraries, Stanford, California.

18. Alan Ternes to Stephen Jay Gould, August 17, 1973, Box 230, Folder 3, Stephen Jay Gould Papers, M1437, Department of Special Collections, Stanford University Libraries, Stanford, California. The other potential contributor was Marvin Harris, an American anthropologist and popular science writer. He wrote twelve articles for the magazine between 1967 and 1979.

19. For a perspective of the rise of special interest magazines in the midcentury period, David Abrahamson, *Magazine-Made America: The Cultural Transformation of the Postwar Periodical* (New York: Hampton Press, 1996). For a study of the role of *Scientific American*

in shaping the place of science in American popular culture from the beginning of the Cold War, see Emma Beintende, "The Scientific American" (senior thesis, Harvard University, 2012).

20. Stephen Jay Gould to Alan Ternes, October 26, 1973, Box 230, Folder 3, Stephen Jay Gould Papers, M1437, Department of Special Collections, Stanford University Libraries, Stanford, California.

21. Stephen Jay Gould to Alan Ternes, October 26, 1973, Box 230, Folder 3, Stephen Jay Gould Papers, M1437, Department of Special Collections, Stanford University Libraries, Stanford, California.

22. Robert Ardrey and Berdine Ardrey, *The Territorial Imperative* (New York: Atheneum, 1966); Konrad Lorenz, *On Aggression* (New York: Harcourt, 1966); Nadine Weidman, "Popularizing the Ancestry of Man: Robert Ardrey and the Killer Instinct," *Isis* 102, no. 2 (2011): 269–299.

23. Stephen Jay Gould to Alan Ternes, October 26, 1973, Box 230, Folder 3, Stephen Jay Gould Papers, M1437, Department of Special Collections, Stanford University Libraries, Stanford, California.

24. Stephen Jay Gould to Alan Ternes, October 26, 1973, Box 230, Folder 3, Stephen Jay Gould Papers, M1437, Department of Special Collections, Stanford University Libraries, Stanford, California.

25. Alan Ternes to Stephen Jay Gould, October 20, 1975, Box 230, Folder 3, Stephen Jay Gould Papers, M1437, Department of Special Collections, Stanford University Libraries, Stanford, California.

26. Jared Haynes, "This View of Writing: An Interview with Stephen Jay Gould," *Writing on the Edge* 2, no. 2 (Spring 1991): 71.

27. Peter Dzwonkoski, *Dictionary of Literary Biography: American Literary Publishing Houses 1900–1980* (Detroit, MI: Gale, 1986); Jay P. Pederson, *International Directory of Company Histories* (Detroit, MI: Gale, 2008), 519.

28. Norton publicity clipping for *Ever Since Darwin*, Box 150, Folder 5, Stephen Jay Gould Papers, M1437, Department of Special Collections, Stanford University Libraries, Stanford, California.

29. Stephen Jay Gould to Edwin Barber, December 20, 1977, Box 149, Folder 8, Stephen Jay Gould Papers, M1437, Department of Special Collections, Stanford University Libraries, Stanford, California.

30. Edwin Barber to Stephen Jay Gould, December 19, 1977, Box 150, Folder 4, Stephen Jay Gould Papers, M1437, Department of Special Collections, Stanford University Libraries, Stanford, California.

31. Jim Buckley, "Review of *Ever Since Darwin*," *American Spectator*, December 1978, Box 150, Folder 4, Stephen Jay Gould Papers, M1437, Department of Special Collections, Stanford University Libraries, Stanford, California. Jim Buckley was a former US senator from New York at the time of this review, and the article was sent to Gould by a colleague with the note, "Did you know Jim Buckley?"

32. Harold Hayes, "Seeking Our Roots: A Few Million Years Ago," *Chicago Sun Times*, January 15, 1978, Box 150, Folder 4, Stephen Jay Gould Papers, M1437, Department of Special Collections, Stanford University Libraries, Stanford, California.

33. Edwin Barber to Stephen Jay Gould, April 25, 1978, Box 150, Folder 4, Stephen Jay Gould Papers, M1437, Department of Special Collections, Stanford University Libraries, Stanford, California.

34. Sokilov, "A Talk with Stephen Jay Gould," 7.

35. This analysis takes its lead from James Secord's study of readership reception. James A. Secord, *Victorian Sensation: The Extraordinary Publication, Reception, and Secret Authorship of Vestiges of the Natural History of Creation* (Chicago: University of Chicago Press, 2003).

36. Stephen Jay Gould, "The Median Isn't the Message," *Discover* 6, no. 6 (1985): 40–42. This essay detailed his experience recovering from cancer after realizing that the median survival time for people with peritoneal mesothelioma was eight months.

37. Janel G. Raney (*NOVA* producer) to Stephen Jay Gould, October 9, 1991, Box 92, Folder 14, Stephen Jay Gould Papers, M1437, Department of Special Collections, Stanford University Libraries, Stanford, California.

38. Ben Shedd, "NOVA: From the Beginning," *WGBH Alumni Records*, accessed February 7, 2012, http://web.archive.org/web/20071005120646/http://www.wgbhalumni.org/essays/1970s -nova.html.

39. David Stewart, "Ambrosino and NOVA: Making Stories that Go Bang," *Current*, May 4, 1998.

40. Shedd, "NOVA," 1.

41. Linda Harrar, "This View of Life," *NOVA*, December 18, 1984.

42. Harrar, "This View of Life."

43. Harrar, "This View of Life."

44. "This View of Life," Film Plan, Box 858, Folder 5, Stephen Jay Gould Papers, M1437, Department of Special Collections, Stanford University Libraries, Stanford, California.

45. Harrar, interview with author, 2012. *NOVA* Audience Statistics, Box 858, Folder 6, SJG Papers, Stanford University, California; "This View of Life," Quarterly Reports, Box 858, Folder 6, Stephen Jay Gould Papers, M1437, Department of Special Collections, Stanford University Libraries, Stanford, California.

46. Harrar, "This View of Life."

47. Harrar, "This View of Life."

48. Harrar, "This View of Life."

49. Evidence of Agnes Pilot's role in Gould's career is dispersed throughout his correspondence and writing. Occasionally, Gould would write asides or corrections on typewritten correspondence indicating that he had had a difference of opinion with his assistant. She typed his manuscripts for him; Gewertz, "Gould Reads from Latest Opus," 2. During his illness from cancer in 1982, she took over most of his correspondence. The best piece on Pilot's role in Gould's career is Benjamin Solomon-Schwartz, "Behind Every Great Harvard Professor," *Harvard Gazette*, April 20, 2000.

50. Details of Gould's papers organization can be found in Jennifer Johnson and Myrna Perez, "Stephen Jay Gould Papers at Stanford University," *Mendel Newsletter* 19 (November 2012).

51. Gary Larson to Stephen Jay Gould, July 11, 1989, Box 698, Folder 3, Stephen Jay Gould Papers, M1437, Department of Special Collections, Stanford University Libraries, Stanford, California.

52. Stephen King to Stephen Jay Gould, "Note on Advanced Copy of *Jurassic Park* by Michael Crichton," DSC 0653, Stephen Jay Gould Papers, M1437, Department of Special Collections, Stanford University Libraries, Stanford, California.

53. Stephen Jay Gould to Mario Bunge, March 11, 1982, Box 114, Folder 8, Stephen Jay Gould Papers, M1437, Department of Special Collections, Stanford University Libraries, Stanford, California.

54. Letter from Bill Clinton to Stephen Jay Gould, December 6, 1996, DSC 0693 in Stephen Jay Gould Papers, M1437, Department of Special Collections, Stanford University Libraries, Stanford, California.

55. Invitation for Stephen Jay Gould to United States Presidential Inauguration, 1993, Box 115, Folder 9, Stephen Jay Gould Papers, M1437, Department of Special Collections, Stanford University Libraries, Stanford, California.

56. Stephen Jay Gould to Peter Stone, November 23, 1979, Box 120, Folder 2, Stephen Jay Gould Papers, M1437, Department of Special Collections, Stanford University Libraries, Stanford, California.

57. Andrew Jewett, *Science, Democracy, and the American University: From the Civil War to the Cold War* (Cambridge, UK: Cambridge University Press, 2012).

58. Jenna Tonn, "Radical Science, Feminism, and the Biology of Determinism," *Lady Science,* December 20, 2018; Myrna Perez, *Criticizing Science: Stephen Jay Gould and the Struggle for American Democracy* (Baltimore: Johns Hopkins University Press, forthcoming 2024).

59. Marian Lowe and Ruth Hubbard to Ethel Tobach and Betty Rosoff, September 2, 1979, Box 2, Folder 5, Records of the Genes and Gender Collection, Schlesinger Library, Radcliffe Institute, Harvard University; Marian Lowe, "Sociobiology and Sex Differences," *Signs: A Journal of Women in Culture and Society* 4, no. 1 (1978): 118–125.

60. Michael B. Shermer, "This View of Science: Stephen Jay Gould as Historian of Science and Scientific Historian, Popular Scientist and Scientific Popularizer," *Social Studies of Science* 32, no. 4 (2002): 489–524; Declan Fahy, *The New Celebrity Scientists: Out of the Lab and into the Limelight* (London: Rowman & Littlefield, 2015), 87–110.

61. The trends that pushed Protestant creationists toward "creation-science" were simultaneously theological and legal, and not all committed creationists were happy with the movement. Numbers, *The Creationists,* 241–248.

62. Michael D. Gordin and W. Patrick McCray, introduction to this volume.

63. John Agar, chapter 4 in this volume.

64. Peter Westwick, chapter 10 in this volume.

65. Cathy Gere, chapter 7 in this volume.

CHAPTER THREE

VisiCalc, Personal Computing, and the Speculative Entrepreneur of 1980s America

Laine Nooney

In a November 1984 article in *Harper's*, tech journalist Steven Levy documented the latest obsession to take over the business world: the computerized spreadsheet, best identified with the landmark software program VisiCalc.[1] In this roving treatise, Levy explains the spreadsheet to the uninitiated through a personalized tour of the church of spreadsheet-ology: "There are corporate executives, wholesalers, retailers, and small business owners who talk about their business lives in two time periods: before and after the electronic spreadsheet." Inspired by the fervor of the power users he interviews, Levy describes the spreadsheet as the new double-entry bookkeeping, the crisp mastery of oil painting, a transcontinental railroad, what horses were to cowboys—in other words, a new climax in Western models of capture and conquest, a blend of high speed and high detail that conferred upon its faithful an "unshakable belief that the way the world works can be embodied in rows and columns of numbers and formulas."[2] Levy tracks this transformation as part of a new "entrepreneurial Renaissance" taking hold in the United States, embodied in "a new breed of risk taker who creates businesses where none previously existed"—and one uniquely reliant on the spreadsheet for its ability to conjure something out of nothing.[3]

Since VisiCalc's release in 1979, journalists and historians have lauded the program as the "killer app" of personal computing. And indeed, there is perhaps no piece of software more infamously associated with the launch of the personal computing industry in the United States than VisiCalc.[4] Walter Isaacson argues VisiCalc transformed expensive, complicated microcomputers into "tools rather than merely toys," and even software historian Martin Campbell-Kelly, bearish on the killer app hypothesis, acknowledges that VisiCalc plausibly accelerated the growth of the personal computing industry by several months. Yet beyond the general knowledge within the history of science and technology that VisiCalc was the first commercial spreadsheet, as well as the biographical accounts from its developers and publisher, we know very little about what VisiCalc did and why it mattered as more than a bellwether for the software industry. In a strange sense, we have taken history at its word: believing those who professed to its transformative power while often underexamining the broader historical conditions that allowed it to be transformative at all.

In the mid-to-late 1970s, microcomputers (or what we would today call personal computers) were still untested and underutilized technologies. A consumer-grade software market barely existed in recognizable form, and what was sold on it had limited utility. Imagine dozens of software packages that were simply hyper-specific calculators: one to formulate your gas mileage, one for running statistics algorithms, another for tracking payroll. Users whose needs were more specific got stuck programming themselves.

But VisiCalc was a paradigm shift in consumer software. With VisiCalc, the program did not prescribe what was to be calculated or how. Instead, VisiCalc allowed users to define their content, as well as the mathematical operations they wished to conduct between numerical categories. In other words, VisiCalc provided a *framework* for mathematical calculation and modeling that was indifferent to its numerical content. Thus, VisiCalc offered remarkably flexible uses across financial analysis, engineering, recordkeeping, budgeting, and other domains—all while ensuring "the user need not know *anything* about computers or programming in order to derive VisiCalc's benefits."[5] As computing industry analyst Ben Rosen put it in his semiregular publication for Morgan Stanley, *The Rosen Electronics Letter*,

"VisiCalc could someday be the software that wags (and sells) the personal computer dog."[6]

While VisiCalc would prove productive for any number of users and industries, its popularity was first driven by an expanding base of non-hobbyist white-collar professionals who reveled in deploying the program for projection, forecasting, modeling, and other modes of professional and financial management. Under the hood of VisiCalc is a transformation in the technics of work that has become so commonplace that today it is hardly noticed: the way "computerizing" one's business became an aesthetic experience unto itself, rife with sensations of transparency, immediacy, and control. The software was thus ideal for those seeking a more immediate visualization of where potential untapped opportunities and externalities might lie—a way of seeing with technology that made "greedy" outcomes all the more likely by accelerating their ease in a business context.

This chapter documents the development, marketing, and consumption of VisiCalc while framing its remarkable success alongside the economic history of the United States during the "long 1980s." Arriving at the sunset of the 1970s, VisiCalc's capacity to energize an "entrepreneurial Renaissance," as Levy put it, stands in stark but important contrast to the economic and global crises that straddled the presidencies of Jimmy Carter and Ronald Reagan, as the United States sagged under spiraling inflation, rising unemployment, ballooning deficit, and a torpid global energy crisis.[7] Yet while some mourned the loss of the traditional symbols of American power—the factory, the steel worker, the assembly line—others would double down on free market ideologies that promoted a neoliberal political economy grounded in the valorization of individual entrepreneurism.[8] (Yulia Frumer, in chapter 13 of this volume, takes up similar themes but unfolding in Japan.) This was a purpose for which VisiCalc was uniquely suited, given its design was predicated on simplifying mathematical calculations to the scale of a single user. As we shall see, themes of the entrepreneurial individual course through VisiCalc's advertising and journalistic coverage, as do anxious references to time, energy, economic surveillance, and the internalization of personal responsibility. VisiCalc's affordances were presented not just as a handy helpmate for business but as a tool for bracing oneself against economic uncertainty: "With VisiCalc, getting your arms around the future seems a trifle easier."[9]

Imagining and Making VisiCalc

Dan Bricklin didn't dream of making software for personal computers, bringing computers to the masses, or disrupting oppressive institutional hierarchies through the power of computing. Like most computer entrepreneurs of his generation, his ambitions were simple: make software, make money. And his pathway into computing was about as predictable as they come. Growing up in Philadelphia in the 1950s and '60s, Bricklin took to electronics hobbyism like many young white men of his generation who showed a proclivity for math and science. Because of family connections, he was allowed to tinker on a local school's mainframe, while the nearby University of Pennsylvania granted him the opportunity to teach FORTRAN to Wharton grad students while he was still in high school.[10] Sticking to the East Coast, Bricklin received a bachelor's in computer science from MIT in 1973, followed by several years' experience working at Digital Equipment Corporation (DEC), the popular minicomputer manufacturer headquartered a few miles to the west, just outside of Boston.[11] When he returned to school in 1977, he would ascend the trellis of the Ivy League, enrolling in the Harvard Business School (HBS) due to concerns that computer programming lacked potential for long-term financial security.[12] VisiCalc would originate where upwardly mobile white men were refashioning themselves for an economy that was leaving many others behind.

At HBS, Bricklin found a world of people who could greatly benefit from computer applications for instantaneous modeling and financial synthesis, especially from software that didn't require much in the way of programming chops. The evidence of this unmet need was everywhere: from the tedious process of writing his own financial projection programs on the university's PDP-10 time-sharing system, to "running the numbers" by hand for his homework case studies, to observing his professors, clouded in chalk dust, onerously recalculate financial tables every time they adjusted a model on the blackboard.[13] As he sat in class, Bricklin's mind wandered. He imagined a heads-up display "like in a fighter plane," a trackball in the back of a calculator, a world of command-and-control hardware interfaces bestowing godlike perspective on the world's numbers. Tech journalist Robert X. Cringely would gloriously embellish the scene as one in which Bricklin, "like Luke Skywalker jumping into the turret of the Millennium Falcon, . . . saw himself blasting out financials, locking on the profit and loss numbers that would appear suspended in space before him"—an unironic reading that only exacerbates the propulsive machismo of it all.

These were, of course, hardware fever dreams. Nothing like this was plausible, not within Bricklin's reach anyway. So he scoped his fantasies downward over the summer of 1978 into the realm of material feasibility. He imagined the general schema of what we would later identify as the *spreadsheet*, a digital, interactive, real-time mathematical matrix that collapsed the typically distinct computational processes of input, calculation, and output into a seamless user experience. His plan for profitability, however, was more in keeping with the industrial spirit of the Death Star than the wily guerilla tactics of the Rebellion: he planned to design his program for a DEC time-sharing system, then hawk it to businesses and institutions. VisiCalc was imagined as what we would today call *enterprise software*: just the logical extension of a market opportunity.

Now, enterprise software wasn't a new concept; computers had been deployed for business and administrative purposes since the early 1950s.[14] By the late 1970s, a software products industry for mainframe and minicomputers was fairly well established, and nearly all of this product would have been oriented to use in business or industrial settings.[15] This included a wide range of systems and applications software, ranging from operating systems, to programming aids and utilities, to industry-specific programs for banking, engineering, insurance, and transportation, to cross-industry applications for handling inventory management, payroll, marketing, and general accounting. By 1980, the cumulative revenue of the US packaged software business reached nearly $2.5 billion.[16] It is unclear to what extent Bricklin thought through the challenge of developing new software in such an established and consolidated market (where the top 15 percent of all software suppliers took 68 percent of all revenue), but with his top-tier educational credentials and unique blend of computational and financial know-how, his schemes may not have seemed entirely outside the realm of possibility.[17]

More intriguing than Bricklin's aspiration, however, is the way his prior educational and professional background shaped how he thought about the relationship between hardware, software, and the people on the receiving end of these technologies: the users. As a student at MIT in the early

1970s, Bricklin had the advantage of working on the cutting edge of new and innovative time-sharing systems—systems that, if not technically "real time," afforded an experience of interactivity and immediacy by virtue of how time-sharing worked. In his post-MIT work at DEC, Bricklin was assigned to the company's Typeset-10 word-processing system, which allowed users to prepare raw text in a format appropriate for a typesetting machine (such systems were increasingly common at newspaper offices, among other places: part of a larger trend in the computerization of labor). Specifically, Bricklin was tasked with the software development for the Typeset-10's editing terminals, which worked in real time—users had to see what they were writing as they wrote it and be able to correct errors as they went.[18] Features he worked with on the Typeset-10, like an embedded on-screen ruler, keystroke minimization, and on-screen scrolling, focused on end-user efficiency and ease of use. Thus, the functionality Bricklin envisioned for his future software product expressed a user-centric philosophy. He designed from the point of view of a working professional, *not* a computing expert.[19]

The impetus to look toward the emerging microcomputer industry came through the privileges of his Harvard network. In the fall of 1978, Bricklin (now accompanied by an old MIT friend turned development collaborator, Bob Frankston) was directed by a Harvard finance professor to get in touch with another recent HBS graduate, Dan Fylstra (also an MIT engineering grad).[20] Fylstra was notable among recent HBS graduates for his fervent faith in the future of personal computing, having founded a microcomputer software mail-order business while finishing his second year at Harvard.[21]

As computing entrepreneurs on the East Coast went, Fylstra was one of the best positioned to understand the vicissitudes of the microcomputer market: he'd been a founding associate editor of *Byte* in 1975, writing the magazine's first reviews of the Commodore PET and the TRS-80.[22] It was the kind of institutional coziness that helped his mail-order racket, Personal Software, grow into one of the nation's most successful microcomputer software publishing businesses by the time he was in conversation with Bricklin and Frankston barely eighteen months later. By that time, Fylstra had even taken on a partner, Peter R. Jennings, a young Canadian developer who had programmed a popular chess game, *Microchess* (by 1979, Personal Software was able to laud *Microchess* as the industry's first

"Gold Cassette," selling more than fifty thousand copies).[23] Yet Fylstra's at-traction to Bricklin and Frankston's finance software project was driven by his own needs for financial diversification: games like *Microchess* made up nearly half of Personal Software's offerings by the fall of 1978. Fylstra desired to move into "more serious software"—code for professional and business applications.[24]

Fylstra was adamant that Bricklin should develop the program for the microcomputer rather than a DEC time-sharing minicomputer. This ad-vice was one part prescience, one part self-interest: software sold directly to corporations went through entirely different distribution channels than consumer software, and Fylstra's business was not structured to support the former. In order to get a cut, Fylstra needed Bricklin to develop some-thing he could sell through the industrial pathways he already held sway in. So in late September 1978, Bricklin learned what he could about the Apple II, then crunched out his first prototype in BASIC over the first week in October on Fylstra's Apple II.[25] VisiCalc's initial prototype was a rudi-mentary thing, lacking many of the advanced features Bricklin hoped for, but Fylstra liked what he saw and conversations continued.[26] In Bricklin's notes, he called his prototype "FINANCE TEST PROGRAM" (the Apple II only wrote in all-caps); other sources refer to it as an "electronic black-board," and Bricklin even wrote a paper for his marketing class about the product in which he called it "Calcu-ledger."[27] The name VisiCalc, short for *visible calculator*, would come later.[28] What Bricklin had written in BASIC was simple, but the foundations of VisiCalc's future were there: columns and rows and the ability to change calculations with a keystroke—an en-tire responsive, interactive mathematical world, glowing on screen.[29]

Publishing and Marketing VisiCalc

No formal agreement governed the early talks between Bricklin, Frankston, and Fylstra. Fylstra's support was at first casual and loosely defined. He provided Bricklin with information on the Apple II, and eventually with the machine itself, and engaged in iterative feedback with Bricklin and Frankston as they developed the idea in the fall of 1978.[30] The unstated commitment between the parties was common enough in this very early stage of industry, where outcomes and markets were still quite taking form. Nonetheless, a verbal royalty arrangement was set early in the development process, before any formal publishing agreement between the two entities, at the substantial rate of 35.7 percent.

No industry best practices existed yet to govern what a royalty arrangement should look like, but Fylstra and Jennings have admitted it was high for the time but not unheard of—although the high rate would bring out everyone's financial self-interest over time and prove the undoing of the entire operation just a few years later.[31] Nonetheless, a formal publishing arrangement emerged in the spring of 1979, sometime after Bricklin and Frankston established themselves as a development studio called Software Arts that January.[32] Aside from laying out royalty stipulations, the contract also established both companies' interdependencies: Personal Software was restricted from developing or marketing any other spreadsheet product, while Software Arts had to follow Personal Software's marketing direction.[33]

While Bricklin and Frankston got to work, Fylstra and Jenning's job, as their publisher, was to prepare the runway for VisiCalc's arrival. This involved a variety of tasks that reflected the range of support early publishers provided for their clients, including industry outreach, manufacturing, distribution, and—perhaps most important—provision of start-up capital. Personal Software sank a reported $100,000 into VisiCalc.[34] Some of this money constituted an advance royalty payment to Software Arts to support development costs, including the monthly lease of a time-sharing system.[35]

Aside from supporting early development costs, Fylstra and Jennings also arranged to demonstrate early versions of the program to key industry players, including Mike Markkula, Steve Wozniak, Ben Rosen, and Carl Helmers. In every case, the goal was to prime interest and solicit feedback, especially with regard to marketing and pricing.[36] Lacking a model for the release of this kind of software, conversations with industry leaders helped Fylstra and Jennings maximize potential profit by adjusting their price to what they thought the market would bear. They had initially imagined a price around $35—more than a game but less than many programming languages or operating systems—but they marked up the price to $99.50 on launch.[37]

Feedback from friends and industry insiders was strong, but would it sell to an audience larger than the then-narrow community of technoliterati?

Fylstra and Jennings played to potential business users' anxieties and desires with a manicured marketing strategy. In an era when most software was sold in Ziploc plastic bags with one sheet of photocopied documentation, Personal Software treated VisiCalc as a "whole product," a marketing concept Fylstra claims to have learned at HBS.[38] The premise was to treat the experiences of handling and learning to use VisiCalc as aspects of the product, as well as make this "whole product" easily understandable to nonprogrammers, at every level of interaction. This meant extensive documentation, reference guides, model spreadsheets, and packaging that made the user feel like they were dealing with a serious product—all of which was a first in the microcomputer software industry. VisiCalc came packaged in a brown vinyl folder, reminiscent of a leather attaché case or executive binder, along with a hundred-page manual divided into four lessons and a quick reference card.[39] Fylstra especially had a sense for the cultural work that had to be done to incorporate the Apple II's unfamiliar technology and VisiCalc's nuanced functionality into the professional imagination; everything about VisiCalc was promoted as easy, helpful, professional, and business oriented.

This sensibility was potently expressed in VisiCalc's first advertisement, produced for Personal Software by Regis McKenna, the same public relations firm Apple used (and a relationship brokered for Fylstra by Steve Jobs).[40] As the first full-page, four-color-process ad for a microcomputer software product ever printed, it was a show of economic force and desire in its own right—a slick and shiny thing telegraphing a new world of expectations for what personal computing software could become.[41] It is also a case study on the financial anxieties of American business practice circa 1979. "Solve Your Personal Energy Crisis," announces the headline, a not-at-all wry nod to the ongoing 1979 crisis over oil supplies. (See chapter 1 for more explication about the nexus of oil, science, and greed.) Unfolding in the wake of the Iranian Revolution, the '79 oil shock was the second energy crisis of that decade, catalyzing panic buying, gas rationing, and generalized economic anxiety throughout the United States (Jennings himself recalls having to take the train from Palo Alto to San Francisco for the 1979 West Coast Computer Faire due to the gasoline shortages).[42] This wasn't software for some kind of countercultural computer hacker utopia, but software that lived in what was real and immediate, even if it was only targeted to the class of people already most privileged in weathering the economic storm—in other words, the white male professional, so prominently featured in the ad itself.

VisiCalc's first ad collapses the personal and the geopolitical, scaling the impact of widespread infrastructural precarity to the level of one man's frantic calculations. In the advertisement's scene, our everyday business-man triples himself: referencing his sums, calculating them, and copying down the results, working at a desk somehow both well lit and void of context. The ad's copy suggests that VisiCalc's power lies not merely in simplifying or removing the tedium of calculation but in redistributing one's labor to a higher realm of meaning: the "what if?" "What if sales dropped 20 percent in March?" "What will happen to our entertainment budget if our heating bill goes up 15 percent this winter?" "What if the oscillation were dampened by another 10 percent?" By redirecting the user's focus toward speculation, VisiCalc proffered an opportunity to displace the uncertainty of the unknown into an infinite number of possibly calculable futures—to replace indeterminacy with probability, to develop strategies for every possible outcome. In the shadow of economic unknowns, VisiCalc was not a technology for time saving but a tool that promised you would always have a contingency plan. Software had never been sold this way: appealing to an individual not on the basis of what made the software good, useful, or easy but with a sensibility for the way it might architect new hopes over uncertain foundations. In such a chaotic world, our calculating hero isn't a self-serving protagonist but rather an enterprising survivalist; as the ad frames it, the hostile outside environment, not his own internal compulsions, forces him to adopt this new regime of creative accounting. As such, the advertisement effectively neutralizes any possible "greedy" interpretation, assuring future buyers that they are simply doing what must be done.

VisiCalc may have provided a rationale for "buying a ten-thousand-dollar computer system to run a one-hundred-dollar program," but *who* was in a position to buy either platform or program points to a marked departure from the traditional microcomputer enthusiast.[43] Many of VisiCalc's earliest customers were wealthy white-collar men interested in microcomputing as a tool to amplify their professional lives. Some would catch the computer bug, yes, but most experienced their new-found computer hobbyism (if they identified as hobbyists at all) through preprogrammed software applications. Stan Veit, owner of the New York

City–based store Computer Mart, witnessed these new computer habits co-agulate in real time, as businessmen of the early 1980s kept coming in asking for a "VisiCalc machine."[44] In Veit's estimation, this marked the beginning of a different kind of user than the traditional hobbyist who knew all about system buses and processor models: "They were more interested in what a computer did than in how it did it."[45]

Putting VisiCalc to Work

As it turned out, a large part of VisiCalc's popularity among white-collar professionals was that it resolved a problem of access to computing power that Bricklin himself had never set out to solve: the executive's interest in tighter control over figures and greater flexibility in economic projection. It wasn't that financial modeling software, or other kinds of programs for dynamic calculation, hadn't existed before. Software packages specialized for the financial, banking, or retail sectors had proliferated within commercial businesses large and small since the 1960s.[46] But within businesses of nearly any size, the access any given employee had to a computer was limited, constrained by the rules and structures of a company's data-processing department (essentially the room where the mainframe computer or time-sharing minicomputer hub was kept, guarded by the employees who managed that computer's use).[47] Access was never direct, and the computer was largely an object of mystery. At most companies, in most scenarios, no rationale existed for an *individual* employee to have exclusive access to a computer.

VisiCalc, and the Apple IIs that it brought through the door, redistributed this specific kind of computing power within the workplace. As *Byte* founder Carl Helmers wrote in the magazine's August 1979 issue, "The techniques used in Visi-Calc [*sic*] are possible only . . . when the concept of 'one user, one processor' is employed, ie: when the computer power is 'personal.'"[48] Consumer microcomputers were the condition of opportunity for white-collar professionals, especially those in the expanding information industries of the late 1970s, to more closely manage their own research, planning, or budgeting analysis, which was preferable to relying on internal, centralized data-processing departments that had little attention to spare for running and rerunning fiscal or scientific scenarios with countless small changes.[49]

So up and down Wall Street, executives went through consumer-facing retail distribution channels to procure microcomputers, creatively accounting their Apple IIs and VisiCalc packages past their data-processing departments and into their offices on the company dime. "The check for a computer installation would have the legend 'Furniture' annotated on the stub," wrote personal computing consultants Barbara and John McCullen, summarizing the history of these transformations in their early 1980s essay "Screen Envy on Wall Street." "In other words, the firm was purchasing a $9000 desk that just happened to have a funny looking machine sitting on top of it at the time of delivery."[50] In the artifact of VisiCalc, a user-centric orientation toward software converged with a hardware platform designed for use by only one person.

For professionals whose job it was to massage numbers, manage data, and scrape bottom lines, the direct access to calculation provided by VisiCalc was revelatory. "Hours of figuring cost projections eliminated," one computing advocate lauded. "Days of waiting for revised estimates reduced to seconds."[51] VisiCalc would be released on other systems in the years following its release, including later models of the TRS-80 as well as some Commodore machines and the Atari 800. VisiCalc's expanded platform compatibility was touted in the second round of advertisements Personal Software ran for the program, along with a litany of user anecdotes underwriting VisiCalc's promise to save "time and money."[52] The types of business users quoted ranged from "the financial VP in Massachusetts" to "a New York dentist" to "a Utah businessman," emphasizing the software's nationwide appeal, while the material of the anecdotes themselves showed the transformation of time itself into money (a mathematics core to VisiCalc's claim of "productivity"). For one user, the time required to produce month end reports dropped from "three days to three hours," while another company was able to use VisiCalc and a personal computer to cut its reliance on time-sharing, a move alleged to save "at least $30,000 the first year." Tellingly, many of VisiCalc's hypothetical nonbusiness uses—such as for scientific calculation or household finances—had dropped out of the marketing appeal as Personal Software ran with VisiCalc's core user base: "besides saving time and money, [users are] simplifying their work and getting more information helps them make better decisions." Across all the major microcomputers of the early 1980s, VisiCalc was honing its edges around the amplifying obsessions of the late twentieth-century business elite: to produce more value with less expenditure, with the hopeful outcome of lining one's pocket with the difference.

The program's initial association with the Apple II remained beneficial, especially among an expanding entrepreneurial class that viewed Apple Computer as a slickly marketed start-up parable, a nimble David facing off against the Goliaths of an outmoded industrial economy.[53] Two years after its release, VisiCalc was still outselling its nearest Apple II software competitor by a two-to-one ratio, and it remained in the top ten of all software products sold month to month for the Apple II until December 1983.[54] As one journalist wrote in December 1981, "The continuing strength of Visi-Calc . . . is a story so often told that it tends to become ho-hum. It should not be so. VisiCalc has validated the personal computer as a useful business tool, and it's the business user who is now flocking into the computer stores everywhere and queuing up for the product."[55]

Business journalism was quick to latch on to the perception of microcomputing as a new trend, with VisiCalc as its central star. The program was feverishly covered in mainstream business periodicals, including *Fortune*, *The Wall Street Journal*, and *Inc.* (which began in 1979 as a "magazine for growing companies," the canary in the coal mine on a rising American obsession with corporate entrepreneurship). And a new wave of profitability would be achieved following the 1981 release of the IBM 5150, better known simply as the IBM PC—the system that would come to standardize the term "personal computer" as the vernacular term for a microcomputer. For business users, the IBM PC was a shot heard round the world. If a company's data-processing department had been reluctant to invest in microcomputers because of their perceived status as computational toys, IBM's seal of approval solidified the trending perception of the microcomputer as the next eagerly awaited disruption to American business. Thus, despite VisiCalc's ongoing advantageous association with the Apple II, it would eventually sell more copies for the IBM PC than for any other platform—a testament to the fact that, no matter what the headlines claimed, most American businesses were fundamentally conservative in their technological orientation and unwilling to experiment with microcomputers until they were normalized through their association with the largest "traditional" computing corporation on earth.

Once VisiCalc had given professionals a taste of what microcomputers could do for them, the scale of the business computing market boomed, developing its own distribution specializations distinct from software development that only targeted consumers at retail. Software publishing for businesses would become the largest segment of the software market by the mid-1980s, dwarfing the consumer and education markets with cumulative unit sales for top products estimated at 2.3 million and revenues near half a billion.[56] The top three business software publishing companies (Microsoft, Personal Software/VisiCorp, and Lotus Development) exceeded the revenue of the top three consumer software companies (Sierra On-Line, Spinnaker, and Brøderbund) by a fivefold margin.[57] Business software would rightfully be considered "*the* major area for software publishers" by the mid-1980s, "the one market where revenues and profits are significant for the leading publishers."[58] Thus, in just a few years, a market that had been starved for "practical, useful, universal, and reliable" software, as industry analyst Ben Rosen had lamented in 1979, was transformed—and it was finally helping to bring microcomputers to the masses.

Thinking Like a Spreadsheet

What VisiCalc offered was not a mere computerization of pen-and-paper practices. Rather, it provided something that had never been possible before: the capacity to *see* the instantaneous transformation of data ripple down the screen as the computer refreshed its screen memory. "If you change any of the numerical data, the electronic worksheet instantly displays a new result. Automatically," promised VisiCalc marketing. "You can play 'what if' as often as you wish to solve thousands of different problems. When finished, you can get a hard copy of all the information on your worksheet from your computer printer. Absolutely no programming is necessary."[59] Whether it was for a Wall Street stock brokerage or a small-town Main Street storefront, VisiCalc promised a way of *assessing* business operations that could be commanded by a single individual.

The sensation VisiCalc bestowed—of not just commanding a Cartesian view of a world of numbers down below but also being able to alter their output—emboldened fiscal manipulators to tweak scenarios to perfection and then execute them in the real world. VisiCalc's fundamental orientation toward instantaneous calculation would encourage new cultures of rapid projection, especially within the financial sector, greasing the operations of Wall Street's most ruthless shock troops. In their hands, tools like VisiCalc displaced responsibility for economic decisions from the person to the program. In other words, no one was greedy; the program just told you what was true. Alongside this was an inherent internalization of economic risk, a sort of financialization-of-the-self that VisiCalc enabled through the presumed transparency of its economic modeling. "In the past, before spreadsheets, people would've taken a guess," Bob Frankston told the tech journalist Steven Levy in 1984. "Now they feel obligated to run the numbers."[60]

What all this amounted to was an epistemology that has sustained into our present moment, that "spreadsheet way of knowledge" Steven Levy identified in the 1984 *Harper's* article that opened this chapter. Returning to Levy's analysis, he also offers a historical arc to the uptake of spreadsheets. While their initial appeal might have been in saving time, their accelerated use soon produced new forms of work. Quarterly updates could be replaced with monthly, weekly, or daily updates, allowing for continual instantaneous assessment. But the real delirium of VisiCalc's appeal ultimately returned to the "what if" factor—the fact that spreadsheets, as one executive put it, allowed them to experiment with "a phantom business within the computer":[61] "All this powerful scenario-testing machinery right there on the desktop induces some people to experiment with elaborate models. They talk of 'playing' with the numbers, 'massaging' the model. Computer 'hackers' lose themselves in the intricacies of programming; spreadsheet hackers lose themselves in the world of what if. . . . The experiments . . . are far-flung attempts to formulate the ultimate model, the spreadsheet that behaves just like an actual business."[62] The presumably god's-eye view afforded by spreadsheets, the ability for users to pinpoint and cast off low-performing assets (or low-performing employees), was fuel for financial operatives hell-bent on finding new pathways for capital accumulation among economic elites in a slumped economy. This was "personal" computing in a new way: computer power harnessed to serve an individual's personal anxieties and ambitions.

Perhaps one of the most iconic exemplars of this tendency was investment banker Michael Milken, the Wall Street "junk-bond king," who made a fortune instigating high-yield leveraged buyouts of companies experiencing depressed stocks during the economic ups and downs of the 1980s. VisiCalc was essential to Milken's process, allowing him to simultaneously stalk multiple companies, calculating which companies would make good targets for buyouts based on their cash flow relative to the amount of debt they could bear. As historian William Deringer has written, "Spreadsheet software like VisiCalc and Lotus 1-2-3 became a durable augmentation for financial agents" like Milken, who exemplified emerging perspectives about the value of "information" as a commodity unto itself.[63] Milken didn't make money by harnessing economies of scale or pursuing new efficiencies but by manipulating relationships of revenue, debt, and tax write-offs to consolidate interests and restructure companies in favor of the investors he represented. Enabled by the electronic spreadsheet, Milken produced profitability for shareholders (rather than employees) where it previously had not existed.

VisiCalc did not somehow single-handedly create the "Deal Decade" of the 1980s, as one contemporary book labeled it, or was responsible for the "shift in emphasis from production to finance as the centerpiece of capitalist class power" that marked American economic trends from the late 1960s forward.[64] It is necessary to understand, however, that VisiCalc, as well as the spreadsheets that followed on its heels, existed in a symbiotic relationship with these larger trends. As Kera Allen has explored in her examination of VisiCalc's use by the Rockefeller Foundation to address agricultural shortfalls in Tunisia, the program encouraged modes of data analysis that were largely only about their own propagation, a scenario in which stakeholders mistook the "amassing data for a more complete analysis, regardless of the outcome" for "improved analytical capacity."[65] VisiCalc itself was not greedy, but it certainly *enabled greed* by turning time-consuming financial assessments into a matter of abstract mathematical comparison—regardless of the material circumstances those numbers represented.

Noncomputable Numbers

For all the advantages of visibility that VisiCalc afforded, there would always be corners of the world it couldn't calculate. For example: nothing in its rows and columns could have predicted a $60 million lawsuit brought by VisiCalc's publisher, VisiCorp (Personal Software had renamed itself in 1982 as part of a "high visibility and aggressive marketing program to establish brand recognition") against VisiCalc's developer, Software Arts, in September 1983.[66] The lawsuit was the culmination of long-brewing tensions, as organizational growth at both companies pulled apart the once-aligned interests between Fylstra and Jennings, Bricklin and Frankston. Communication crumbled; accusations grew bold. VisiCorp was overshadowing Software Arts in taking credit for VisiCalc. Software Arts wasn't responding quickly enough to improve the product, causing VisiCorp to lose market share on the IBM PC. Neither party was satisfied with the royalty deal struck four years prior.

At the core of it all was, indeed, greed—or, to put it in capitalist parlance, the imperatives of growth and profit. Bricklin and Frankston's company had been paid over $22 million in royalties by early 1984. "Running the numbers" through VisiCalc would have made it readily apparent how much VisiCorp was potentially losing on this deal—a hypothetical financial scenario that became an inescapable source of friction between the two companies. This was especially true since, no matter how many complementary products VisiCorp released, VisiCalc brought in the majority of VisiCorp's revenue.[67] Without a diversified revenue stream, VisiCorp's profitability rested on a single piece of software. The testy machinations between VisiCorp and Software Arts, full of suits and countersuits, became a case study on "how a software winner went sour," as one *New York Times* headline put it.[68]

In the shadow of all this upheaval, something else VisiCalc couldn't have predicted was changing the stakes of business software: Mitch Kapor, a former VisiCorp employee, had developed a not-so-little program called Lotus 1-2-3.[69] Released in the spring of 1983 and customized especially to the new 16-bit IBM PC, Lotus 1-2-3 harnessed capabilities VisiCalc had yet to implement. With better features and product integration, Lotus 1-2-3 set itself apart from VisiCalc in the eyes of corporate information technology services executives eager to make the right choice for their company. VisiCorp and Software Arts were bickering away in court as Kapor ate the business software market for lunch. In the end, it was all cannibalized: Software Arts' remaining assets were sold to Kapor's company in the spring of 1985—a final bid to stave off bankruptcy. With no reason to keep a prime competitor alive, Lotus shuttered VisiCalc. As for VisiCorp: despite being the world's fifth-largest microcomputer software company, bringing in $43 million in sales in 1983, the fall of VisiCalc sent the company into a rapid decline.[70] By November 1984, VisiCorp was on the auction block.[71] There is something poignantly ironic about the fact that the company that unleashed the power of hypothetical projections into the hands of individual users was unable to leverage that same software to protect its own interests—a failure that is suggestive of the unpredictable hypervolatility resting at the core of late-stage capitalism.

But even with VisiCalc gone, the "spreadsheet way of knowledge" was here to stay. More than merely a "killer app" that made the computer a productive tool, VisiCalc showcased, in a way no program had before, how computational power might be leveraged for individual gain. The added value was in black and white (or green and black or orange and black, depending on your monitor). Computation in the form of individual computers and individual units of software brought a presumed transparency, a hypothetical objectivity to the financial world, transforming the problem of any given business into just another numbers game. Like the "what ifs" VisiCalc proposed for individual users, the very existence of the program invited a different kind of "what if": what if the failure of the American economy was not a failure of America per se but an opportunity for continued innovation, a time for necessary upheaval and the shedding of old skin? What if greed wasn't greed at all, but just survival? Such, at least, was the party line of the cult of entrepreneurship, which found ready circulation for new ideas among an American mass media eager for a hopeful spin on uncertain times.

NOTES

1. Steven Levy, "A Spreadsheet Way of Knowledge," *Harper's*, November 1984. Reposted on *Wired*, October 24, 2014, https://www.wired.com/2014/10/a-spreadsheet-way-of-knowledge/.

2. Levy, "A Spreadsheet Way of Knowledge."

3. Levy, "A Spreadsheet Way of Knowledge."

4. The "killer app" hypothesis, as Martin Campbell-Kelly explains, argues that "a novel application, by enabling an activity that was previously impossible or too expensive, causes a new technology to become widely adopted." Martin Campbell-Kelly, *From Airline Reservations to Sonic the Hedgehog: A History of the Software Industry* (Cambridge, MA: MIT Press, 2003), 212. Many popular press and retro enthusiast writings on VisiCalc maintain this mythological stance, claiming the software is the reason we have a personal computing industry. Such claims, of course, are overwrought. As Campbell-Kelly writes, "The personal computer revolution would have happened with or without VisiCalc. However, it is plausible that VisiCalc accelerated the process by several months" (212). For further dissection of the "killer app" hypothesis, see Campbell-Kelly, *Airline Reservations*, 212–214.

5. Benjamin M. Rosen, "VisiCalc: Breaking the Personal Computer Software Bottleneck," *Morgan Stanley Electronics Letter*, July 11, 1979, 2.

6. Rosen, "VisiCalc," 2.

7. Peter N. Carroll, *It Seemed Like Nothing Happened: America in the 1970s* (New Brunswick, NJ: Rutgers University Press, 1990), 207–232.

8. David Harvey, *A Brief History of Neoliberalism* (New York: Oxford University Press, 2005), 2.

9. Steven Levy references this as a primary observation from Allerton Cushman's pamphlet *Confessions of an Apple Byter*. Levy, "Spreadsheet Way of Knowledge."

10. Robert Levering, Michael Katz, and Milton Moskowitz, *The Computer Entrepreneurs: Who's Making It Big and How in America's Upstart Industry* (New York: New American Library, 1984), 130.

11. For more information on Bricklin's background, see Dan Bricklin, *Bricklin on Technology* (New York: Wiley, 2009), 423–426; Levering, Katz, and Moskowitz, *The Computer Entrepreneurs*, 129–131.

12. Martin Campbell-Kelly, "Number Crunching without Programming: The Evolution of Spreadsheet Usability," *IEEE Annals of the History of Computing* 29, no. 3 (July–September 2007): 7.

13. For references to these various anecdotes, see Bricklin, *Bricklin on Technology*, 424, 426; Campbell-Kelly, "Number Crunching," 7; Robert X. Cringely [Mark Stephens], *Accidental Empires: How the Boys of Silicon Valley Make Their Millions, Battle Foreign Competition, and Still Can't Get a Date* (New York: Harper, 1996), 64–66; Burton Grad, "The Creation and the Demise of VisiCalc," *IEEE Annals of the History of Computing* 29, no. 3 (July–September 2007): 21; Levering, Katz, and Moskowitz, *The Computer Entrepreneurs*, 131.

14. The use of computing in business settings traditionally starts with the 1951 UNIVAC. Initially conceived to aid the automation of census calculations for the US government, the system was soon thereafter bought by a market research firm, an insurance company, and other corporate operations. Martin Campbell-Kelly, William Aspray, Nathan Ensmenger, and Jeffrey R. Yost, *Computer: A History of the Information Machine*, 3rd ed. (Boulder, CO: Westview Press, 2014), 99–103.

15. For an overview of the midcentury software products industry in the United States, see Campbell-Kelly, *Airline Reservations*, chaps. 4 and 5.

16. Roughly two-thirds of that revenue derived from industry-specific and cross-industry applications. Campbell-Kelly, *Airline Reservations*, 126.

17. Campbell-Kelly, *Airline Reservations*, 126–127.

18. Grad, "The Creation and the Demise of VisiCalc," 21.

19. Burt Grad provides a deeper overview of the influence the Typeset-10 and the Harris 2200 had on Bricklin's design ethos. "The Creation and the Demise of VisiCalc," 21–22.

20. Campbell-Kelly, "Number Crunching," 7. While most sources claim Bricklin and Fylstra were introduced by an HBS professor, Paul Freiberger and Michael Swaine claim Frankston already knew Fylstra, having converted a bridge program to the Apple II for Personal Software. Paul Freiberger and Michael Swaine, *Fire in the Valley: The Making of the Personal Computer* (New York: Osborne/McGraw-Hill, 1984), 229.

21. Dan Fylstra, "Personal Account: The Creation and Destruction of VisiCalc," May 2004, 4, https://www.computerhistory.org/collections/catalog/102738286.

22. Dan Fylstra, "The Radio Shack TRS-80: An Owner's Report," *Byte*, April 1978, 49–60; Dan Fylstra, "User's Report: The PET 2001," *Byte*, March 1978, 114–127.

23. For more information on the development of *Microchess*, see Freiberger and Swaine, *Fire in the Valley*, 134; Peter Jennings, interview by Sellam Ismail, "Oral History of Peter Jen-

nings," *Computer History Museum*, February 1, 2005, 10–19. For "Gold Cassette" reference, see ad in *Byte*, June 1979, 142.

24. Allan Tommervik, "Exec Personal: The VisiCalc People," *Softalk*, October 1980, 7.

25. While not addressed in this chapter, the selection of the Apple II as the initial platform was a choice with important implications for both the program and the platform. For more information on this, see Laine Nooney, *The Apple II Age: How the Computer Became Personal* (Chicago: University of Chicago Press, 2023), chap. 3.

26. For a summary of VisiCalc's design and implementation, see Grad, "The Creation and the Demise of VisiCalc," 24–26.

27. Dan Bricklin and Bob Frankston, interview by Martin Campbell-Kelly and Paul Ceruzzi, Charles Babbage Institute, May 7, 2004, 15; Dan Bricklin, "Special Short Paper for the HBS Advertising Course," 1978, http://www.bricklin.com/anonymous/bricklin-1978-visicalc -paper.pdf; Grad, "The Creation and the Demise of VisiCalc," 23.

28. Both Frankston and Fylstra have taken credit for coming up with the name "VisiCalc." See Bricklin and Frankston, interview, 40; Fylstra, "Personal Account," 8.

29. Bricklin discusses his development process on the Apple II in *Bricklin on Technology*, 427.

30. Fylstra, "Personal Account," 6.

31. According to Jennings, the rate was informed by the royalties he himself had on *Microchess*. Peter Jennings, "VisiCalc—The Early Days," *Benlo Park* (Peter Jennings's personal website), http://www.benlo.com/visicalc/index.html, accessed December 13, 2023.

32. Grad, "The Creation and the Demise of VisiCalc," 27.

33. Grad, "The Creation and the Demise of VisiCalc," 27.

34. Tommervik, "Exec Personal," 7.

35. The time-sharing system was a Prime 350 minicomputer, which had a PL/1 compiler similar to the time-sharing system Bricklin and Frankston had used during their time at MIT. It was not unheard of to develop microcomputer software in a custom minicomputer environment, which allowed for more deployment of programming utilities and tools. See Bricklin, *Bricklin on Technology*, 447; Grad, "The Creation and the Demise of VisiCalc," 26; Jennings, "VisiCalc—The Early Days."

36. Jennings, "VisiCalc 1979," Benlo Park, http://www.benlo.com/visicalc/visicalc2.html and http://www.benlo.com/visicalc/visicalc3.html. There is no archival evidence that Personal Software conducted any user testing beyond the development team (although apparently Frankston and Jennings both used VisiCalc to compile their 1978 taxes).

37. See ad in *Byte*, September 1979, 50.

38. Fylstra, "Personal Account," 9. "Whole product" was a concept developed by Regis McKenna. For more information on Regis McKenna's marketing strategies, see Regis McKenna, *The Regis Touch: New Marketing Strategies for Uncertain Times* (Reading, MA: Addison-Wesley, 1985).

39. According to Bricklin, the manual was drafted three times: first by Bricklin, then by a freelancer, and finally by Fylstra. Personal Software handled everything except the printing of the reference card; that was reproduced by Bricklin's father, Baruch Bricklin, who owned a print shop in Philadelphia.

40. Fylstra, "Personal Account," 10.

41. This image can be found in the September 1979 issue of *Byte*.

42. Jennings, "VisiCalc 1979," 3.

43. Bricklin, *Bricklin on Technology*, 452.

44. Stan Veit, *Stan Veit's History of the Personal Computer* (Asheville, NC: Worldcomm, 1993), 99.

45. Veit, *Stan Veit's History*, 100.

46. For more on the development of the mainframe and minicomputer software industry, see Campbell-Kelly, *Airline Reservations*, chap. 5.

47. For further explanation of the structure of data-processing departments and the challenges microcomputers posed to those employees, see Jeffry Beeler, "Personal Computing Is Big Business: No Turning Back," *Computerworld*, December 17, 1982–January 3, 1983, 21–24.

48. Carl Helmers, "Returning to the Tower of Babel, or . . . Some Notes about LISP, Languages and Other Topics . . . ," *Byte*, August 1979, 156.

49. Barbara E. McMullen and John F. McMullen, "Screen Envy on Wall Street," in *Digital Deli: The Comprehensive, User-Lovable Menu of Computer Lore, Culture, Lifestyles and Fancy*, ed. Steve Ditlea (New York: Workman, 1984), 277–278.

50. McMullen and McMullen, "Screen Envy on Wall Street," 278.

51. Peter A. McWilliams, *The Personal Computer Book* (Los Angeles: Prelude Press, 1982), 60.

52. Advertisement can be found in *Byte*, January 1981, 31.

53. Apple was so esteemed as an up-and-coming entrepreneurial company, it was featured in the 1979 premiere issue of *Inc.* See Norman Sklarewitz, "Born to Grow," *Inc.*, April 1979.

54. "Softalk Presents the Bestsellers," *Softalk*, January 1984, 272. This edition of *Softalk*'s bestsellers list reported that *VisiCalc* slipped from number four (its previous month's ranking) to number eleven. *Softalk* attributed the slippage to several factors, including the strong showing of entertainment software in the holiday season, competition from other products such as *Multiplan*, and what *Softalk* identified as the "changing profile of new owners" who were "more directed toward word processing than toward number crunching" (267).

55. "Softalk Presents the Bestsellers," *Softalk*, December 1981, 205.

56. Efrem Sigel and Louis Giglio, *Guide to Software Publishing: An Industry Emerges* (White Plains, NY: Knowledge Industry Publications, 1984), 49–50. While Sigel and Giglio document, in table 3.11, that the *value* of consumer purchases for the year was $905 million, publishers would not have captured more than 50 percent due to the prevalence of discounts and OEM (original equipment manager) bundling (placing publisher revenues closer to half a billion).

57. Sigel and Giglio, *Guide to Software Publishing*, 47.

58. Sigel and Giglio, *Guide to Software Publishing*, 46.

59. Personal Software, "VisiCalc," https://ia600803.us.archive.org/32/items/TNM_Visicalc_-_Personal_Software_Inc_20170922_0447/TNM_Visicalc_-_Personal_Software_Inc_20170922_0447.pdf, accessed December 13, 2023, 2.

60. Levy, "Spreadsheet Way of Knowledge."

61. Levy, "Spreadsheet Way of Knowledge."

62. Levy, "Spreadsheet Way of Knowledge."

63. William Deringer, "Michael Milken's Spreadsheets: Computation and Charisma in Finance in the Go-Go '80s," *IEEE Annals of the History of Computing* 42, no. 3 (2020): 56.

64. Harvey, *A Brief History of Neoliberalism*, 2.

65. Kera Allen, "'The Official Response Is Never Enough': Bringing VisiCalc to Tunisia," *IEEE Annals of the History of Computing* 41, no. 1 (2019): 42.

66. Dan Fylstra, quoted in "Tradetalk," *Softalk*, March 1982, 43.

67. For sources on the lawsuit between VisiCorp and Software Arts, see Andrew Pollack, "How a Software Winner Went Sour," *New York Times*, February 26, 1984, F1–F12; Denise Caruso, "VisiCorp, Software Arts Battle for VisiCalc Rights," *Infoworld*, March 5, 1984, 13–15.

68. Pollack, "How a Software Winner Went Sour," F1.

69. Kapor had developed VisiCorp's complementary VisiCalc products, VisiPlot and VisiTrend. When Kapor decided to leave VisiCorp, he was bought out of his royalties on these products for $1.2 million—a price VisiCorp hastily paid, and without challenging Kapor's

written addendum to his buyout agreement, which permitted him to work on his own spreadsheet product. More information on Kapor and Lotus 1-2-3 can be found in Levering, Katz, and Moskowitz, *Computer Entrepreneurs*, 188–195; Campbell-Kelly, *Airline Reservations*, 216; Amanda Hixson, "Lotus 1-2-3, Integrated Program for the IBM PC," *Infoworld*, August 1, 1983, 39–41.

70. Denise Caruso, "Software Gambles: Company Strategies Boomerang," *Infoworld*, April 2, 1984, 80–83.

71. Andrew Pollack, "Visicorp Is Merging into Paladin," *Infoworld*, November 3, 1984, 29.

Part II / Privatization

CHAPTER FOUR

Thatcherism, Science, and Greed

Jon Agar

> **Douglas Keay, *Woman's Own*:** . . . When I first interviewed you six or
> seven years ago you used almost the same words. Government
> statistics show divorce rate under 35 is nearly 50%, abortions
> have nearly doubled. We seem to have more violence, we have the
> yuppies of the City sort of violent with money. We have competi-
> tion and free enterprise and it seems somehow to go together
> with greed.
>
> **Prime Minister Margaret Thatcher:** No, it does not go with greed at
> all.[1]

The interview that appeared in *Woman's Own* magazine in 1987 was
controversial at the time and has been remembered ever since (not least by
sociologists) as the source of one of Margaret Thatcher's most infamous
remarks: "There is no such thing as society." But in it she also rejected the
public's tendency to associate the social problems and visible moneymak-
ing of the 1980s with greed. She was asked about this issue more than
once. To the popular BBC radio host Jimmy Young the following year, she

not only rejected the association of Thatcherism with greed but also expanded on the point to say it was a matter of individual qualities:

> **Jimmy Young:** Roy Hattersley [leading Labour politician], the other day, said: "The thug with too much money and too little conscience is the monster which has mutated from Mrs. Thatcher's open advocacy of selfishness and greed!" How do you answer the charge that your policies do in fact encourage selfishness and greed?
>
> **Prime Minister:** Not in any way! My policies encourage personal responsibility.[2]

In this chapter, I explore what the interconnections might be between Thatcherism, greed, and science. One point I argue is that Thatcher's broader emphasis on personal responsibility (which she viewed as the very opposite of greed) also shaped how she saw the need for scientists to take more responsibility as individuals and as entrepreneurs. It was part of what I see as a long-sought and late-coming formation of a distinctively Thatcherite science policy. Personal responsibility was also central to how Thatcher thought about environmental issues, as I show in the last section of this chapter, on climate change. So what we might take as an instantiation of "greedy science" (the individualistic scientist embracing wealth creation) became an ideal for Thatcher, but she, as the above quotations suggest for the general case, would have rejected the label of "greedy." To make sense of Thatcherism, science, and greed, we have to say something about what these three terms mean, and meant, in Britain in the 1980s.

Thatcherism

Margaret Thatcher served three consecutive terms as the prime minister of the United Kingdom, from 1979 until 1990, throughout our entire "greedy" decade of the 1980s. She was born Margaret Roberts in a lower-middle-income, Methodist, Liberal-voting, small-town family—hence the class-inflected "grocer's daughter" jibe. She studied chemistry at Oxford University, starting in the midst of war in 1943, completing a fourth-year project under the future Nobelist Dorothy Hodgkin. Not only was she to become the first British prime minister with a science degree, but also and, in a way more importantly, she gained a working knowledge of employment in science. Her first two jobs were in industrial science: at the materials firm BX Plastics (BX stands for British Xylonite; xylonite was a very early plastic, a contemporary of Bakelite) and the second, deliciously for "greedy science," at the large and innovative food and teashop firm Lyons & Co. While working for Lyons, Margaret, soon also to be married to the businessman Denis Thatcher, stood as a Conservative candidate in parliamentary elections in 1950 and 1951. Her campaign featured photographs of her in a white lab coat, posed with chemistry glassware. Margaret Thatcher eventually won the constituency of Finchley, North London, in 1959, a seat she held until she was forced to resign as prime minister over three decades later.

In 1970, Edward Heath appointed Thatcher to a ministerial position, as secretary of state for education and science. While most people recall her period as a minister, if at all, for the withdrawal of the free, publicly funded provision of milk for schoolchildren ("Thatcher, Thatcher, Milk Snatcher!" was a chant that endured), it was also the time for a key, perhaps symbolic, decision in science policy. The issue in 1971 was the framing and funding of science conducted for government departments, and the choice was between tradition (a relatively hands-off provision of funding) and something new and distinctly market oriented, in the shape of the "customer–contractor principle." In the words of its architect Lord Rothschild, the customer–contractor relationship meant that "the customer [a government department] says what he wants; the contractor [the scientist, the provider of research and development] does it (if he can); and the customer pays."[3] Thatcher went into the crucial meeting with Heath, Rothschild, and a pair of the most senior servants, briefed to the gills by her department to support and protect the traditional mechanism but emerged from the room convinced that the market route was right.

The moment is telling because at some point during the early 1970s, Thatcher became a Thatcherite. Previously by no means exceptional in terms of ideology, by the time of the fall of the Heath government, in 1974, Thatcher was within a small circle of ideologues on the right of the Conservative party, some of whom in turn were fully immersed in the economics of Austria and Chicago. Although prodigious at commanding ministerial paperwork, there is little evidence that at the time Thatcher read Hayek and the like deeply at source (although she certainly had later). It is just as likely, I have argued, that Thatcher was led into thinking as a Thatcherite by confronting decisions such as that on science policy in 1971 and other moments of practice.[4] A conviction politician, for Thatcher it was moments, not ideas alone, that convinced.

What do we mean by "Thatcherism"? We can distinguish a parochial and an international meaning, both of which seemed more distinct to commentators in the 1980s than they do to scholars in the 2020s. The parochial version refers to policies made and implemented in Britain under Thatcher. Contemporary critics such as Stuart Hall and political science academics such as Andrew Gamble would agree that Thatcherism encompassed and aimed for a "free economy and the strong state" but also the reduction of the size and role of the public sector, opposition to the public provision of public services, privatization of nationalized industries, a championing of market mechanisms over social democratic alternatives and consensus, "family values," and an ideological opposition, it almost goes without saying, to the socialist left.[5] The international meaning overlaps with the parochial one but would stress Thatcherism as a national variety of neoliberalism. Thatcher is often bracketed with Reagan here.[6] Later historians have been skeptical about both Thatcherism and neoliberalism as revolutions, noting continuities and contradictions aplenty, in particular arguing that the framing of Thatcherism as a smashing of, and indeed liberating from, a postwar consensus is to repeat without questioning the rhetorical move of Thatcher and her insurgent circle of the mid-1970s.[7] "Thatcherism," argues David Edgerton, "did not transform the rate of growth of the British economy, nor did it unleash a radical new British entrepreneurialism, or indeed consumer sovereignty in a national market," but rather is best seen as a "ruler's revolt, a radical strengthening of the power of wealth, and the wealth of the powerful, and a new economic form, where, for example, privatized infrastructure was a means of extracting profit on an unprecedented scale," an internationalization enabled by the contingencies of unprecedented food security, the sale of public capital, and a relatively generous welfare system.[8]

The best of contemporary historiography has argued that historical investigation of Thatcherism should look beyond the contemporary first takes of journalists to see what "Thatcherism" meant to ministers and especially civil servants.[9] If we do so, then it is striking how rare the term is. Indeed, my research into Whitehall documents relating to science and technology, revealed one, albeit interesting, appearance.

In October 1986, after ten years of piecemeal construction, Margaret Thatcher opened the M25, a 117-mile orbital motorway encircling London.[10] In the previous month, a fierce argument had raged across Whitehall on how a final segment might be built and, more significantly, funded. The M25 must cross the Thames twice, and east of the city, the river is at its widest. The vast, continuous circling of traffic was likely to back up at the pinch point as it had to use the old, existing tunnel at Dartford. What was needed was a new East London crossing. In the view of the Number 10 Policy Unit, it was a "precedent setting case with wider implications—a way to harness the efficiency and enterprise of the private sector, and to provide useful infrastructure earlier than [traditional] constraints on public expenditure . . . would allow. In this context the Dartford project is seen as the touchstone of the Government's resolve to press forward and secure the benefits of a new aspect of privatisation."[11] Removing the Dartford motorway bottleneck by a private-finance initiative was explicitly described, to Thatcher, as the "Thatcherite solution"[12]—"a touchstone" (again) of whether Thatcherism is to succeed against conventional Treasury thinking."[13] Indeed, given the choice between tunnel and bridge, it should be the latter, as it would be visible, all the better as a lasting Thatcherite symbol.[14] The Dartford bridge (an artifact with politics) is a rare example of the term "Thatcherite" being applied to a technology by civil servants and even then only by special advisers in the Number 10 Policy Unit, the part of the civil service closest to Thatcher both physically and ideologically.[15] To find Thatcherite science, we will not see the term being used, but I argue below that the label fits.

"Science"

What was science in the United Kingdom in the 1980s? A very brief overview might look like this. Through the 1960s and 1970s, the United Kingdom devoted a relatively high, globally speaking, or relatively average (when compared to other rich, industrial nations) portion of its economy to science, gently declining from about 2.3 percent to 2.1 percent gross domestic product between 1964 and 1975.[16] Just over half of the funds for science came from the government, although nearly two-thirds was performed by industry. The difference reflected the presence of an active industrial strategy, including large programs, such as the £350 million Alvey initiative discussed below, of support for research and development of interest to high-tech and science-based industries. Key sectors were pharmaceuticals, chemicals, aerospace, and electronics. About half of research and development was military and half civil. In 1983/1984, the funding of science through the Ministry of Defense almost exactly matched government funding channeled through civil routes.[17] Scientific advice to government was essential in formulating responses to the great challenges of the 1980s, including AIDS, Chernobyl, acid rain, and nuclear missile defense.

Science was therefore done in a huge number and variety of sites. Universities would have both teaching and research laboratories, enabled by a "dual support" system that funded a teaching base through a university grants committee and specific research programs through several research councils. Scientists' participation in international networks and projects (such as fundamental physics at CERN) was supported through public funds. Government scientific sites included both large and small defense research labs, as well as centers for a range of subjects that might be of public service (such as meteorology for weather forecasting or metrology for standards), support nationalized projects (such as nuclear energy), or provide scientific monitoring necessary for regulation and enforcement (such as water quality).

The prominence of government funding meant that choices over the government's science policy were highly consequential for scientists working in Britain. Likewise, the scale of funding meant that science had to make a case for its support against opposing claims on the public purse. For both of these reasons, Margaret Thatcher, who came to power promising a new politics, was met with trepidation. One of the very first experimental privatizations, testing the water for the immensely consequential transfer of assets from public to private sectors—the sale of the Radiochemical Centre to become Amersham International Ltd—was of a science-centered body.[18] Thatcher also promised cuts to public expenditure, and the cuts as they affected the university dual-support system in the early 1980s (some of which, it is now clear in retrospect, had unintended consequences for science) seemed an existential threat to academics—a Save British Science movement was launched in 1985. By other measures, science in Britain was doing well: there were five science Nobel Prize winners (duly invited to Number 10) during Thatcher's prime ministerial years, including two born outside the United Kingdom, and a science book (*A Brief*

History of Time) by a British scientist (Stephen Hawking), published in 1988, was a publishing phenomenon.

"Greed"

What did "greed" mean in the 1980s? There are clearly long-established meanings. First, greed can refer to excessive individual wealth. In the 1980s, at least in perhaps clichéd commentary and popular culture, greed as the cause of wealth could be valorized (as we are reminded in figure 4.1, which pairs Michael Douglas and Margaret Thatcher in a cartoon from the Labour-supporting, left-wing, red-top British tabloid *The Daily Mirror*).

"YOU'RE REALLY CHEERING US UP, MR DOUGLAS. COULD WE HEAR THAT BIT AGAIN ABOUT GREED IS GOOD?"

Figure 4.1   Charlie Griffin, "You're really cheering us up, Mr. Douglas, could we hear that bit again about greed is good?" *Daily Mirror*, April 28, 1988. © Mirrorpix, supplied by British Cartoon Archive, University of Kent, https://archive.cartoons.ac .uk/Record.aspx?src=CalmView.Catalog&id=41420

Second, greed has also meant excessive consumption. (As an aside, we might under the greedy science banner consider 1980s developments in food science and technology.[19]) The greedy consumer could be corporate. Margaret Thatcher, in 1983, considered Big Science, especially the particle accelerator facility at CERN, to be greedy consumers of resources for science.[20] One of the largest companies in the United Kingdom, the electronics, transport, and defense conglomerate GEC, was frequently accused in the 1980s of sitting on a "cash mountain," money that could have been freed up and distributed to support research and development.[21] "The companies have money coming out of their ears—about £8000 million worth at the last count," noted science journalist Tom Lloyd in 1985, in an article subtitled "Greed Foils Britain's Attempt to Encourage High-Tech Businesses," about big companies applying for Alvey program support for fifth-generation computing, adding "GEC's notorious 'cash mountain' was once a special case; it is now merely a prime example."[22] This cash mountain was a fund built up for the purposes of acquisitions, the prevailing style of 1980s greedy capitalism. GEC's CEO, Arnold Weinstock, was one of the few businessmen who could call on Number 10 Downing Street almost when he wished.

If the big companies were greedy, they were contrasted to the hungry, lean entrepreneur. "There are signs now that a new breed of 'scientist/entrepreneur' is emerging in Britain, based on the American model," Lloyd wrote. And it was the "small companies" of these individuals that "depend for their survival on making the most of their limited financial resources. Large companies have grown used to, and have become fat on, the luxury-funding traditions of the Ministry of Defence."[23]

Greed could be corporate but was more often, like wealth, an individual vice. It is noticeable that the older generation of entrepreneurs (such as Robert Maxwell, who had built a science publishing business from the 1960s and by the 1980s owned the *Daily Mirror* and was one of the United Kingdom's most prominently rich businessman) were portrayed as gluttonous, in ways that the next generation (say Richard Branson in the mid-1970s, Clive Sinclair in the early 1980s) generally were not. How would the emerging "scientist/entrepreneur" be seen?

In what ways might science intersect with greed? Greed in science, or at least in representations of science, can work on one of at least three, broad ways. First, greed might be the *target* of science, its opponent. At the end of the 1941 film *Dr. Ehrlich's Magic Bullet*, Edward G. Robinson, who played the titular scientific hero, utters the following dying words: "There will be epidemics of greed, hate and ignorance. We must fight them in life as we fought syphilis in the laboratory. We must fight, fight. We must never, never stop fighting."[24] The film, one of the peaks of Hollywood science, left enough of an impression for the *New Scientist* to quote the lines, somewhat erroneously, thirty-five years later, on the eve of our greedy science era, although by then the sentiment must have felt quaint, redolent of a past era.[25]

Second, greed can be a motivation of scientists because scientists are human like the rest of us, or at least once they are represented as such (Jim Watson's *Double Helix*, his controversial, all-too-human account of the discovery of the structure of DNA, published in 1968 and discussed by Robin Wolfe Scheffler in chapter 8 of this volume, was a significant moment). Theodore Roszak picked out these lines while reviewing William Broad and Nicholas Wade's history of fraud in science, *Betrayers of the Truth* (1983): "Science, they say, is 'a human process governed by the ordinary human passions of ambition, pride and greed.' In every example they offer, it is competitive egoism that has subverted the ideal of objectivity."[26] Or consider the view that cold fusion was an acrimonious mess because it was "a combination of glory and greed."[27] In these cases, greed was evoked as an ordinary human characteristic that could lead science astray. But greed could also be the ordinary human characteristic that drove ordinary science. The suggestion might be disturbing. In July 1985, *New Scientist* published a letter from a concerned reader: "I have noted with growing distaste the advertisements placed in New Scientist by Scientific Staff Consultants, illustrated with unpleasant comic strips representing young scientists as envious, greedy careerists. Is this joke in bad taste, or a calculated insult to the readers? Or worse yet, is it true?"[28] SSC (Scientific Staff Consultants) were regular advertisers in the 1970s and 1980s in *New Scientist*, which in turn was one of the main carriers of job adverts for scientists. Some SSC ads had cartoons. Only one I've found looks like a candidate for what upset the reader above.

The advertisement carried in the March 28, 1985, issue showed a man looking enviously at the shiny new car of an ex-colleague and wondering about seeking a better-paid job (figure 4.2). What is striking about this

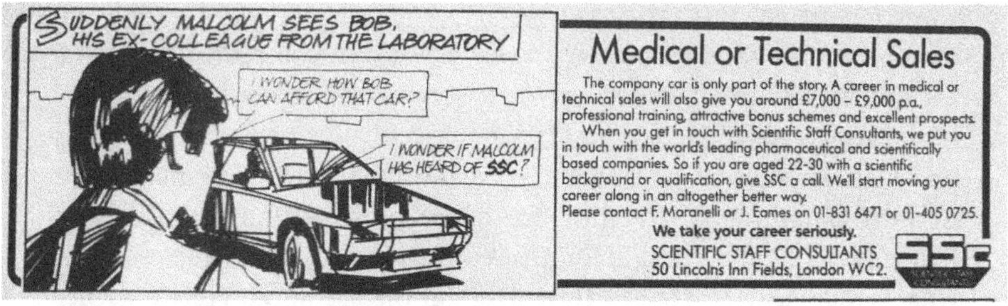

Figure 4.2   Advertisement for scientific staff consultants. *New Scientist*, March 28, 1985, 77

incident is that the (presumably older) reader objected to this mildly ava-ricious advertisement and saw it as a sign that a younger generation of scientists might be "envious, greedy careerists."

The third position would be that greed is integral to flourishing science. How might this work? The motivations for (good) science were once a cen-tral concern of the sociology of science. Merton's mid-twentieth-century CUDOS norms (communal holding of knowledge, universalism in the sense of equality of participation, disinterestedness, and organized skep-ticism) had no place for greed. Neither, really, did the "counter-norms," offered in the countercultural 1960s and 1970s, although "miserism," the hoarding or "protective control over the disposition of one's discoveries," the opposite of communal ownership, came close.[29] More recently, the distinctive epistemic value of scientists has been curiosity, again un-greed-like, although I note below it has greedy origins. Good scientists might have an avaricious appetite for knowledge. But we are missing, as far as I can find, an explicit and articulated 1980s defense of the greedy scientist along the lines of Gordon Gekko in a white coat; there are no trickle-down justifications of overconsumption and personal pursuit of vast quantities of certified, natural knowledge.

A Thatcherite Science Policy

When Margaret Thatcher was considering who to promote, or who to ex-clude, from her Cabinet, a question she would ask her closest advisers would be: is he or she "one of us"? Inclusion meant that the person in ques-tion subscribed to Thatcher's particular brand of Conservativism and shared her values. *One of Us* was used as the title of one of the first substantial biographies of the prime minister, by the political journalist Hugo Young. The phrase captured something distinctive about Thatcher's approach to people and politics. In 1981, still in the fast-moving early years of her administration, Thatcher made a point of inviting a clutch of independent inventors to Number 10 to discuss their issues. The inventors were individuals who worked in garage workshops or at most ran very small companies; their economic significance to the British economy was negligible. So what was it about them that merited symbolic inclusion?

Thatcher opened the meeting with a speech: "I am very pleased to have the opportunity of meeting you tonight. As you know, a principal aim of the Government's economic policies has been to stimulate individual initiative by encouraging the formation of new businesses and enabling their owners to retain more of the wealth that they have created."[30] She expressed hope that the inventors might "take up much of the labour now being shed because of the rundown of our older industries," if only barriers to exploitation of new ideas could be removed. Yet it would be hard to imagine that the employment capacity of the steel, coal, and shipbuilding industries might be replaced by the growth of businesses represented by the inventors in the room—inventors such as M. A. Hiles, who offered "sorbothane, a polymer that simulates the energy-absorbing properties of human flesh" suitable for orthopedic shoes; P. C. Dowles, inventor of a "new type of hotel safe"; or H. Calvert, a schools design prize competition winner who offered a "portable gymnasium." Their "individual initiative" might have appealed to Thatcher, but no amount of encouraging them to retain more of the wealth they created would lead Britain to a bright, new economy.

In the room with the "inventors/innovators" were people from "government" (civil servants and politicians), "industrialists" (representing big industry), and a third category, "entrepreneurs/financiers." It is striking that this category described sources of capital: venture capitalists, bankers, and pension fund holders. It did not encompass, as yet, the scientist-entrepreneur. A month after the inventors came to Number 10, Thatcher reflected on what she had learned, telling the special interest group for science of members of Parliament:

> We have a long way to go before we will be giving proper recognition to our inventors. Of course scientific excellence is admirable, but so is engineering

competence and technological capability. So is founding a new successful business and creating a lot of wealth.

But most to be admired are those *rare, gifted persons who can do all of this.* The supercilious attitude that some sections of our society had towards our engineers and entrepreneurs was always a ridiculous affectation now it is positively dangerous [my emphasis].[31]

Biotechnology is the story here. Just as in the United States, the exploitation of genetic engineering techniques had generated excitement and an increasing vortex of investment in the 1970s and 1980s. But the scientist-entrepreneur, of the type we associate with the commercialization of recombinant DNA techniques and the biotech boom (the "academic entrepreneur" discussed by Cyrus C. M. Mody in chapter 1 of this volume), was not yet in evidence in Britain. Patents generated from university research were often placed in a national corporation, the National Research and Development Corporation (NRDC). Indeed, the NRDC had the right of first refusal on patents from work funded by the research councils. The monopoly had only been partially broken by the establishment in 1980 of Celltech, significantly a limited company, not a corporation, supported by venture capital and a deal with the Medical Research Council for exclusive, though time-limited, rights to its patentable research in the areas of recombinant DNA and cell fusion.[32] Celltech was hardly the creation of greedy science; it was guided into being by an instrument of big government intervention and planning, the National Enterprise Board, and built on its own monopoly of a niche of publicly funded research.

In 1980, Margaret Thatcher returned from a visit to Cambridge, United Kingdom, extremely angry. She had heard two stories of scientists not receiving commercial backing. One involved image-intensifying technology at the Mullard radio astronomy observatory. The other concerned monoclonal antibodies. In some ways, it was already an old story, although it seems to have been new to the prime minister. In the mid-1970s, building on work going on in universities across the world, Georges Köhler and César Milstein at the Medical Research Council (MRC)–funded laboratory of molecular biology had succeeded in creating a "hybridoma technology," an immortal cell line capable of producing an endless supply of monoclonal, or in other words identical, antibodies with known specificity. Monoclonal antibodies would prove to have many, and many lucrative, uses in diagnostics and therapeutics. Before publication, they offered the patent to the NRDC, which did not take it up. Seeking to patent such a technique was not, in the mid-1970s, yet routine or expected, although attitudes to commercialization were fast changing.[33] Thatcher heard a more simplistic telling of the story: Milstein did not patent.

She demanded an explanation from her chief scientific adviser. He replied hastily but in detail. The "three parties in this debacle—the NRDC, the MRC and the individual research worker," he wrote, "must be treated separately." The NRDC sat on its monopoly rights and was "cautious and risk averse," although this was being shaken up by the formation of Celltech. The MRC was only now undergoing a "growing realisation . . . that commercial, as well as medical, objectives must become acceptable aims of MRC policies," a "change in attitude . . . greatly helped by the acceptance of the so-called 'Rothschild' principle of a customer-contractor relationship." And third and finally, the individual research worker was "now much more conscious than they were a decade ago of the need to take a personal interest in the commercial fate of their discoveries"; new policies pioneered for biotechnology would "publicise the benefits that can accrue to both Universities and their employees from entrepreneurial activities. Such publicity helps, as would <u>greater financial rewards to academic entrepreneurs</u> . . . as would <u>discrimination against those who do not become entrepreneurial</u>" [the underlining is Thatcher's].[34]

In 1985, Keith Joseph, the secretary of state for education, drafted a new policy statement "reflecting [Thatcher's] wish that he should go further towards devolving rights in research to the individual researcher"; specifically, a key paragraph exhorted universities to give individual researchers the opportunity of exploiting their own work. This paragraph also notes that universities should in these circumstances share in royalties because public funds are involved.

While Thatcher declared herself "delighted" with this specific move, her anger did not subside.[35] She would remain intensely sensitive to any mention of the monoclonal antibody case, and the moral she drew from it would be remembered. Something, for her, was not right in British science and science policy.

Until 1987, there was much about UK science policy that Thatcher didn't like. She thought British scientists weren't entrepreneurial enough. She was deeply unenthusiastic about the industrial strategy programs of pumping research money into promising areas of high technology, such as the flagship Alvey program for computing, or into defense research. She thought the UK contribution to the European big science centers, such as CERN, was wasted and wasteful.

The change in policy in 1987 came from her interaction with two people, each of whom pointed her toward a way her policies for science could be more Thatcherite: Max Perutz, who insisted that the individual scientist must take responsibility, and George Guise, who showed her how and why public money should be withdrawn from near-market research in order to unleash science as a wealth-creating force.

Max Perutz, born in Vienna in the year of the outbreak of World War I, arrived in Cambridge on the eve of World War II. He established, within the Cavendish, what would become the MRC Laboratory of Molecular Biology. His 1950s investigations of hemoglobin using X-ray crystallography would win a share of a Nobel Prize in 1962, and by the 1980s, he was a senior, highly respected scientific figure developing a more public reputation for essays and reviews. In 1987, he wrote an intemperate (or, if you would rather, passionate) article for the *New Scientist*, "How to Stifle Innovation," a response to a plan floated by the Advisory Board of the Research Councils to rank institutions by research quality and channel more money to the higher-ranked ones.[36] Perutz argued that the best science came from independent individuals, or at best small partnerships, working without higher direction and taking personal responsibility for their choices. In "How to Stifle Innovation," Perutz wrote,

> My own laboratory . . . is often held up as a model of a centre or excellence, but this is not because I ever "managed" it. I tried to attract talented people by giving them independence, listening to them and taking an interest in their work, helping them to get what they needed for it and making sure they got credit for it afterwards. I also followed the tradition that I had learnt at the Cavendish Laboratory of letting young people publish their work independently, because I knew this to be one of the most important stimuli to originality. Had I tried to direct people's work, the mediocrities would have stayed and the talented people left. The laboratory was never "mission oriented."

Thatcher was shown the article, and read it, as we can tell from the blue ink underlining. Perutz's mention of monoclonal antibodies as an example of such research probably fired her up further.

She had been shown Perutz's article by the man now her closest adviser on science, the businessman interned at the Number 10 Policy Unit, George Guise. For him, it was one more piece of evidence that supported his argument for a new science policy. As he wrote to Thatcher, "Government should fund basic science, but very little technology except where it is the user."[37] Specifically, government could and should "be quite generous" funding basic science, "with the money concentrated intellectually and geographically on specific centres of excellence," until it was "sufficiently well understood to make some machine based upon it" but no further. "Beyond that point," argued Guise, "we are into the realm of commercial development."

So the new science policy would have two sides, one concerning applied research relevant to industry and the other side concerning basic research.[38] On the first, Guise persuaded Thatcher that her government should cut funding from near-market research, because only then would profitable companies respond and fund more of their own R&D. As Guise pointed out, this science policy would truly resonate with Thatcherite policy; it was the one "consistent with our long term goal of minimising State intervention."[39] The "proper role for Government," he said, for industrial research, would be restricted to "co-ordination, information dissemination and the reduction of bureaucratic impediment." "Basic research," on the other hand, was "essential to long term national prosperity and its funding is a primary function of Government."[40] It was "organically part of the national interest and . . . the route to success is to back individuals and teams"—and to do so in an unashamedly elitist way, to spread the jam widely was "a diluted form of socialism"—and not set "remote goals which pre-judge the outcome of work." Perutz's testimony was useful here in forbidding mission-oriented approaches to basic science. Furthermore, such basic science was rebranded "curiosity-driven research."[41] Guise employed a large quantity of quite tendentious examples from the history of science to try and show that the independent, individual work of curious scientists had large economic rewards in unanticipated ways.

The shift in policy, with its immense implications for the support of industry, was never formally announced but began shaping reports and choices. It was fully present, for example, in Thatcher's most significant speech on science, to the Royal Society in 1988, where she made a clarion call for action on climate change but also spoke of the "benefits

of curiosity-driven research," again recalling Perutz's view on the independent researcher:

> I have always had a great deal of sympathy for Max Perutz's view that we
> should be ready to support those teams, however small, which can demon-
> strate the intellectual flair and leadership which is driven by intense curios-
> ity and dedication.
>
> A good researcher is keenly competitive and wants to be first.
>
> The final stage of the race for the DNA structure was as exciting as any
> Olympic marathon. The natural desire of gifted people to excel and gain the
> credit for their work must be harnessed. It is a great source of intellectual
> energy.[42]

In a 1989 speech on the occasion of the fiftieth anniversary of the Parliamentary and Scientific Committee (essentially a group of members of Parliament interested in science), she not only credited Perutz but also set
out the Guise argument about the central importance of curiosity-driven
science for technology and economic prosperity. She reminded the members of Parliament that when their committee had been formed, in the
1940s, "much of the basic science underlying our modern economy was
already known. . . . Electromagnetic theory, relativity, and the quantum
mechanical basis of modern chemistry were already understood, but we
did not have things like digital watches or computers or cellular telephones
or word-processors or fax machines or polythene," and it had taken "many
years for industry to appreciate and develop the range of new products that
could flow from this knowledge and every bit as important was the imagination to design and market them effectively."[43] "The truth," concluded
Thatcher, was "that the greatest economic benefits of scientific research
have always resulted from advances in fundamental knowledge rather
than the search for specific applications."

Perutz, although he elsewhere expressed views nonaligned and even
critical of Thatcher's,[44] continued to be source of guidance. We also know
they met at least once. In another 1989 speech at the *Good Housekeeping*
awards dinner, which contains many comments on science, including climate change, she recalled,

> I had recently to lunch in Number Ten all of the Nobel Prize winners. It was
> a great joy because we've got rather a lot in our country, more in proportion

to our population than almost any other country. And sitting next to me was one of the most remarkable scientists called Max Perutz, who came over from Germany during the dark days, became a Nobel Prize winner and has done so much for us. And I was quite startled when he said, "Do you know women's liberation could not have succeeded if science had not provided women with household technology and many other scientific advances."[45]

Thatcher then went to on expound her theory that the British empire would have been impossible after the invention of the telephone: "It grew up because we didn't have the telephone and people had to make their own decisions on the spot and not refer them to headquarters." In other words, like the independent, individual scientist, they took personal responsibility for decisions.

Thatcher on the World Stage: Responsible, Not Greedy?

In the late 1980s, Margaret Thatcher, quite startlingly, became the foremost advocate among world leaders for action on climate change. This prominence has been seen as evidence of Thatcher's clear-sighted capacity for leadership, taking responsibility by urging difficult decisions at an international level, and the influence of Thatcher's background competence in science. Only her prepared mind, among international leaders, could grasp the implications and necessity for action in such a complex, science-based issue. Neither account is persuasive in their simple forms. To do better, we need a greater understanding of what responsibility meant to Thatcher.

The path to climate change leadership was full of contingencies and interests that were national (and nationalistic) and party political as much as they were international. In the early 1970s, a combination of environmental concern, futurism, and excitement about the predictive possibilities of computer modeling had opened a space within British governmental circles for attention to long-term trends; climate change, as I have shown elsewhere, became a topic of attention as a result.[46] A report on climate change was written and ready to be published in 1979. But the new Conservative government declined to prioritize it, or even at one stage to publish it—with prime ministerial "coolness" mentioned by civil servants as a reason. First, then, it certainly was not inevitable that climate change would become a major issue for Thatcher. In her introductory meeting with her chief scientific adviser, she had reacted "incredulously, 'Are you telling me I should worry about the weather?'" when he raised the issue.[47] Second, however, there existed a deep commitment to the United Kingdom as a nuclear power. The pursuit of a British nuclear bomb, notoriously "with a Union Jack on it," was part and parcel of maintaining a seat at the highest tables and a basis for a less unequal relationship with the United States in postwar decades of greatly diminishing Great Power status. Thatcher, like most British politicians, embraced this position. Furthermore, nuclear was not coal. Coal, to the Tories, meant strikes and union power. If action on greenhouse gases meant more nuclear and less coal, then party interests were aligned. Thatcher in climate change debates would frequently remind listeners that Labour's interest in coal came from union influence.

Third, across Europe there were gathering anxieties about environmental degradation, and by the late 1980s, in West Germany for example, the issue had become influential at the polls again. European politicians wanted to be seen to take action. Performing this action more visibly, better and on a world stage, had attractions to Thatcher.[48] Fourth, there was the contingency of the war in the South Atlantic. The defeat of Argentina in the Falklands War was critical to Thatcher's political fortunes, and Thatcher subsequently emphasized that she had supported increased funding of science through the British Antarctic Survey, overriding advisers.[49] (Conducting science was essential, under the 1959 Antarctic Treaty, to diplomatic claims for national interests in the South Atlantic.) When, again quite contingently, the depletion of the ozone layer was discovered in 1985, three years after the Falklands, partly on the basis of British Antarctic Survey data, Thatcher saw vindication and a narrative for action on an issue of international atmospheric pollution. The removal of damaging industrial gases, chlorofluorocarbons used as refrigerants, envisaged by the Montreal Protocol gave a model of what international action on greenhouse gases might look like. Thatcher grasped this accidental opportunity.

Climate change can be seen as the result of unbearable industrial appetites for energy. Emission of carbon dioxide, the primary greenhouse gas, is integral to coal- and gas-powered industrialized economies. "Responsibility" is a fraught term here: responsibility in the past, present, and future; responsibility under conditions of knowledge and ignorance of the facts of climate change; responsibility as it lies unevenly among the rich and the poor. How did Thatcher project leadership and finesse responsibility on an issue that was, as much as any other, her country's legacy as the first industrialized nation, the first country greedy for coal? What did Thatcher mean by responsibility and personal responsibility in particular? First and foremost, for Thatcher, personal responsibility was what was denied by socialism. In her speech on "The Principles of Thatcherism" in Korea in 1992, after citing Hayek and Karl Popper, she stated that collectivism "denies the spirit and substitutes 'state' judgment for personal responsibility."[50] Likewise, in an early post–Cold War speech at the Jagiellonian University in Krakow, she argued that "socialism is a dangerous, but undeclared, enemy of freedom—and I do not just mean the socialist command economy but all those attitudes which socialism fosters: dependence, passivity, conformity, envy, lack of personal responsibility and lack of initiative."[51] This bracketing of personal responsibility among other consequences could also be heard in her speech on the event of Reagan's eighty-third birthday, in which she praised the ex-president in the following terms: "When others spoke of the fear of war, you spoke of the need for warriors and peace through strength. When others bewailed the failure of big government to provide for the collective good, you spoke of self-reliance, of personal responsibility, of individual pride and integrity."[52] Ideologically, then, Thatcher viewed responsibility as a virtue suppressed by socialism, by collective visions of society. Talk of "social responsibility" would have been a contradiction in terms for her. (It is worth recalling here that "social responsibility" was the left's framing of science and society issues in the United Kingdom, dating from the launch of the British Society for Social Responsibility in Science in the late 1960s.[53]) On one rare occasion, after stating that politicians should be "very very careful to speak about religion because it is a personal thing," she said that she had "aways taken the view" that the "personal responsibility" as a foundation for democracy and human rights rested on Christianity (she had had a Methodist upbringing).[54] Finally, she thought her government's policies were bringing about a "rebirth" of personal responsibility and stated so explicitly in her New Year's message of 1985, looking back at a year that had seen her near assassination by an IRA bomb, close to record unemployment, and the miners' strike.[55]

Now let's turn to Thatcher on climate change action and responsibility. Her first major statement of the seriousness of climate change came in the speech to the Royal Society in 1988, discussed above: "we have unwittingly begun a massive experiment with the system of this planet itself."[56] It was followed, this time on the world stage, by a speech to the United Nations (UN), in which she said all countries must act together on the basis of "sound science and sound economics." In the drafting of the UN speech, Thatcher stripped reference to responsibilities to "society," replacing it with "the way we live"—personal, not social, responsibilities.[57] Globally, then, her message to nations was that "none of us can do it alone,"[58] but the correction in the draft points toward a significant tension.

Thatcher saw a chain of responsibilities once the "shock" of the scientific revelation of climate change had been accepted. First, she recognized a duty to act but also the need to recognize that action had financial costs. One cost was investment in nuclear: "we have to do the things on environment because we have a duty to do so and most of us wish to improve the environment in any event. It cannot be done without a cost. . . . The answer to the greenhouse effect is, of course, to have more nuclear and if we have more nuclear, all the technology is known to look after the residual nuclear waste, that too costs money but you do not get the greenhouse effect from that."[59] Second, the responsibility was intergenerational: "We, who have inherited so much, must hand on a safe, secure future to our children and to their children; to all who come after us. . . . 'No generation has a freehold on this earth. All we have is a life tenancy—with a full repairing lease.' "[60] She saw intergenerational responsibility as a facet of conservatism: "My whole sort of political philosophy is that what you inherited from your forefathers, it is your duty to add to it, to give to next generations. I belong to a conservative party, conserving. The problems science has created science in fact can solve and we are setting about it." The tension over responsibility is, however, most apparent when it is noticed that Thatcher, when asked about environmental issues, would frequently answer questions on global scale issues with points about local ones, especially litter.[61] At a press conference at the Scott Polar Institute, Thatcher was asked twice, directly, about the responsibility of government and both times turned it into a point about the responsibility of all of us, as individuals, and again gave littering as an example. Litter was disorder and visible evidence of lack of personal responsibility.[62]

The rhetorical drift from global to local is significant, I suggest, because at the local level, the ideological argument about personal responsibility and the role of government was clearer: "Again, I have to say to people: 'Look! You must work this out with local people! Freedom incurs responsibility, but you must make those decisions and not just all put them upwards!' and some of them are now making them."[63]

In summary, I would argue that Thatcher's brief period as world leader for climate change action came from a mixture of shallow, contingent, political reasons and responsiveness to the deeper revelation of climate science. She was able to raise the issue on the global stage and had a model to hand, in the Montreal Protocol, of what international action might look like. But her gut instinct, ideological and even religious, was that environmental action came from personal responsibility, the individual, an ultimately inadequate response. In her post–prime ministerial years, Thatcher moved toward climate skepticism. In the three decades since her speeches on climate change, international political action has been agonizingly slow, even as the planet boils and the air fills with the smoke of wildfires.

Conclusion

Well this morning I am cooking some red cabbage. There it is boiling away. You can see it. Let us just turn it off; it is getting too much smoke.

What I am going to do is pour some of the juice out into this jug so that we can see it more clearly. There! Goodness me, a little splash there. Let me just tidy that up. I do hate an untidy kitchen. There we are!

Now you can see it is really a rather dark purpley colour. That is because, as you and I know, there is something called anthocyanin in it, which is the colour that gives . . . the dye that gives red colour to rose petals, berries, as well as to red cabbage, and to black grapes. I am sorry. We are going to start again. We are stumbling too much! Stumbling too much. (CUT).[64]

Margaret Thatcher's attempt to demonstrate chemistry in the kitchen to a TV audience was ill-judged. She had been invited onto the BBC2 science show *Take Nobody's Word for It* early in the election year of 1987. It was meant to show the prime minister as competent, practical, science trained, as much in command of domestic as national economies. She gave a recipe for bread. The result was inedible, tasted like chewing gum, and generated mailbags of complaints. Her people went unfed.

Thatcherism was accused, by its opponents, of being a politics of greed. When the charge was put directly to her, she rejected it, as the two opening quotations from interviews with the prime minister show. I have noted in this chapter some 1980s anxieties that the younger generation of scientists might be motivated by greed. I have also traced how Thatcher came to adopt, after years of tensions, an approach to science policy that aligned with what we understand as Thatcherism in at least two ways. The first was an aim, if not often realized, to reduce public expenditure in the name of freeing, or rather forcing, (large-scale) private industry to make its own bigger investments in science. The second was the valorization of the independent, ideally often individual, curiosity-driven scientist. Government funds for basic science increased, and basic science centers such as CERN were viewed much more warmly.

In one version of the consequences of this policy shift, the connection with wealth creation was left incredibly sketchy if supposedly profound and relied if anything on moral tales from the history of science that said that fundamental science led to great wealth creation down the line. In another version, these independent scientists might take the responsibility onto themselves to become entrepreneurs. Examining Thatcher's remarkable, if brief, leadership on climate change, I showed again that there was a marked tendency, even when urging global action, to return to personal, individual responsibility, as if global climate change was continuous with local littering. In the literature on neoliberalism, there is the notion of "responsibilization," the taking of previously collectively held burdens and pushing them onto the individual.[65] The pushing down of the responsibilities for wealth creation onto the individual researcher (and the simultaneous withdrawal of some of the collective, large-scale responsibilities), I suggest, is an analogue found in changing policies toward science during our "greedy science" period. The pushing down of responsibilities for global environmental action is an inadequate response to catastrophe.

NOTES

1. Interview with Margaret Thatcher for *Woman's Own*, September 23, 1987, https://www.margaretthatcher.org/document/106689.

2. Radio interview of Margaret Thatcher for BBC Radio 2, *Jimmy Young Programme*, July 27, 1988.

3. "Framework for Government Research and Development, HMSO," HMSO Cmnd. 4814, London (1971). See also Miles Parker, "The Rothschild Report (1971) and the Purpose of Government-Funded R&D—A Personal Account," *Palgrave Communications* 2 (2016), article 16053.

4. Jon Agar, "Thatcher, Scientist," *Notes and Records of the Royal Society* 65 (2011): 215–232.

5. Andrew Gamble, *The Free Economy and the Strong State: The Politics of Thatcherism* (Basingstoke, UK: Macmillan Education, 1988); Stuart Hall, "The Great Moving Right Show," *Marxism Today*, January 1979, 14–20.

6. Quinn Slobodian, *Globalists: The End of Empire and the Birth of Neoliberalism* (Cambridge, MA: Harvard University Press, 2018), 23, groups the governments of Reagan and Thatcher together as having "neoliberal policies."

7. Aled Davies, Ben Jackson, and Florence Sutcliffe-Braithwaite, eds., *The Neoliberal Age? Britain Since the 1970s* (London: University College London Press, 2021), 8, 10–11. They also find at least three distinct understandings of "neoliberalism": political, economic (mostly from Marxist analyses), and governmentality (starting with Foucault's claim that the neoliberal should be seen as the entrepreneur of the self). See also this volume's introduction, by Michael Gordin and W. Patrick McCray, and chapter 6 by Angela Creager, for further comments on neoliberalism.

8. David Edgerton, "What Came between New Liberalism and Neoliberalism? Rethinking Keynesianism, the Welfare State and Social Democracy," in Davies, Jackson, and Sutcliffe-Braithwaite, *The Neoliberal Age?*, 46. "Neoliberalism is a term which should . . . not be used," he adds.

9. Richard Vinen, *Thatcher's Britain* (New York: Simon & Schuster, 2009); Davies, Jackson, and Sutcliffe-Braithwaite, *The Neoliberal Age?*, 10.

10. Iain Sinclair called her "Margaret Scissorhands." Iain Sinclair, *London Orbital: A Walk around the M25* (London: Granta, 2002), 4.

11. The National Archives (hereafter TNA) of the United Kingdom, PREM 19/1477, Wybrew to Thatcher, September 5, 1986.

12. TNA PREM 19/1477, Wybrew to Thatcher, September 5, 1986.

13. TNA PREM 19/1477, Wybrew to Thatcher, September 9, 1986.

14. TNA PREM 19/1477, Wybrew to Thatcher, September 9, 1986. "A bridge is much better than a tunnel. It's a visible capital investment," underlined in Thatcher's blue ink.

15. The great scholar of contemporary British government Peter Hennessy says the Number 10 Policy Unit was "*hers* to the last paperclip" (his emphasis). It made no pretense to traditional generalist civil service neutrality. Peter Hennessy, *Whitehall* (London: Macmillan, 1989), 653.

16. Philip Gummett, *Scientists in Whitehall* (Manchester, UK: Manchester University Press, 1980), 55.

17. £2,073m through the Ministry of Defense compared to £1,150m for the "science base" (research councils and universities) and £947m through civil departments. NA PREM 19/1769, Advisory Council for Applied Research and Development (ACARD) and Advisory Board for the Research Councils (ABRC), "The Science Base and Industry," joint report, 1986.

18. For Amersham, see "Radioactive Privatisation," in Jon Agar, *Science Policy under Thatcher* (London: University College London Press, 2019), chap. 5, esp. 172–178.

19. Both the microwave oven and the frozen ready meal might have origins in the 1950s and 1960s but only become widespread in Britain in the late 1970s and 1980s. The first frozen pizza in the United Kingdom, an experiment of Northern Foods, also dates from our period of greedy science. The main anecdote relating to Thatcher's career in science that widely circulated in the 1980s concerned her alleged involvement in the innovation of "Mr. Whippy"–style ice cream from when she worked as a food chemist at Lyons.

20. Jon Agar, *Science Policy under Thatcher*, 49.

21. GEC was the third largest founding member, by market capitalization, of the FTSE100 in 1984 and employed around 250,000 people.

22. Tom Lloyd, "To Have and to Have Not: Greed Foils Britain's Attempt to Encourage High-Tech Businesses," *New Scientist*, February 7, 1985, 37–38, 37.

23. Lloyd, "To Have and to Have Not."

24. Interestingly, the actual Paul Ehrlich (probably) once said that science depended on *Geld* (money), *Geduld* (patience), *Geschick* (skill), and *Glück* (luck).

25. David Robinson, "Scientists of the Silver Screen," *New Scientist* 23/30 (December 1976): 732–734, 734. The article has the dying actor gasping of the need "to go on fighting, to rid men's hearts of the diseases of hatred and greed."

26. Theodore Roszak, "Does Science Need a Vice Squad? Review of Broad and Wade, *Betrayers of the Truth*," *New Scientist*, February 24, 1983, 537.

27. Nina Hall, "Of Fusion, Nobels and Nobility: The Pursuit of Kudos and Wealth Has Clouded the Excitement in the Search for Cold Fusion," *New Scientist*, May 6, 1989, 28–29. Cold fusion as "inauspicious" greedy science born in the 1980s is discussed by Michael D. Gordin and W. Patrick McCray in the introduction to this volume.

28. Letter, David Bowler (Rhynd, Perth) to the Editor, *New Scientist*, July 26, 1985, 64.

29. Ian I. Mitroff, "Norms and Counter-Norms in a Select Group of the Apollo Moon Scientists: A Case Study of the Ambivalence of Scientists," *American Sociological Review* 39, no. 4 (August 1974): 592.

30. TNA PREM 19/585, "Prime Minister's Reception for Inventors—26 January 1981. Opening Remarks," 1981.

31. Margaret Thatcher, Speech to Parliamentary and Scientific Committee, February 25, 1981. This speech, and others cited below, can be found in the repository maintained by the Margaret Thatcher Foundation at https://www.margaretthatcher.org/speeches.

32. G. H. Fairtlough, "Exploitation of Biotechnology in a Smaller Company," *Philosophical Transactions of the Royal Society of London. Series B, Biological Sciences* 324 (1989): 589–597.

33. Soraya de Chadarevian, "The Making of an Entrepreneurial Science: Biotechnology in Britain, 1975–1995," *Isis* 102, no. 4 (2011): 601–633. For the changing attitudes, see also Susan Wright, "Recombinant DNA Technology and Its Social Transformation, 1972–1982," *Osiris* 2 (1986): 303–360.

34. TNA PREM 19/585, Ashworth to Sanders, September 2, 1980.

35. TNA PREM 19/1931, Addison to Thatcher, April 9, 1985. "Delighted" is handwritten by Thatcher on the letter.

36. Perutz, "How to Stifle Innovation," *New Scientist*, November 10, 1987, 57; ABRC/ACARD, A Strategy for the Science Base, 1987.

37. TNA PREM 19/2478, Guise to Thatcher, September 17, 1987.

38. See Agar, *Science Policy under Thatcher*, for a full account.

39. TNA PREM 19/2477, Guise, "Public Expenditure on Science," July 24, 1987. Thatcher underlined "minimising State intervention."

40. One might ask: Why fund basic research at all? Why not rely on other countries? Guise dismissed this suggestion, since without well-funded basic research in the United Kingdom, "there would be nobody here who understood what was being discovered or [would be] invited to participate in the idea flow. . . . The argument that we can leave everything to the United States is as fallacious on research as it is on defence."

41. See Jon Agar, "2016 Wilkins-Bernal-Medawar Lecture: The Curious History of Curiosity-Driven Research," *Notes and Records of the Royal Society* 71, no. 4 (2017): 409–429.

42. Margaret Thatcher, Speech to the Royal Society (took place in Fishmongers' Hall, City of London), September 27, 1988.

43. Margaret Thatcher, Speech to the Parliamentary and Scientific Committee (50th Anniversary), December 6, 1989. The speech ended with a vision of a third industrial revolution: "The third Industrial Revolution will be knowledge-driven, science-driven, and enterprise-driven. In this new world, we will need the Parliamentary and Scientific Committee as never before to bring together the scientists who open up the possibilities of the future, the men of enterprise who harness their discoveries, and the legislators who must enable both to flourish for the betterment of the people."

44. Two examples can be found in *Is Science Necessary?*: "Like Faraday, Rutherford never took out a patent and would strongly have disapproved of the gene technologists" present scramble for money" (170), and, out of the blue during a review of a biography of a Portuguese woman scientist, "But Anna is deaf to their well-founded objections and adheres to her faith in Hodgkin's disease being primarily a defect in the white blood cells' handling of iron—rather as another iron lady sticks to monetarism despite all proofs that it is making the patient worse" (239).

45. Margaret Thatcher, Speech at Good Housekeeping Awards Dinner, September 25, 1989. Perutz's claim about the technological causes of women's liberation can be found early on, in his *Is Science Necessary?*: "Women's liberation could not have succeeded if science had not provided women with contraceptives and household technology" (4).

46. Jon Agar, " 'Future Forecast—Changeable and Probably Getting Worse': The UK Government's Early Response to Anthropogenic Climate Change," *Twentieth Century British History* 28 (2015): 602–628.

47. John Campbell, *Margaret Thatcher*, vol. 2: *The Iron Lady* (London: Jonathan Cape, 2003).

48. Such rivalry is clearly visible, for example, in dismissive comments within Number 10 about French international diplomacy over the climate issues in the late 1980s.

49. See, for example, Margaret Thatcher, TV Interview for BBC1 Nature, March 1, 1989.

50. Margaret Thatcher, Speech in Korea ("The Principles of Thatcherism"), September 3, 1992.

51. Margaret Thatcher, Speech at Jagiellonian University in Krakow, October 3, 1991.

52. Margaret Thatcher, Speech on President Reagan's 83rd birthday, February 3, 1994.

53. Alice Bell, "The Scientific Revolution That Wasn't: The British Society for Social Responsibility in Science," *Radical History Review* 127 (2017): 149–172; Jon Agar, "What Happened in the Sixties?" *British Journal of the History of Science* 41, no. 4 (2008): 567–600.

54. "I have always taken the view that the great human rights come and the great significance of the individual in the democratic system relied for its birth upon the kind of personal responsibility and accountability which you get both in the Old Testament and the New, which you do not get in the other religions." Margaret Thatcher, interview for *The Times* (with Robin Oakley), October 25, 1988.

55. Margaret Thatcher, New Year Message, December 31, 1984.

56. Margaret Thatcher, Speech to the Royal Society (took place in Fishmongers' Hall, City of London), September 27, 1988.

57. Charles Powell, draft speech (United Nations General Assembly event on environment) with Thatcher's corrections. Thatcher MSS, Churchill Archive Centre: THCR 5/1/5/684 f4.

58. Margaret Thatcher, TV Interview for TV-AM Frost on Sunday, December 30, 1988.

59. Margaret Thatcher, TV Interview for TV-AM Frost on Sunday, December 30, 1988.

60. Margaret Thatcher, Speech to the Per Cent Club (Royal Academy), December 8, 1988.

61. For example: Margaret Thatcher, TV Interview for BBC1 Nature (Michael Buerk), March 1, 1989: "I come back from The Hague, I come back from Frankfurt, and say: 'Look,

how can we persuade people not to throw litter out of car windows, how can we persuade them if they go on a picnic to take the litter back with them?'"

62. Margaret Thatcher, Press Conference after Visiting Scott Polar Research Institute, August 31, 1989.

63. Margaret Thatcher, TV Interview for TV-AM Frost on Sunday, December 30, 1988.

64. Transcript of take one of the recording of Margaret Thatcher, with Professor Ian Fells, for BBC2 *Take Nobody's Word for It*, January 12, 1987. Program broadcast January 28, 1987. The program title is a translation of the Royal Society's motto "nullius in verba."

65. See, for a discussion of the "responsibilization of the working mother," Helen McCarthy, "'I Don't Know How She Does It!': Feminism, Family and Work in 'Neoliberal' Britain," in Davies, Jackson, and Sutcliffe-Braithwaite, *The Neoliberal Age?*, 135–154.

CHAPTER FIVE

Kids, Commerce, and Communists

Access to Space in the 1980s

Margaret A. Weitekamp

On January 28, 1986, millions of schoolchildren across the United States had a searing generational experience: witnessing the loss of the Space Shuttle *Challenger* and its seven-person crew. Inspired by the inclusion on the mission of Christa McAuliffe, the teacher in space, hundreds of schools had arranged for students to watch the launch on televisions wheeled into gymnasiums, cafeterias, and classrooms. The outsized attention to McAuliffe that morning overshadowed the presence in that same crew of another nonprofessional astronaut: engineer Gregory Jarvis, representing Hughes Aircraft. Jarvis was one of two Hughes representatives who had been chosen to represent his company aboard space shuttle missions as payload specialists, a different kind of space travelers. These temporary astronauts embodied the changed position of the National Aeronautics and Space Administration (NASA) vis-à-vis aerospace companies. In the midst of a decade characterized by financialization, monetization, neoliberal politics, and popular culture that glorified acquisitiveness, directing national resources to support regular human spaceflights to orbit and back could be considered greedy. More so, Jarvis and other payload specialists had access to spaceflights aboard the reusable spacecraft because of one of the other greedy trends of the 1980s, the increasing dissolution of the traditional lines between corporate interests and the state.[1]

No longer serving simply as contractors executing NASA's vision, aerospace and technology companies also vied to propose experiments and other payloads to be carried into space, which would then be tended to by a representative of their choice (approved by NASA). The inclusion of McAuliffe and Jarvis in the crew of STS-51L represented a strategic democratization of who could go into space, a reflection of a larger turn in US human spaceflight in the 1980s that was facilitated by the reusable launch vehicle itself as well as the political rhetoric surrounding NASA's Space Transportation System (the formal name of the space shuttle).

For both the United States and the Soviet Union, the 1980s were a period that featured the development of infrastructures and programs for more frequent and longer-term spaceflights, for the purposes of facilitating financial and political partnerships. In the United States, the goal of the commercialization of spaceflight (seeking to develop a new market-driven field) both overlapped and coincided with the continued allure of spaceflight as an advertising opportunity (capturing and capitalizing on the excitement associated with spaceflight for profit). Thus, at the same time that payload specialists brought corporate representatives on board missions as cooperative partners, an educational program called the Young Astronaut Program also created branding partnerships intended to promote education by teaching about the shuttle. For the Soviet Union, the evolving Cold War context led the Soviet program to pursue affiliations using the infrastructure and capacity created with the Salyut space stations to provide opportunities for affiliated space travelers to demonstrate comradeship through shared spaceflights. Rejecting capitalistic greed, the Soviet space program craved instead the spread of socialism. The Interkosmos program used the cachet of spaceflight to strengthen alliances and make inroads in global geopolitics, even with allies of the United States. In the 1980s, the erasures of established lines, whether between corporate interests and state-run programs or between a highly centralized socialist state and its satellites, illustrated permeabilities that were characteristic of the era.

History of Commercial Ties

In the 1980s, economic relevance and corporate branding came to be seen as signs of a new era in human spaceflight in the United States. Unlike its predecessor programs, the shuttle was fundamentally shaped by economic considerations throughout its development. But the rhetoric around those issues took on new energy as a result of President Ronald Reagan's embrace of spaceflight and his administration's eagerness to marry the technological project to probusiness interests. Especially before the loss of the *Challenger* in 1986, NASA's space shuttle program became directly and indirectly affiliated with corporations, companies, and brands—a history recorded in media accounts, official documents, and physical artifacts.

In the first decade after NASA's creation in 1958, the economic ramifications of spaceflight programs were not tracked systematically by the agency. Having been built on the administrative structure of the National Advisory Committee on Aeronautics (NACA), itself a federal agency conducting aeronautical research, NASA was fundamentally configured as a research and development (R&D) agency. That, coupled with the tremendous pressures created by executing US President John F. Kennedy's lunar landing mandate, meant that socioeconomic analyses did not factor primarily into NASA's policy decisions. As the cover letter for an analysis commissioned by NASA in 1965 concluded, attention to such issues had been, to that point, "separate, uncoordinated, and almost incidental activities."[2]

Rather, the goals and execution of human spaceflight programs in the 1960s were shaped by geopolitical concerns,[3] and any relationship with corporations or industry was limited to engaging them as suppliers. The Mercury, Gemini, and Apollo programs were carried out with carefully managed relationships with industry partners as contractors. To offer just two specific examples: McDonnell Aircraft Corporation of St. Louis, Missouri, built the crewed spacecraft for Project Mercury and the Gemini program; North American Aviation was the prime contractor for the command and service modules as well as the second stage of the Saturn V launch vehicle for the Apollo program.

Contractors took great pride in their association with NASA projects. Photos of the Gemini spacecraft being prepared show the vehicle surrounded by McDonnell workers, each wearing a white coverall with the

McDonnell name emblazoned across the back (figure 5.1). Likewise, satellite development was often done in partnerships between NASA centers and private companies. For instance, Project Echo, which constructed two passive communications satellites, was a project shared by Bell Telephone Laboratories, NASA, and its Pasadena-based center, the Jet Propulsion Laboratory. Private companies touted their connections to NASA in the industry and popular press through advertisements. Just as advertisers sought to associate their companies and products with the space program, looking to capture some of the public excitement about spaceflight, NASA also worked to sell the American public on space exploration as well. Both sides of these relationships have been explored well by researchers.[4]

Even as the Apollo program came to its peak, however, NASA officials started looking to the future to develop its next generation vehicle, which included coping in new ways with budgetary limits and economic impact. By the time that firm decisions were being made in the early 1970s about the Space Transportation System or space shuttle, which would become the defining driver of American space activities in the 1980s and beyond, economic considerations were integral to the process. Certainly, NASA officials framed their arguments in terms that they hoped would be received favorably in an environment in which business-style decisions about budget had become paramount. As NASA administrator James Fletcher conveyed in 1972 to Casper Weinberger, then at the government Office of Management and Budget (OMB), "From the budgetary point of view, perhaps the most important consideration is that we have selected the configuration [of the space shuttle system] which, for a given payload size and weight, entails the lowest development cost."[5]

The complex story of how the space shuttle developed as a vehicle of compromise has been well developed by other historians.[6] Simply put, the size, composition, and functionality of the final design required balancing the Department of Defense's needs with NASA's own goals, while assessing development costs against operational ones. Moreover, these complex decisions were carried out in a constrained economic environment governed by the complexities of congressional and presidential approval. Originally scheduled to launch for the first time in 1979, the space shuttle languished due to production delays with specific components (including the space shuttle main engines and the thermal protection system of tiles and woven fireproof blankets that protected the orbiter). Advocates

Figure 5.1 The prominence of McDonnell's name as contractors prepare Gemini-3 in this photo from 1965 illustrates NASA's long history with industry. Image courtesy of the National Air and Space Association, https://images.nasa.gov /details/s65–21090

defended the prospective vehicle by appealing to national pride and the need for the United States to retain a homegrown human spaceflight capacity—as well as with promises that the versatile system would reduce flight costs even as it carried out many different kinds of missions. By the time the system was ready for launch in 1981, a new administration was in office and the Reagan White House's probusiness stance dovetailed with the economic arguments for the new spaceflight vehicle. Indeed, the Reagan administration's first National Security Decision directive in space policy, issued in 1981, affirmed the importance of the shuttle program for "broad commercial, civil, and national security needs." Commercial was mentioned first.[7]

As a federal agency well aware of the larger turn toward neoliberalism, NASA refined its public messaging, which was then amplified by the media, to illustrate the enthusiasm with which the dream of the space shuttle was figured as an economic driver. In the runup to the inaugural launch of the Space Shuttle *Columbia*, initially planned for April 10, 1981, the press trumpeted the potential of the untested system. According to one of the articles published in a five-part series in the *Chicago Tribune*, "The shuttle is intended to open a highway into space where orbiting factories, permanently manned space stations, huge communications satellites, and other innovations may produce wonderous products and services for people on Earth." The *Tribune*'s writers concluded, "Backers of the world's only reusable rocketship shuttle have hung out their shingle for business. And it reads, 'Space for rent.'" They speculated that some businesses might rent an entire mission for a cost of approximately $30 million. Or, finding that too expensive, customers could choose to purchase the use of small canisters, appealingly dubbed "getaway specials," which allowed the buyer to fly up to 60 pounds in a 2.5 cubic foot container, having the contents returned to them postflight. No longer defined primarily as missions of exploration undertaken to achieve geopolitical gains in international prestige, state-sponsored spaceflight was instead a new kind of business.[8]

From the beginning, the space shuttle was touted as an economically viable space launch system. In 1981, the chief of technical planning at the NASA Johnson Space Center "equated the first shuttle launch with the first train west out of St. Louis," as a driver of economic development. Initially, the program had the goal of recouping the fleet's operating costs—projected to be $30 million in 1975 dollars—over the course of twelve years. Two ancillary arguments also existed from the beginning of the space shuttle era. The first was the suggestion by NASA that a national commitment to building a crewed space station should follow quickly upon the demonstrated success of the shuttle. The second was the idea that the government project could become so economically successful that the booming world of corporate America might take note or even take over. Once NASA demonstrated the capabilities of the shuttle, the argument went, the government agency could get out of the business of operating it and transition it to other management, perhaps even to a private agency. As NASA administrator James Beggs said in an interview with *Omni* magazine in 1982, "My personal predilection would be to try making it a commercial enterprise." He hoped that the increase in business interest would relieve NASA of the "large operational responsibility [of the shuttle program], because it is dilutive of the basic reason for this agency's existence, which is research and technology."[9]

Business concerns, as well as a belief in the efficacy and promise of market-oriented solutions, were therefore tied into discussions of all kinds of other developments regarding the space shuttle orbiter systems. The White House hosted a major one-day meeting on August 3, 1983, inviting aerospace executives to a lunch to discuss space commercialization and how to get the private sector involved in space efforts. Even the hoped-for expansion of the shuttle launch facilities to a West Coast site at Vandenberg Air Force Base in California, which would have enhanced the vehicle's national security applications by allowing polar orbits to be flown, got tied to commercial interests. In an article directed at military engineers explaining the merits of developing the West Coast launch and landing facility for the shuttle, the last line summed up the impact: "The construction of a space launch facility at Vandenberg Air Force Base will not only benefit the defense community, but will also provide ready access to space for the business community."[10]

If greed can be defined as the base avarice of desiring more than is reasonably attainable, then the space shuttle program represented a voracious overselling of the next steps in spaceflight. The reusable vehicle system could never achieve the hoped-for but ultimately unrealistic goals of economic relevance and financial independence that shaped its early hype. By 1982, only one year into flights, the government's General Accounting Office (GAO) was already projecting that NASA could lose up to

$1.2 billion by 1985. Hardware and operational costs greatly outstripped projected fixed pricing. Moreover, early decisions to lower developmental costs resulted in larger operational costs. Ultimately, the Space Transportation System would never achieve the pace of turnaround and relaunch that would be required to meet the ambitious launch schedules initially speculated. As successful as the space shuttle orbiters were, they would never fulfill the overinflated budgetary promises that had been made to persuade decision-makers to build them in the first place.[11]

During the first years of the Space Shuttle program, in the Reagan era, spaceflight advocates leaned into commercial affiliations in new ways, touting them as evidence of progress toward the goal of shifting spaceflight from being a government-sponsored effort into a new realm of space business. At the same time, an ancillary group intended to promote the educational benefits of the Space Shuttle program for schoolchildren explicitly embraced corporate sponsors.

Young Astronaut Program

The development of space-themed educational initiatives for schoolchildren in the United States took place in the context of profound shifts in global industrial structures. American dominance in traditional industrial sectors, including coal, steel, oil, and the auto industry, collapsed while microelectronics charged to the fore, with the centers of these new industries in Asia and elsewhere. The celebration of a new American launch system offered a counternarrative to these global shifts. Moreover, astronauts seemed the epitome of this new kind of skilled, postindustrial, knowledge labor, going to work in low Earth orbit—an aspirational new workplace that was far above and apart from the Rust Belt.

Efforts to prepare students for these new kinds of work also addressed the profound concerns raised in April 1983 by the report of the US National Commission on Excellence in Education. Titled "A Nation at Risk: The Imperative for Educational Reform," the report called out American schools for failing to prepare students in fundamental mathematics, science, and other subjects, resulting in a "rising tide of mediocrity" that threatened the nation's ability to move successfully into this new future. In this historical moment, several different innovators began educational initiatives to educate schoolchildren about spaceflight or to use spaceflight to excite students about their education. One well-known example, Space Camp, the spaceflight-themed experience based at the US Space and Rocket Center in Huntsville, Alabama, was founded in 1982. Another was the Young Astronaut program.[12]

In October 1983, Pulitzer Prize–winning journalist Jack Anderson approached President Reagan with his idea to get children excited about science, math, and technology, and the president authorized the project the next summer. The Young Astronaut Council, which Anderson chaired, coordinated between NASA, the White House, and the private sector. The goal was to prepare young people for life in the space age. By 1992, the organization was "the largest, space-related, educational youth organization in the world." In keeping with the Reagan administration's belief in private initiatives rather than public or government programs, the council was supported by private corporations. Sponsors included Adidas, Commodore International computers, Martin Marietta Aerospace, M&M Mars, Pepsi-Cola, Safeway Stores, and Xerox.[13]

The new organization was announced at a White House ceremony on October 17, 1984. With the seventeen newest astronauts just selected by NASA in attendance, Reagan greeted an audience of dignitaries and children on the South Lawn. His speech echoed the formal letter written to dedicate the group: "In order to maintain our position of leadership in the world of high technology, we need to rekindle the spirit of scientific adventure and help nurture it in our nation's schools." New members of the Young Astronaut program pledged to raise their grades in math and science, learn about space, and support other members of the group. At the ceremony, child actress Drew Barrymore, well known for her role in *E.T.: The Extra-Terrestrial*, attended as a special guest and read the membership pledge aloud for the group. Barrymore then handed Reagan a baseball cap bearing the Young Astronaut program logo. Reagan wore the cap not only during the event but also "in leaving the ceremony to reemphasize his support for the educational initiative."[14]

Within the first month after Reagan's kick-off event at the White House, invitations to form chapters went out to 77,000 elementary and junior high schools. The Young Astronaut Council also distributed curricular materials to "all categories of schools, including rural and inner city, gifted and handicapped, public and private." Many responded. Within a year of its founding, 50,000 inquiries had been received and more than 3,700 chapters were operative. Even without any explicit outreach outside the United States, sixty chapters were founded in seventeen different nations during the organization's first year.[15]

The Young Astronaut program became another example of the dissolution of traditional lines between government projects and corporate interests in the 1980s. To do the work that would prepare students to "operate the computers, robots, and lasers that will be standard equipment in just a few years," council executive director Wendell Butler sought business sponsors using the model offered by Peter Uberoff's 1984 Summer Games in Los Angeles, the first profitable Olympic Games since 1932—and one achieved with significant corporate sponsorship. The Young Astronaut program created a partnership among educators, parents, and the business community. Although several of the participating organizations were federal (including NASA, the Department of Education, and the Smithsonian Institution's National Air and Space Museum), private companies such as Commodore computers also took a leading role. The Young Astronaut Council allowed participating sponsors to use Young Astronaut program logos on their product. Such partnerships allowed the group's organizers to tout the program's independence from government support. The emphasis on private enterprise dovetailed with the Reagan administration's hopes for the American space program itself.[16]

Some of the corporate affiliations connected easily with the program's core content. Monogram, a company that produced boxed model kits, rebranded its existing Snap-Tite "Space Shuttle" model kit to carry the Young Astronaut's logo. Marvel Comics planned a series of branded comic books. CBS planned an animated Saturday morning television show. Each episode of *The Young Astronauts* would conclude with an "Astro-minute" in which real astronauts would explain scientific concepts. Some companies were still welcomed to associate themselves with the program even if the connection between their products and spaceflight enthusiasm was tenuous. For instance, the Safeway chain of grocery stores found a way to participate in the program by printing Young Astronaut registration forms on its brown paper grocery bags.[17]

Participation in the program offered some extraordinary opportunities for a few children. In 1985, two Young Astronaut members went on location with ABC Motion Pictures for the filming of *SpaceCamp* (1986). The movie's plot featured a group of Space Campers who accidentally get launched into space on a real shuttle mission—and, in true 1980s teen flick fashion, come together as a team without losing their quirky individuality. Monologues extoled the inspirational and educational value of human spaceflight. Ultimately, the plot's resolution rested in each character's ability, when tested, to rise to a challenge. Although the Young Astronaut Council's literature did advertise, "Eventually, we hope to send a Young Astronaut on a Space Shuttle mission," including a child on a future launch was not the group's purpose. In contrast, flying a teacher was the centerpiece of the Teacher in Space program. Both the privately funded Young Astronaut Council and NASA's Teacher in Space program were announced in close succession during the summer of 1984.[18]

The association of the Young Astronaut program with corporate sponsors sometimes caused complications. As a Young Astronaut program partner, NASA flew memorabilia for the group on both STS-51F and STS-61A[19] to build excitement about the project. On the first of the two missions, however, NASA also had to negotiate a delicate truce in the contemporary pop culture beverage rivalry. In 1984, Coca-Cola modified a soda can in a way that would allow astronauts to drink a Coke while in orbit. They proposed to fly it in space to test its efficacy. Coca-Cola, however, was the major competitor of Young Astronaut program sponsor Pepsi-Cola. Caught between the two rivals in the 1980s cola wars, NASA also allowed Pepsi to provide its own modified can. The space agency called the lighthearted but otherwise serious experiment the "Carbonated Beverage Dispenser Evaluation." Although the experiment largely failed (neither unrefrigerated soda had much appeal to the astronauts who got to sample them), Pepsi nonetheless celebrated. Commemorative cans bearing the Young Astronaut program logo appeared in stores across the nation.[20]

In the prelaunch press conference for STS-51F, NASA officials tried to redirect the attention from the two rival beverage cans to Spacelab, the collaborative science laboratory specially designed to be flown in the space shuttle orbiter's payload bay. The *Washington Post* quoted Dr. Jeffrey Rosendahl, NASA's associate administrator for space science, informing reporters, "We're carrying some very sophisticated instruments on this flight and possibly the best telescopic aiming equipment ever built." Working in Spacelab during that mission, along with two NASA astronauts, were two payload specialists: Dr. John-David Bartoe, a civilian Navy scientist, and

Dr. Loren Acton, a physicist representing the Lockheed Palo Alto Research Laboratory, which had an agreement with NASA for their participation.[21]

Payload Specialists

The existence of payload specialists as a category of space traveler personi-fied the cooperation encouraged during the Reagan administration be-tween NASA's big government/big science expenditures and the agency's industry, research, and corporate partners. The first Spacelab mission (STS-9 in 1983) included Byron Lichtenberg, an American engineer from the Massachusetts Institute of Technology (MIT) who is credited as the first astronaut to serve as a payload specialist, along with Dr. Ulf Merbold, a West German physicist. They flew in space for ten days, conducting mul-tiple experiments. The very first nongovernment individual to fly in space was McDonnell Douglas engineer Charles D. "Charlie" Walker. Walker launched into space on August 30, 1984, aboard STS-41D, tending a device for which he shared a patent.

Even before the advent of payload specialists, the Space Shuttle program had already seen the division of the NASA astronaut corps into two differ-ent kinds of astronauts.[22] Beginning with the recruitment for the Space Shuttle program in the late 1970s, NASA created two separate tracks: pilot astronauts and mission specialists. Pilot astronauts applied based on their flight experience. Mission specialists applied based on their credentials as researchers with professional degrees such as PhDs or MDs, for instance. Both tracks were considered "real" NASA astronauts. The new designation, payload specialist, was used for researchers or international astronauts who were added to specific missions to tend to particular payloads. Described in a press kit from 1985 as "NASA's newest breed of workers in space," pay-load specialists were "career scientists and engineers . . . identified and selected by their peers to fly into space and conduct experiments. After the mission, they return to their previous position at the institution sponsor-ing their research." Such cooperation with private companies supported efforts to encourage business interests in space. Their inclusion in missions put substance to the promise that the space shuttle orbiters would make spaceflight and low Earth orbit research within the reach of more people, whether that meant participants from cooperating nations or representa-tives of various corporate partners.[23]

Close readings of artifacts and images reveal how the first corporate payload specialist, an employee of a major American aerospace manufacturing corporation and defense contractor, was rendered almost indistinguishable from NASA's regular astronauts, government employees working as civil servants. Scholarship in museum studies and material culture offer analytical structures for unpacking how artifacts can reveal underlying structures of meaning—and how the close examination of things both enhances and complicates those meanings. As museum curator Martin Collins has argued, artifacts "illuminate a culture through the details of their creation and use (materials, craft skills, exchanges, rituals) as well as through their associated symbolism." There is much to be learned, then, by employing careful readings of literal things as evidence. In this case, examining operational flight clothing and mission patches supports broader analyses of interpersonal relationships, organizational hierarchies, and professional functions.[24]

The Smithsonian's National Air and Space Museum holds a flight spare of Walker's name badge, which notably does not include anything that visually differentiates it from those worn by NASA's regular astronauts, whether pilot astronauts or mission specialists. The simple black leather rectangle has Walker's name as he chose it[25]—Charlie Walker—embossed in silver on the surface of the Velcro-backed piece. When seen in the context of one of his official NASA portraits, one notices that Walker wore the same uniform as any shuttle astronaut, the standard light blue fabric flight suit used during launch and landing. (NASA only introduced protective launch and landing partial-pressure suits after the *Challenger* disaster resulted in the loss of the crew and vehicle on January 28, 1986.) Like the other shuttle astronauts, he wears two patches on his right chest: one with the NASA "worm" logotype and a triangular one for the Space Shuttle program (figure 5.2). His uniform includes an American flag patch on his left shoulder and a McDonnell Douglas patch with the company name and logo on his right. Below his name badge, Walker wears a patch for the EOS (electrophoresis operations in space) project that he designed to process pharmaceuticals on board a shuttle flight in microgravity.

In the official crew photo for Walker's first mission, STS-41D in 1985, one would have to look very closely to discover that Walker was not an official NASA astronaut in the employ of the federal government. In the

Figure 5.2 Civilian engineer Charlie Walker flew in space three times in 1984 and 1985 as a payload specialist representing McDonnell Douglas. Image courtesy of the National Air and Space Association, https://images.nasa.gov/details/S84-27269

group picture, Walker stands in the second row in such a way that his EOS and McDonnell Douglas patches are almost completely obscured behind NASA astronaut Steve Hawley. The image conveys the impression of a uniform(ed) crew, with the only distinguishing elements being called out through two ship models included in the picture. The inclusion of the sailing ship symbolized the first flight of OV-103 (orbital vehicle 103), named *Discovery* after several different ships of exploration that bore the same name. To the left of Walker stands a model of the orbiter deploying the OAST-1 solar array. The miniature solar array all but covers the absence of the NASA logotype on Walker's shoulder.

Even as Walker was integrated into the crew, however, McDonnell Douglas was very excited to promote their alignment with the Space Shuttle program as distinctive, a corporate victory in the competitive field of aerospace manufacturing. For instance, McDonnell Douglas created a photo button made for its employees to wear to celebrate Walker's first flight (figure 5.3). The example in the museum's collection came from a donation by John Bickers, who started his career as a public affairs representative when the company was still just the McDonnell Company. The button features a main, central photograph—an official astronaut portrait—showing Walker in his official light blue astronaut flight jacket. Walker is posed with a US flag behind his right shoulder and the top of a solid rocket booster (SRB) from a space shuttle model poking out behind his left shoulder. Both of those elements remain in the frame, reinforcing his role as an American space traveler, even though the photo has been cropped into a circle to fit the button's form. White type on the red border of the button calls out the mission designation, 41-D, and the date, June 1984. With McDonnell Douglas's name on the lower border, the clever combination of uppercase and lowercase letters on the upper border declares both McDonnell Douglas as a "space manufacturer" and Walker as a "SPACE MAN."

Internally at NASA, the question of Walker's inclusion was initially a very different story. As he recalled in an oral history recorded in 2006, although NASA's agreement with McDonnell Douglas included an agreement for seven flights of the electrophoresis experiment that he had developed—and his presence on a mission to tend to it—nonetheless, the career astronauts initially enforced a clear distinction between their roles in the space program and those of the payload specialists. As Walker recalled, "It was made clear to us [payload specialists] from the beginning

Figure 5.3   This button from 1984 celebrated Walker's overlapping role as a "SPACE MAN" and McDonnell Douglas as a "space manufacturer." Smithsonian National Air and Space Museum

that they [career astronauts] didn't expect to see us over on the fourth floor in Building 4 [the Astronaut Office at Johnson Space Center] except for scheduled meetings. We were just outsiders who would become crew members for a short period of time and would train mostly on our own, but when there was necessary crew combined training, certainly we would be there."[26]

Historian Matthew Hersch has analyzed the ways that the early NASA astronaut corps defined itself like a guild, building its sense of inclusiveness through shared skills, experience, and brotherhood. Beginning with the introduction of scientist astronauts in the late 1960s, the membership in the astronaut corps underwent another shift with the addition of mission specialists in 1978. At least at the beginning of the payload specialist program, the definition of the astronaut guild did not stretch enough to include short-term corporate representatives. Walker recalled that this exclusionary stance moderated over time. As the individuals worked together, they came to feel less need to enforce a separation. In fact, late in 1985, the payload specialist office was moved from its location in Building 39 over to Building 4 with the astronauts, just before the loss of the Space Shuttle *Challenger* changed the landscape for spaceflight.[27]

Payload specialists carried out their assigned duties during spaceflight missions but also functioned as embodiments of relationships. In many ways, the changes in crew assignments that led to Greg Jarvis's assignment with the ill-fated STS-51L crew illustrated how NASA officials juggled and prioritized the agency's relationships. In the mid-1980s, NASA officials made several changes to mission assignments in order to strengthen the agency's position with its congressional funders. Jarvis had been scheduled to fly in April 1985 but was bumped in favor of US senator Jake Garn, a veteran naval combat pilot and member of the Senate Appropriations Committee. Jarvis's mission to space was again delayed in January 1986, when he was removed from the flight roster to provide a seat for US Representative Bill Nelson (who later became NASA administrator in 2021), then chair of the Space Subcommittee of the House Science, Space, and Technology Committee. That fateful change moved Jarvis to the ultimately fatal flight. Irrespective of the unintended consequences, the reordering of flight crews reflected how crew cohesion and mission goals could be rendered secondary to other agency priorities, including valuing the favor of federal-funding gatekeepers over new relationships with corporate sponsors. Even in an era of neoliberalism, NASA remained first and foremost a federal agency, a part of the state. The shift to new understandings and goals about business-style decisions was in some ways more an aspirational attempt to capture the zeitgeist than a practical reordering of the core funding of spaceflight operations.

Interkosmos and Soviet Space Partnerships

The Union of Soviet Socialist Republics (USSR) also used the inclusion of external space travelers on missions to solidify relationships. Between 1978 and 1988, the Soviet Interkosmos program flew fourteen different non-Soviet space travelers on Soyuz missions, buttressing the USSR's relations with nations including Czechoslovakia, Poland, the German Democratic Republic (GDR), Bulgaria (twice), Hungary, Vietnam, Cuba, Mongolia, Romania, France, India, Syria, and Afghanistan. This program had its technological origins in the mid-1970s with the decisions to cycle the Soyuz spacecraft servicing different Soviet space stations—thus opening up a seat for a non-Soviet occupant to accompany the Soviet cosmonaut on a flight—but its political roots can be traced to a 1965 meeting about possible space collaborations held between Soviet officials and representatives from eight fellow Communist nations.[28]

The capacity for the new geopolitical outreach grew from the success of second- and third-generation Soviet space stations. Although the first crew sent to *Salyut 6* failed to dock, once the first crew entered the station in December 1977, the orbiting station had the capacity to host both long-term resident crews and short-term visitors. Those shorter missions allowed for seats to be allocated to cosmonauts who trained for their missions but were not necessarily members of the professional Soviet cosmonaut corps. Interkosmos missions were flown to *Salyut 6* between 1978 and 1981, and to *Salyut 7* in 1982 and 1984. After 1987, the Interkosmos missions docked with the *Mir* space station.[29]

The Interkosmos project involved two different representatives from the chosen nation being selected for each flight, with both training for up to two-and-a-half years before the mission made only one of them an actual space traveler. The potential cosmonauts selected in 1976 represented Poland, the GDR, and Czechoslovakia. In 1978, pairs from Bulgaria, Cuba, Hungary, Romania, and Mongolia joined. Two candidates from Vietnam were selected in 1979, followed by representatives chosen to train for spaceflight from India in 1982, from Syria in 1985, from Bulgaria a second time in 1987, and from Afghanistan in 1988. With each Interkosmos mission, the Soviet program sought to buttress relationships with the research cosmonaut's home country and to illustrate its capabilities to the world. Notably, Vietnamese cosmonaut Phạm Tuân, who flew aboard Soyuz 37 in 1980, was the first Asian person in space. The flight of Cuban cosmonaut Arnoldo Tamayo Mendez in 1980 made him the first person of African descent in space, three years before Guy Bluford became the first African American in space aboard STS-8 in 1983.[30]

As the 1980s ended, the Interkosmos program chose guest cosmonauts not only to buttress political relationships but also to help infuse the Soviet space program with funding. In 1989, as the Soviet Union faltered, officials at the Tokyo Broadcasting System (TBS) decided to celebrate its upcoming fortieth anniversary with an out-of-this-world publicity stunt: sending someone into space. Dozens of employees applied for the opportunity. Japanese news correspondent Toyohiro Akiyama was chosen, in exchange for a reported payment of ¥1.5 billion from TBS to the Soviet space program, Akiyama trained for fourteen months before he flew to the *Mir* space station in December 1990. As such, Akiyama was not only the first Japanese national in space but also the first journalist to report from space. US payload specialists may have flown in space as representatives of private companies, but those companies did not pay for the flights directly. In contrast, Akiyama is considered the first privately sponsored individual to fly in space. The televised broadcasts from the mission featured him doing some science experiments, talking about his experience (he missed being able to smoke), and addressing his family from space.[31]

In the United Kingdom, an effort called Project Juno set out to follow the TBS model by raising money from a private consortium to send the first Briton to space. Thirteen thousand people applied for the chance, with four finalists being selected in 1989. In the end, chemist Helen Sharman trained at Star City for eighteen months before launching to *Mir* in May 1991. Her mission included her performing experiments that had been designed by British schoolchildren. Although the consortium backing Project Juno failed to raise all of the money needed, Soviet premier Mikhail Gorbachev approved the flight nonetheless. The international relations goals of the cooperation carried enough weight to balance the lack of funds.

The coda to these stories of the corporate sponsorship is the episode that resulted in two Russian cosmonauts wearing MTV (Music Television) baseball hats and logo T-shirts, carrying a dismantled Video Music Awards "Moonman" statuette aboard the *Mir* space station. MTV debuted in 1981 with spaceflight footage followed by an astronaut planting an MTV logo flag—an avante-garde marketing image for a new kind of network taking a giant leap. When the cable network developed an awards show, the Video Music Awards or VMAs, the Moonman became the icon awarded to winners. In 1996, Pepsi, one of MTV's corporate sponsors, paid to have a VMA award statue flown to the Russian *Mir* space station. The cosmonauts had the statuette on hand during the downlink with host Dennis Miller on the live broadcast.

Ultimately the pressures of live television caught Miller by surprise when their unscripted interaction went very differently than expected, but the resulting event fit the VMAs' reputation for unconventional and unpredictable programming. On a night during which Smashing Pumpkins, Alanis Morrisette, and Coolio won awards, sardonic comic Miller contended with several unwieldly portions of the live program, which was punctuated by a brief striptease done onstage by the bassist for California alternative rock band Red Hot Chili Peppers, known professionally as Flea. But even more than that incident, the inherent complications of the downlink from space left Miller unnerved. Transmission delays, coupled with additional pauses for language translations, meant that the short conversation consisted of long moments of awkward silence punctuated by both host and cosmonauts speaking at once. "Has anything ever gone this poorly?" he asked out loud, "I will be haunted by this for the rest of my life." The nakedly commercial advertisement from space only added to the awkwardness. Regardless of whether it made for good television, the private funding was crucial for Russia's human spaceflight effort at the time and dovetailed with US government support provided via the Shuttle-*Mir* program that ran from 1993 to 1998.[32]

Conclusion

Spaceflight creates meaning in many different ways. It can be simultaneously symbolic, aspirational, and practical, with cultural, social, political, and economic implications. For the US civil human spaceflight program, the 1980s seemed to herald a new era of regular spaceflight that might even fulfill ambitious goals for financial independence. At the same time, the Young Astronaut program sought to excite young people about spaceflight and education in equal measure through a program that was designed to be self-sustaining through the embrace of corporate sponsorships. For its part, the Soviet Interkosmos program flew fourteen different non-Soviet cosmonauts to space between 1978 and 1988. As for so many things in the parallel yet separate histories of the US and Soviet superpowers in this era, however, the overlapping timelines defy simple conflation. US efforts explored the ways that corporate interests could overlap with and reinforce state interests in human spaceflight. Soviet efforts used their increased crew capacity aboard orbiting space stations to offer opportunities that would increase the strength of political alignments—or offer access to much-needed infusions of cash. Despite these differences in intentions and execution, it is not perhaps too oversimplified to say that both space-going powers seized opportunities to use their maturing space efforts to strengthen important relationships in ways that built on the inherent symbolism of spaceflight.

The idea that the future of spaceflight lay in partnering with private businesses coalesced in the 1980s. But the rhetoric of the time was never matched by real operational shifts that would change the fundamental institutional structures of spaceflight. Recent examples of commercial spaceflight, such as the missions flown by private companies including SpaceX, Virgin Galactic, and Blue Origin, owe their origins, in part, to dreams first dreamt in the 1980s. Ultimately, however, the underpinnings of today's commercial human spaceflight companies can be traced to a very different greedy era when the speculative dot-com bubble allowed tech moguls to amass the outsized fortunes necessary to bankroll entirely new spaceflight firms.

NOTES

1. An estimated 2.5 million students watched via satellite dishes, with more watching via cable television. Lynn Olsen, "TV Brought the Trauma to Classroom Millions," *Education Week*, February 5, 1986, https://www.edweek.org/education/tv-brought-the-trauma-to-classroom-millions/1986/02.

2. Good histories of NASA's founding include Roger Bilstein, *Orders of Magnitude: A History of the NACA and NASA, 1915–1990* (Washington, DC: NASA, 2013); Roger D. Launius, *NASA: A History of the U.S. Civil Space Program* (Malabar, FL: Krieger Publishing Company, 1994); and Michael J. Neufeld, *Spaceflight: A Concise History* (Cambridge, MA: MIT Press, 2018). Jack G. Faucett, President, Jack Faucett Associates, Inc., to Willis H. Shapley, Associate Deputy Administrator, NASA, November 22, 1965, in John M. Logsdon, ed., with Roger D. Launius, David H. Onkst, and Stephen J. Garber, *Exploring the Unknown: Selected Documents in the History of the U.S. Civil Space Program*, vol. 3: *Using Space*, The NASA History Series, NASA SP-4407 (Washington, DC: National Aeronautics and Space Administration, 1998), 402.

3. For the ways that diplomacy and soft power shaped US goals with the Apollo program, see Teasel Muir-Harmony, *Operation Moonglow: A Political History of Project Apollo* (New York: Basic Books, 2020).

4. Megan Shaw Prelinger, *Another Science Fiction: Advertising the Space Race 1957–1962* (New York: Blast Books, 2010); David Meerman Scott and Richard Jurek, *Marketing the Moon: The Selling of the Apollo Lunar Program* (Cambridge, MA: MIT Press, 2014).

5. Underlining in the original. James C. Fletcher, Administrator, NASA, to Caspar W. Weinberger, Deputy Director, Office of Management and Budget, March 6, 1972, in John M. Logsdon, ed., with Ray A. Williamson, Roger D. Launius, Russell J. Acker, Stephen J. Garber, and Jonathan L. Friedman, *Exploring the Unknown: Selected Documents in the History of the U.S. Civil Space Program*, vol. 4: *Accessing Space*, The NASA History Series, NASA SP-4407 (Washington, DC: National Aeronautics and Space Administration, 1999), 255.

6. For the history of the shuttle's development, see Roger D. Launius, "NASA and the Decision to Build the Space Shuttle, 1969–72," *The Historian* 57, no. 1 (Autumn 1994): 17–34; T. A. Heppenheimer, *The Space Shuttle Decision: NASA's Search for a Reusable Space Vehicle*, NASA History Series, NASA SP-4221 (Washington, DC: National Aeronautics and Space Administration, 1999). For a complete operational history of the space shuttle, see Dennis R. Jenkins, *Space Shuttle: Developing an Icon, 1972–2013* (North Branch, MN: Specialty Press, 2017). For a broader social, cultural, and political analysis, see Valerie Neal, *Spaceflight in the Shuttle Era and Beyond: Redefining Humanity's Purpose in Space* (New Haven, CT: Yale University Press, 2017). For NASA's relationship with the Department of Defense regarding the shuttle program, see James E. David, *Spies and Shuttles: NASA's Secret Relationships with the DoD and CIA* (Gainesville: University Press of Florida, 2015).

7. Brigadier General Robert Rosenberg, National Security Council, "Why Shuttle Is Needed," undated but November 1979, in Logsdon, *Exploring the Unknown*, 4:292–293; The White House, National Security Decision Directive 8, "Space Transportation System," November 13, 1981, in Logsdon, *Exploring the Unknown*, 4:333–334.

8. Jon Van and Ronald Kotulak, "Shuttle Could Open 'Truck Highway' to Space," *Chicago Tribune*, April 7, 1981, 1.

9. John Noble Wilford, "The Industrialization of Space: Why Business Is Wary," *New York Times*, March 22, 1981, F1; M. Mitchell Waldrop, "NASA Struggles with Space Shuttle Pricing," *Science* 216 (April 1982): 278–279; "Space Shuttle Good Business: NASA Chief," *Chicago Tribune*, March 28, 1982, A1.

10. Issue paper, Office of Policy Development, February 5, 1983, ID# 142305PD, OS, WHORM: Subject File, Ronald Reagan Library; Executive memorandum, Heritage Foundation, April 11, 1983, ID #137066, OS, WHORM: Subject File, Ronald Reagan Library. For records of the invitations issued, the agenda, and the resulting white papers from the one-day White House meeting on space commercialization, see ID #073513, OS 001 Space Flight, WHORM: Subject File, Ronald Reagan Library; Joseph V. Link, "Vandenberg's Space Shuttle: The Why and the How," *The Military Engineer* 76 (January–February 1984): 70–71.

11. Waldrop, "NASA Struggles with Space Shuttle Pricing"; "Space Shuttle Good Business: NASA Chief," *Chicago Tribune*, March 28, 1982, A1.

12. For other such efforts, see Margaret A. Weitekamp, *Space Craze: America's Enduring Fascination with Real and Imagined Spaceflight* (Washington, DC: Smithsonian Books, 2022).

13. Poster, n.d.; Jack Anderson, "Young Astronauts Program Under Way," *Washington Post*, November 12, 1984, D19, s.v. "Young Astronaut Program," 008926, NASA Historical Reference Collection, NASA Headquarters, Washington, DC. Hereafter, the NASA Historical Reference Collection will be cited as NASA.

14. Ronald Reagan to "Dear Friends," October 17, 1984, s.v. "Young Astronaut Program," 008926, NASA; "Young Astronaut Program Launched by the President," *NSI/Space World*, December 1984, n.p., s.v. "Young Astronaut Program," 008926, NASA; Ellen I. Kelly, "The Young Astronauts," *Aerospace: Official Publication of the Aerospace Industries Association*, Summer 1985, s.v. Young Astronaut Program, 008926, NASA. Frank Johnston of the *Washington Post* published the next day a close-up head shot of Reagan wearing the cap.

15. Jack Anderson, "Young Astronaut Program Under Way," *Washington Post*, November 12, 1984, D19, s.v. "Young Astronaut Program," 008926, NASA; T. Wendell Butler, "The American Young Astronaut Program: A First Year Progress Report," 1985, 2, 5–6, s.v. "Young Astronaut Program," 008926, NASA.

16. Kelly, "Young Astronauts"; Advertisement, "MARS OR BUST!" *Omni*, December 1985, n.p., s.v. Young Astronaut Program, 008926, NASA; Advertisement, "Will You Be Ready for the 21st Century?" *Omni*, February 1986, n.p., s.v. Young Astronaut Program, 008926, NASA; Letter, Ronald Reagan, October 17, 1984, s.v. "Young Astronaut Program," 008926, NASA; "American Space Movement," *Space Calendar*, November 19–25, 1984, 7, s.v. "Young Astronaut Program," 008926, NASA; "Young Astronaut Program," *Sky & Telescope*, March 1985, s.v. "Young Astronaut Program," 008926, NASA; T. Wendell Butler, "The American Young Astronaut Program: A First Year Progress Report," 1985, 5, s.v. "Young Astronaut Program," 008926, NASA.

17. Sandra Salmans, "Television: Why Saturday Morning Is One Big 'Cartoon Ghetto,'" *New York Times*, August 25, 1985, H21; John Carmody, "The TV Column," *Washington Post*, April 11, 1985, B11; Jack Anderson and Dale van Atta, "Space Can Help Kids Rise above Times," *Washington Post*, June 17, 1985, B8, s.v. "Young Astronaut Program," 008926, NASA.

18. T. Wendell Butler, "Young Astronauts Blast Off into New School Year," *Young Astronaut: The Official Newsletter of the Young Astronaut Council* 2, no. 1 (September/October 1985): 1, s.v. "Young Astronaut Program," 008926, NASA; Jack Anderson, "Frontier Facing Today's Youth Is the Universe," *Washington Post*, September 10, 1984, n.p.; Flyer, "Join the Young Astronauts!" n.d., s.v. "Young Astronaut Program," 008926, NASA.

19. NASA ceased using the straightforward numbering system for missions with STS-9. The new system combined the usual prefix, "STS," with a two-digit number based on the last digit of the fiscal year for the launch, combined with a digit indicating the site of the launch and a letter indicating its order in the original flight plan. So, STS-51F was initiated in 1985 and would launch from the Kennedy Space Center in Florida (1), and 2 would have been used for Vandenberg Air Force Base in California, had it ever been used. The "F" denoted its place as the sixth mission scheduled. Flight delays sometimes caused missions to be flown out of alphabetical order. Such designations ended in 1988 with STS-26, the return to flight mission after the loss of the Space Shuttle *Challenger* in 1986.

20. Thomas O'Toole, "Cola Clash Is Taking to Heavens: High-Tech Cans Fly on Shuttle Today," *Washington Post*, July 12, 1985, A3.

21. O'Toole, "Cola Clash Is Taking to Heavens."

22. NASA recruited the first scientist astronauts in 1965 for the Apollo program but created the designation of Mission Specialist for the Space Shuttle program.

23. NASA, STS 51-F Press Kit, https://www.nasa.gov/wp-content/uploads/2023/05/sts-51f-press-kit.pdf?emrc=68b394, accessed October 30, 2021. For a complete history of the payload specialist program, see Melvin Croft and John Youskauskas, *Come Fly with Us: NASA's Payload Specialist Program* (Lincoln: University of Nebraska Press, 2019).

24. Bill Brown, "Objects, Others, and Us (The Prefabrication of Things)," *Critical Inquiry* 36, no. 2 (Winter 2010): 183–207; Martin Collins, "Introduction," in *Showcasing Space*, ed. Martin Collins and Douglas Millard (London: Science Museum, 2006), 1–6.

25. Astronauts could have more than one removable nametag. Some astronauts chose nicknames or even just first names on their nameplates. For instance, in addition to having name badges that included their full names, first American woman in space Dr. Sally K. Ride had a nametag that just read "Sally," and space shuttle pilot and commander Robert Crippen had one that simply said, "Crip."

26. Edited Oral History Transcript, Charles D. Walker interviewed by Sandra Johnson, Springfield, Virginia, November 7, 2006, NASA Johnson Space Center Oral History Project, https://historycollection.jsc.nasa.gov/JSCHistoryPortal/history/oral_histories/WalkerCD /WalkerCD_11-7-06.htm, accessed November 15, 2021.

27. Regarding the astronauts' evolving definition of themselves as a professional group, see Matthew H. Hersch, *Inventing the American Astronaut* (London: Palgrave Macmillan, 2012).

28. Officially, the first Interkosmos project was the launch of satellite DS-U2-GK, known as *Kosmos-261*, in 1968. Colin Burgess and Bert Vis, *Interkosmos: The Eastern Bloc's Early Space Program* (Chichester, UK: Springer-Praxis Books, 2016), 1–4.

29. For the history of second- and third-generation Soviet space stations, see Wayne R. Matson, ed., *Cosmonautics: A Colorful History* (Moscow, CIS: Cosmos Books, 1994), 60–90.

30. Cathleen S. Lewis, "Arnoldo Tamayo Mendez and Guion Bluford: The Last Cold War Race Battle," in *NASA and the Long Civil Rights Movement*, ed. Brian S. Odom and Stephen P. Waring (Gainesville: University Press of Florida, 2019), 145–166.

31. Tom Hale, "The Bizarre Story of Japan's First Astronaut," *IFL Science*, April 5, 2017, https://www.iflscience.com/the-bizarre-story-of-japans-first-astronaut-36284.

32. Mark Sheerer and the Associated Press, "Party On: Flasher, Profanity, Cosmonauts Highlight MTV Awards," CNN.com, September 5, 1996, www.cnn.com/SHOWBIZ/9609/05 /mtv.awards/index.html.

CHAPTER SIX

Neoliberal Mutations

Angela N. H. Creager

> And what is important and decisive in current neo-liberalism can, I
> think, be situated here. . . . Can the market really have the power of
> formalization for both the state and society? This is the important,
> crucial problem of present-day liberalism and to that extent it
> represents an absolutely important mutation with regard to tradi-
> tional liberal projects, those that were born in the eighteenth
> century. It is not just a question of freeing the economy. It is a
> question of knowing how far the market economy's power of
> political and social information extend.
> —MICHEL FOUCAULT, *THE BIRTH OF BIOPOLITICS*, 117–118

In my domain, the phrase "greedy science" usually calls up the biotech industry's rapid ascent in the late 1970s and 1980s. When Genentech had
its initial public offering on October 14, 1980, the stock prize doubled in
one day, setting off a series of other "heady biotech IPOs that defined the
era."[1] Of course, molecular biology had become a hot area of science for
some time before the commercialization of genetic engineering. When Erwin Chargaff quipped in 1963, "We now have DNA tycoons and others
have 'made a killing' in RNA," he was being metaphorical, skewering the
raw, uninformed ambition of young biologists.[2] Fifteen years later, it was
no longer a joke. High-profile molecular biologists partnered with venture
capitalists in a flurry of start-ups. Faculty with expertise in gene cloning
could moonlight as consultants to industry without giving up their positions, but many molecular biologists and biochemists abandoned universities altogether for the seemingly freer and certainly more freewheeling
atmosphere of biotech.[3]

The initial public alarm about the safety of recombinant DNA was over-
taken in the 1980s by ethical concerns about the corruption of academic

biology by investors.[4] The launch of the Human Genome Project (with its own funding stream for Ethical, Legal, and Social Implications, ELSI) prompted further criticism from scholars and public interest groups of the possible abuses arising from the commercialization of genetic knowledge, above all human DNA sequence data.[5] Many journalists and science studies scholars (some with ELSI grants) portrayed biology in the 1980s as a story of the commodification of scientific knowledge and of life itself, enabled by new intellectual property provisions for patenting organisms and DNA sequences.[6] This picture of "greedy biology" captures many important developments in the life sciences and in American capitalism but misses others. For example, changing understandings of how mutations cause cancer reflected and reinforced political trends, with a shift away from government action protecting consumers from damaging chemicals to an emphasis on market mechanisms and individual risk management.

This chapter uses the theme of mutation to compare the regulatory and public health responses to cancer in the 1970s to the deregulatory turn in the 1980s. The central role assigned to genetic mutation in cancer management was part of the legacy of the atomic age, which put the spotlight on effects of radiation.[7] Ionizing radiation, a known carcinogen, was shown to damage DNA, inspiring prominent scientists to attend to the public health threat of other environmental carcinogens.[8] In 1971, Nixon's Council on Environmental Quality called for more stringent regulatory controls of chemicals to prevent the cancers, birth defects, and heritable mutations from long-term, low-dose exposure.[9] The passage of the Toxic Substances Control Act in 1976 was a direct outgrowth of this regulatory approach to controlling cancer.

Biologists in the 1980s did not abandon the centrality of mutations in causing cancer but began seeing DNA damage as an inevitable by-product of metabolism and aging rather than predominantly due to exogenous carcinogens, such as chemicals.[10] The oncogene theory of cancer, which originated in tumor virology, came to implicate specific cellular genes in controlling growth and suppressing tumors.[11] If only a handful of mutations in critical genes could turn a cell rogue, inheriting one or two from a parent could dramatically increase one's personal susceptibility to cancer. The Human Genome Project provided cancer biologists and subsequently clinicians with new tools, particularly through sequence diagnostics, to evaluate individual predisposition.[12] In conjunction with this changing picture of cancer, public health messaging shifted away from chemical culprits to encouraging individuals to adopt healthy lifestyles and manage their hereditary risk. This was part of a broader trend, especially at the Food and Drug Administration, of using nutritional labeling and direct-to-consumer drug advertising as part of a new reliance on "informational" regulation, trusting informed individuals to make rational (hence healthy) market choices.[13] A shift that might be dubbed neoliberal.

The term "neoliberalism" gestures to a set of sweeping political and economic changes in the 1980s and 1990s associated with deregulation, financialization, and globalization.[14] Daniel Rodgers has tried to give the term some analytical grip by pulling apart four distinct strands conflated in the term "neoliberalism": (1) finance capitalism, (2) market fundamentalism, (3) business-friendly policy measures, and (4) the Foucauldian "neoliberal reason," the subjugation to all human experience and culture to economic rationality.[15] The trajectory of cancer biology and prevention in this period resonates particularly with Rodgers's third and fourth aspects of neoliberalism—and Foucault's own metaphor of neoliberalism as a mutation in the liberal project. To wit, if a key element of neoliberal governance consists of entrusting informed consumers and their market choices to reach policy outcomes, rather than "command-and-control" regulation, then the story of the migration of cancer-causing mutation from environmental pollutants to hereditary risk could hardly be more emblematic. This transition can be attributed to either the effectiveness of corporate actors in reshaping public policy or the emergence of new scientific evidence about DNA damage. Rather than adjudicating between these explanations, I seek to integrate the history of cancer biology with that of political economy and environmental regulation in the 1980s, situating "neoliberal mutations" as the product of this specific historical moment.

Carcinogens in the Environment

After World War II, thousands of new synthetic chemicals entered the marketplace and, eventually, the environment.[16] Protection against exposure to toxic chemicals had mainly developed in the realm of occupational health, through both industrial hygiene and government oversight. But the perils of contamination from the petrochemical revolution extended beyond the workplace, contributing to the rise of environmental science and environmentalism.[17] In 1962, Rachel Carson's *Silent Spring* called attention to the potential hazards of these contaminants to both wildlife and human health. She drew on public concerns generated during the debates over radioactive fallout from atomic weapons testing in pointing to the additional dangers from chemicals: "In this now universal contamination of the environment, chemicals are the sinister and little-recognized partners of radiation in changing the very nature of the world—the very nature of its life."[18]

Efforts to protect ordinary people and the environment from radioactivity shaped the response to chemicals of concern in three important ways. First was the enduring politicization of contamination from the fallout debates. As the quote above illustrates, Carson and others channeled the public opposition to atomic weapons testing (and, less visibly, nuclear waste) to encompass synthetic chemicals as well.[19] Second, the establishment in 1958 of national standards for population exposure to ionizing radiation was an important precedent for regulations controlling chemical exposures, a point to which I will return.[20] Third, a group of geneticists, drawing largely on studies in radiation biology and medicine, advanced the idea that human cancers typically arose from mutations to somatic cells.[21] The so-called somatic mutation theory, which had been first proposed in the early twentieth century, focused attention on the role of environmental factors (including exposure to radioactivity) in cancer incidence.[22] Since that time, this theory has been so fully assimilated into current thinking about cancer that we often fail to recognize the novelty of its underlying assumption equating mutagenicity with carcinogenicity—that is, the mutagenizing ability of an agent such as radiation was responsible for inducing cancer.

Because there appeared to be no threshold for genetic damage, even low-level exposures might induce cancer-producing somatic mutations. In addition, the "multistage model" of carcinogenesis proposed by Peter Armitage and Richard Doll in 1954 treated cancer as the result of not just one, but several, cellular events.[23] These events (particularly the initial trigger of carcinogenesis) could be envisioned as mutations, such as those induced by radiation. In 1972, the National Academy of Science's Advisory Committee on the Biological Effects of Ionizing Radiation stated that the carcinogenic effects of ionizing radiation should be regarded as linearly dose dependent, with no threshold below which damage did not occur.[24] More

complex nonlinear (e.g., quadratic) models of the dose–response for ionizing radiation are now utilized by some regulatory bodies, but the non-
threshold assumption for mutational carcinogenesis remains.[25]

This understanding of the carcinogenicity of even low-dose radiation
informed the growing public resistance in the United States to the development of nuclear power—and suspicions about other low-level contaminants. As the emerging environmental movement focused attention on
the dangers of chemical residues, the somatic mutation theory provided a
way to link synthetic chemicals to cancer risk. Chemicals, like radiation,
were presumed to act at the level of DNA in causing mutations and potentially cancer.[26] Not only did this have implications for the regulation of
chemicals in the workplace, but as there were no safe thresholds for mutagens, this analogy opened the question of protecting the population at
large from chemical exposures in pollution, pesticides, and consumer
products, just as government regulation specified population exposure
limits for ionizing radiation.

Testing Chemicals

The idea that environmental mutagens induce cancer by damaging DNA
gained ground steadily through the 1960s. Scientists working for the US
Atomic Energy Commission in the Biology Division at Oak Ridge National
Laboratory contributed to the base of knowledge by establishing a computer registry on the mutagenicity of chemicals (known as the Environmental Mutagen Information Center). The National Institutes of Health
provided funding for this work at Oak Ridge and established a National
Institute of Environmental Health Sciences in 1968. As Scott Frickel has
argued, both these institutional developments reflected the collective action of scientists in response to environmentalism and in concert with
government efforts to improve environmental health.[27] Many biologists
turned their attention to harnessing their knowledge and methods to detect carcinogens, with an eye toward their control.[28]

Scientists discussed possible approaches to these issues at the Conference on Evaluating the Mutagenicity of Drugs and Other Chemical
Agents, which took place in Washington, DC on November 4–6, 1970. At
this gathering, six months after the first Earth Day, biologists drew on their
familiarity with mutagenesis as a laboratory tool to consider the parallel
hazards of ionizing radiation, drugs, and chemical mutagens in everyday

life. New detection methods adapted from molecular genetics could be used to assess the dangers of synthetic chemicals. As a reporter for *Science* noted, "Many workers believe that chemical damage is now a more important problem than radiation hazard."[29]

The most influential tool to come out of these concerted efforts was a rapid bacterial test to identify mutagenic chemicals, developed by Bruce Ames at the University of California, Berkeley. The "Ames test" relied on four mutant strains of *Salmonella typhimurium*, adapted from their prior application in studies of bacterial metabolism and mutagenesis. All four strains were deficient in their ability to synthesize a particular amino acid, histidine, so they required this supplement in the growth media. Each of the four strains could be used to genetically screen compounds inducing a specific kind of mutation in the DNA sequence. These registered as reverse mutations, or revertants, which compensated for the histidine deficiency. In other words, on petri dishes plated with these bacteria and the substance to be tested, cells that grew represented new mutations. If the number of colonies on a test plate was greater than that of the control (representing the spontaneous mutation rate), the test substance was mutagenic.

The Ames test registered specific types of mutation at the DNA level. Three of Ames's test strains (originally TA1531, TA1532, and TA1534) were designed to detect different kinds of frameshift mutagens (adding or deleting a nucleotide). The fourth strain, TA1530, contained a base-pair change, so it would detect mutations that involve base-pair substitutions. The four *Salmonella* strains were further customized with additional genetic changes that made the cells more permeable to large molecules and eliminated some kinds of DNA repair.[30] Last but not least, an extract of rat liver was added to the cells and chemical being tested. In some cases, researchers had realized, the active carcinogen was not the original chemical but a breakdown product generated in the liver through the body's detoxification pathways. In the Ames test, the enzymes in the liver extract produced these metabolic derivatives of the chemical being assayed.[31]

The value of Ames test, which was embraced by industry and environmentalists alike in the 1970s, relied on two key assumptions.[32] First, as Ames put it, a carcinogen is a mutagen.[33] Human cancer, in this view, was usually triggered by exposure to environmental mutagens. Chemical compounds that do not induce mutations were presumed to not initiate cancer, either. There were known nonmutagenic carcinogens, notably hormones; these were envisioned as *promoting* tumor formation after the *initiating* genetic mutations.[34] Second, Ames assumed that a microbe was a suitable model organism for detecting genetic damage as it occurred in human (or animal) cells. Bacterial tests were much quicker and cheaper than rodent bioassays, which between 1940 and 1960 became the standard method to evaluate the safety of chemicals in foods, drugs, and cosmetics.[35]

The regulatory importance of this new bacterial test became clear when Ames and Arlene Blum showed that the most commonly used flame retardant in children's pajamas, tris(2,3-dibromopropyl) phosphate, commonly called Tris, was a mutagen.[36] Toxicologists subsequently found Tris caused kidney cancer in rodents.[37] In April 1977, the US Consumer Product Safety banned the sale of Tris-treated garments, citing Blum and Ames's work.[38] The compound had only been introduced a few years prior, to meet 1973 federal regulations for decreased flammability in children's sleepwear. In screening over 300 chemicals, Ames and collaborators demonstrated that 90 percent of *Salmonella* mutagens were also rodent carcinogens and that 89 percent of animal test carcinogens were also bacterial mutagens.[39] Of more than 109 "noncarcinogens" tested, none proved to be mutagens. Even if the Ames test was not a perfect predictor of carcinogenicity, its ability to identify such a large percentage of cancer-causing chemicals meant it could be used to address the bottleneck of rodent carcinogenicity testing, which took two years and cost at least $150,000 (over $800,000 adjusted to 2023).[40]

Both scientists and government officials were interested in using short-term bioassays like the Ames test in the regulation of chemicals, to identify and control (if not ban) potentially carcinogenic substances. This approach was exemplified in the Toxic Substances Control Act (TSCA), which was introduced as a bill in 1971 and debated by Congress until 1976, when it was finally signed into law. The main avenue for regulation in TSCA was toxicological testing and government review of test data. Concessions to industry in the language of the bill before it was passed weakened the Environmental Protection Agency's ability to require tests or data submission for either existing or new commercial chemicals.[41] Nonetheless, the legislation still represented an approach to regulation that charged the government with making decisions about consumer health and safety. Beginning with the Jimmy Carter administration, and intensified under Ronald Reagan, this approach to regulation was challenged by an emphasis on consumer choice, product labeling, and a turn to market mechanisms to achieve environmental and public health objectives. Cancer prevention was no exception.

Tumor Viruses to Cancer Genes

By the 1980s, the role of mutation in carcinogenesis was rarely contested. But *which* genes were being mutated, to enable cells to escape the regulatory controls over cell division, or to circumvent apoptosis, the programmed cell death after a certain number of mitotic events? Molecular biologists were well poised to pursue these lines of investigation. The turn to cancer genes, however, came through a rather circuitous route, namely, the study of a different set of external cancer agents, tumor viruses. In the 1960s and beyond, US federal funding had poured into research on tumor viruses. In 1964, the National Cancer Institute established a Special Virus Leukemia Program (soon renamed Special Virus Cancer Program), with the ambition of identifying human tumor viruses in order to produce cancer vaccines. Despite massive funding and use of the most up-to-date administrative measures, this effort was deemed a failure and—by some—a boondoggle.[42] Yet the burst of intensive research produced experimental systems involving tumor viruses in cultured cells that were repurposed to other ends. The fact that some small viruses such as polyoma and SV40 could transform animal cells (i.e., causing uncontrolled growth) gave molecular biologists reason to think that just a few genes, or even just one, might suffice in tripping the switch to malignancy, and tumor viruses could be used to find them.[43]

Important clues came from work being done with Rous sarcoma virus (RSV), which had first been identified by Peyton Rous in 1910 in chickens. There was an early connection between work on RSV and the mutation theory of cancer; in 1931, James Murphy had suggested that fowl tumor agents might be referred to as "transmissible mutagens."[44] In 1956, researchers at Rutgers showed that RSV infection produced discrete foci in cultured chicken embryo cells; these were cells that lost their usual morphology as well as the ability to stop multiplying when adjacent to other cells, creating a characteristic pile-up. This cell transformation provided a useful in vitro proxy for cancer.

In Renato Dulbecco's laboratory, postdoc Harry Rubin and graduate student Howard Temin used transformation in cultured cells to study the dynamics of RSV infection. RSV contains RNA, not DNA; Temin showed that the viral-derived DNA was synthesized off RSV RNA, an apparent violation of Francis Crick's "central dogma" that DNA makes RNA, and RNA produces protein.[45] (RSV was thus called a retrovirus and its enzyme that could generate a DNA copy of RNA "reverse transcriptase.") The synthesized viral DNA was integrated into the host cell DNA, creating the genes that triggered cell transformation. In Rubin's laboratory at Berkeley, Hidesaburo Hanafusa and colleagues showed that a single gene was responsible for this transformation of cells from RSV infection.[46] This gene was soon mapped to a region of the RSV genome called *src*, shorthand for sarcoma. The biology was complex; for instance, RSV needed a "helper virus" (Rous-associated virus, or RAV) to replicate in cells, but it appeared that the key step of transformation was due to one gene, a viral oncogene, as it was called.

Several groups were hot on the heels of the oncogene *src*. At the University of California, San Francisco (UCSF), Dominique Stehelin, Harold Varmus, and Michael Bishop wondered how RSV had acquired a cancer-causing gene, absent in related viruses. They isolated a probe for *src*, labeled it with radioactive phosphorus, and hybridized it to cellular DNA not only from chicken but also from other birds, to light up any complementary sequences. Stehelin showed that there were *src* sequences in chicken cells uninfected with RSV, as well as variants in other species of birds.[47] Two years later, Deborah Spector, working with Varmus and Bishop, found cognate genes in human, calf, mouse, and salmon.[48] It remained unclear how *src* caused cell transformation, but other oncogenes (*ras*, *mos*, *myc*, *mil*, etc.), some with viral variants, began to be identified in other species. Varmus and Bishop hypothesized that these cellular genes were proto-oncogenes, normal residents of cellular genomes that could be disrupted by mutation to cause cancer.[49] Cell division was normally regulated by a whole network of proteins, encoded by a variety of genes, each of which, if mutated, might contribute to cancer.[50]

Bruce Ames's conviction that all carcinogens were mutagens inspired these tumor virologists to search out all cellular genes involved in cancer. As Robert Weinberg recalls, "To maximize our chances of a hit [on an oncogenic gene], we looked for cancer cells that seemed most likely to carry mutated genes. It was here that Bruce Ames and his sermon came to mind. Ames had drummed into our heads the message that potent carcinogens were potent mutagens. It followed that chemically induced cancers would carry mutant genes, and that these genes would operate as oncogenes that drove the growth of these cancers."[51]

Weinberg's lab showed that chemical carcinogens could activate oncogenes, "just as Ames's work had predicted."[52] The group's experiments between 1981 and 1984 demonstrated that the same oncogenes were responsible for cancer whether they were induced by viruses or environmental carcinogens, and they were involved in noncancerous cell growth as well.[53] In fact, about forty different proto-oncogenes were identified. As Michel Morange describes the state of the field circa 1985, "most biologists agreed that studying this small group of genes was the best way to understand cancerous transformation. Alterations in these genes and their functioning were considered to be the cause of cancer."[54]

In this way, a line of research originally aimed at finding out how tumor viruses caused cancer led to the identification of a whole class of cellular genes that, if mutated, could cause malignancy. Retroviruses had led cancer biologists to "the enemy within."[55] The range of genes that can betray the organism expanded further. By the end of the 1980s, it was clear that a *second* class of genes, tumor suppressor genes, were also implicated in carcinogenesis. These genes encode proteins that put the brakes on cell proliferation, and in diploid cells, both copies of the gene must be mutated for cancer to result.[56]

The best-known tumor suppressor gene goes by the name p53, after the estimated molecular weight (53 kilodaltons) of the protein it encodes.[57] This protein was initially identified as one of the two produced in response to infection of cells by a polyomavirus, simian virus 40 (SV40).[58] Arnold Levine and his colleagues found that p53 was present in uninfected cells as well as infected cells, indicating that it was derived from the host, not the virus.[59] For years, virologists assumed that this must be another oncogene, but Levine's lab used two copies of the gene, one of which was mutated, to demonstrate that the wild-type copy could *prevent* cell transformation, even in the presence of the mutant copy that usually caused transformation.[60] Mutations in this cellular gene could disrupt its ability to serve as a brake on cancer. In 1989, Bert Vogelstein, his doctoral student Suzanne Baker, and a number of other collaborators published a paper documenting a strong genetic link between colon cancer and p53. In 75 percent of the colorectal cancers they analyzed, the gene for p53 was either deleted or mutated in both chromosome 17 copies.[61] Vogelstein hypothesized that for colon cancer, at least six consecutive genetic mutations were required, including one in the *ras* gene and another disabling p53.[62] Mutations in the gene for p53 are now thought to be involved in more than 50 percent of human cancers.[63]

Testing People

The Human Genome Project, which began in 1987, reinforced the preoccupation with genes (and their mutations) as determinants of cancer. In some cases, the mapping and sequencing of genes brought greater precision to hereditary determinants that had been identified already using pedigree analysis and chromosomal genetics. For instance, childhood retinoblastoma (a rare eye cancer) had long been observed to run in families, with a high percentage of children having tumors in both eyes. Family histories suggested inheritance in a dominant Mendelian fashion; about half of the children of affected parents developed the tumors themselves. In a 1971 paper, Alfred Knudson presented an analysis of 48 cases.[64] Those children who inherited a germline mutation were far more likely to have bilateral (two-eye) tumors and presented at an earlier age; those children with unilateral tumors, whose average age at diagnosis was older, did not inherit either mutation. Knudson showed that the data were consistent with the cancer being due to *two* mutations, using a two-hit model: children with a hereditary mutation developed cancer after acquiring a second somatic mutation. The gene associated with retinoblastoma (RB) susceptibility was mapped to human chromosome 13q14.[65] In 1988, a research group at the University of California, San Diego cloned the segment of chromosome 13 that corresponded to deletions in this region among individuals affected with familial RB. Inserting the normal copy of this gene into cultured retinoblastoma cells reversed their oncogenicity.[66] Thus, RB was considered the first human oncogene identified (although the gene product worked as a tumor suppressor).

Retinoblastoma is a rare cancer, but scientists had also recognized familial dispositions to more common cancers. In the 1960s, physician Henry Lynch traced the elevated incidence of hereditary nonpolyposis colorectal cancer (HNPCC) to certain "cancer families."[67] This particular type of colon cancer was distinctive in that it was not preceded by polyps, which were already being identified and removed by gastroenterologists in cancer prevention. By the mid-1980s, molecular biologists were attempting to identify genetic mutations associated with what Lynch had called "cancer family syndrome."[68] This took most of a decade; in the 1990s, mutations in two human DNA mismatch repair genes were implicated in most cases of HNPCC.[69] While mutations in these genes have proven strongly predictive of this type of colorectal cancer, they are not prognostic for most forms of the disease, which emerge from polyps. Mutations in a different gene, called *APC*, have been associated with sporadic colorectal tumors (both benign and malignant).[70]

The search for gene mutations associated with breast cancer was even more influential, reinforcing popular perceptions that tumors arise because individuals inherit "cancer genes."[71] In the mid-1970s, Mary-Claire King at Berkeley started a project to identify mutations associated with hereditary breast cancer. In the early 1980s, she began working on data and blood samples from almost 1,600 patients at UCSF diagnosed with breast cancer. Using this data set, she determined that 4 percent possessed a hereditary predisposition to breast cancer.[72] Focusing on DNA from these patients enabled King and her colleagues to identify linkage of early-onset breast cancer with a specific allele on human chromosome 17, band 21.[73] Her announcement of this finding, along with the dramatically improved technologies for mapping, cloning, and sequencing human genes, meant she soon had rivals.[74] Although she joined forces with Francis Collins, who had identified the gene associated with cystic fibrosis, they were beat in the race to clone the breast cancer gene (dubbed *BRCA1*) by Mark Skolnick, a professor at the University of Utah who had a start-up company called Myriad Genetics.[75] Identification of a second gene, *BRCA2*, soon followed. Myriad obtained patents for both of the *BRCA* genes, which became the basis for a diagnostic kit the company developed and licensed for the most common breast cancer–related mutations in these genes.[76] Hundreds of thousands of women have undergone testing for mutations in *BRCA1* and *BRCA2*, although the availability and cost of testing is highly dependent on national context.[77] Strikingly, the DNA testing company 23andMe, popular for its ancestry services, obtained authorization in 2020 from the US Food and Drug Administration to test for certain mutations in the *BRCA* genes.[78]

Many scientists, policy analysts, and scholars have criticized the patenting of the breast cancer–associated genes, the high cost of the test, and the uncertainty of extrapolating susceptibility to the general population from families showing very high rates of breast cancer.[79] I want to underline another feature of this mode of cancer prevention: the way in which a new generation of genetic diagnostics made patients responsible for managing their personal health risk, such as mutations predisposing them to cancer. An educational booklet from the National Cancer Institute depicts cancer as resulting from several successive mutations in the same cell (figure 6.1). The legend says, "Persons with hereditary cancer already have the first mutation." The other three mutations are presumed to be spontaneous, perhaps due to exposure to DNA-damaging chemicals, whether they originated outside or within the body.

Conclusions

To be sure, cancer prevention was already moving toward an emphasis on individual responsibility before the Human Genome Project—not least because of public health campaigns against smoking. In 1981, Richard Doll and Richard Peto published an influential review of cancer causation, in which they estimated that most of the "avoidable" causes of cancer were linked to individual decision-making, particularly smoking, diet, and alcohol use.[80] By contrast, they attributed only 2–5 percent to industrial pollution.[81] (It should be noted that Doll had begun consulting for the chemical industry.[82]) Ongoing research into DNA repair led many scientists to regard cancer more as a breakdown of genetic maintenance (the defense) than the results of an assault by environmental mutagens (the offense). This could account for the long-standing correlation of cancer incidence with age.

Note that the basic picture of carcinogenesis remained unchanged during the decades under discussion: cancer was fundamentally a disease of mutation. What had changed was the target of prevention. In the 1970s, the development of mutagenicity tests was motivated by a policy goal of limiting exposure to carcinogens, effectively putting the responsibility for cancer prevention on the government and industry. In the 1980s and 1990s, consumers were "empowered" to take charge of their own cancer risk, with public health messaging about healthy behaviors and the new availability of biomarkers and other diagnostics for individual disease

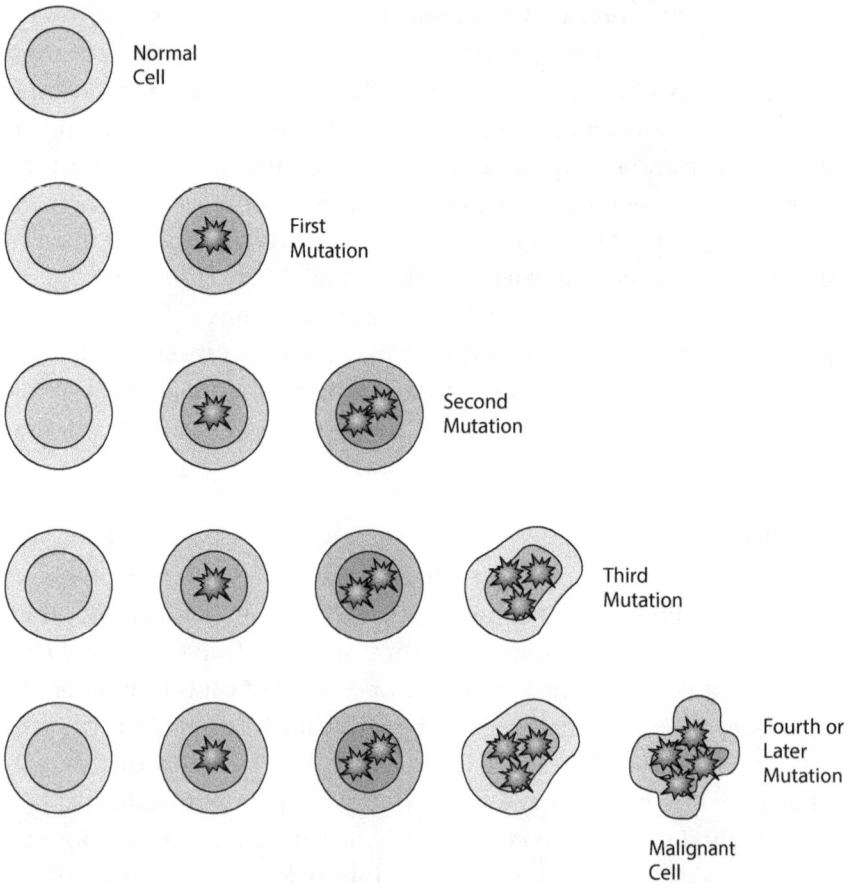

Cancer usually arises in a single cell. The cell's progress from normal to malignant to metastatic appears to follow a series of distinct steps, each controlled by a different gene or set of genes. Persons with hereditary cancer already have the first mutation.

Figure 6.1   An educational booklet from the National Cancer Institute depicts cancer as resulting from several successive mutations in the same cell. From National Cancer Institute, *Understanding Gene Testing,* NIH Publication No. 96–3905, December 1995, 13

susceptibility. Some patients, notably those mobilized around breast cancer and AIDS, were eager to be active participants in such decision-making, in part due to frustration with regulators and the pharmaceutical industry.[83] But as Wendy Brown has noted, the devolution of responsibility from government to individuals—and market choices—was a key facet of the thorough-going economization of politics and society.[84] As she puts it, "Responsibilization tasks the worker, student, consumer, or indigent person with discerning and undertaking the correct strategies of self-investment and entrepreneurship for thriving and surviving."[85] One might add to her list of subjects the patient, whose clinical encounters became pervaded by a managerial language of predisposition, risk, lifestyle choice, and informed decision-making.[86] In this world of the 1980s (and beyond), cancer mutations became neoliberal.

NOTES

1. Laura Fraser, "Genentech Goes Public," *Genentech*, April 28, 2016, https://www.gene.com/stories/genentech-goes-public; Robin Wolfe Scheffler, chapter 8 in this volume.

2. Erwin Chargaff, "Amphisbaena," in *Essays on Nucleic Acids* (New York: Elsevier, 1963), chap. 11, 176.

3. Paul Rabinow, *Making PCR: A Story of Biotechnology* (Chicago: University of Chicago Press, 1996), makes the case, though Steven Shapin, *The Scientific Life: A Moral History of a Late Modern Vocation* (Chicago: University of Chicago Press, 2008), is also relevant.

4. Sheldon Krimsky, *Genetic Alchemy: The Social History of the Recombinant DNA Controversy* (Cambridge, MA: MIT Press, 1982). However, as Doogab Yi has shown, the commercialization of biology was driven as much by the federal government as by corporate funding: Doogab Yi, "Who Owns What? Private Ownership and the Public Interest in Recombinant DNA Technology in the 1970s," *Isis* 102, no. 3 (2011): 446–474.

5. Sheldon Krimsky, "Science and Wall Street: Academic Entrepreneurship in Biology," in *Biotechnics & Society: The Rise of Industrial Genetics* (New York: Praeger, 1991), 59–79.

6. Martin Kenney, *Biotechnology: The University-Industrial Complex* (New Haven, CT: Yale University Press, 1986); Sheldon Krimsky, *Biotechnics & Society: The Rise of Industrial Genetics* (New York: Praeger, 1991); Daniel J. Kevles and Leroy Hood, eds., *The Code of Codes: Scientific and Social Issues in the Human Genome Project* (Cambridge, MA: Harvard University Press, 1992); George J. Annas and Sherman Elias, *Gene Mapping: Using Law and Ethics as Guides* (New York: Oxford University Press, 1992); David Magnus, Arthur L. Caplan, and Glenn McGee, eds., *Who Owns Life?* (Amherst, NY: Prometheus Books, 2002); Philip Mirowski, *Science-Mart: Privatizing American Science* (Cambridge, MA: Harvard University Press, 2011); Elizabeth Popp Berman, *Creating the Market University: How Academic Science Became an Economic Engine* (Princeton, NJ: Princeton University Press, 2012), esp. chap. 4; Nicolas Rasmussen, *Gene Jockeys: Life Science and the Rise of Biotech Enterprise* (Baltimore: Johns Hopkins University Press, 2014); Doogab Yi, *The Recombinant University: Genetic Engineering and the Emergence of Stanford Biotechnology* (Chicago: University of Chicago Press, 2015).

7. Angela N. H. Creager, "Radiation, Cancer, and Mutation in the Atomic Age," *Historical Studies in the Natural Sciences* 45, no. 1 (2015): 14–48.

8. John Higginson, "Present Trends in Cancer Epidemiology," *Proceedings of the Canadian Cancer Conference* 8 (1969): 40–75; Philippe Shubik, "Current Status of Chemical Carcinogenesis," *Proceedings of the National Academy of Sciences of the United States of America* 69, no. 4 (April 1972): 1052–1055.

9. Council on Environmental Quality (US), "Toxic Substances" (Washington, DC: US Government Printing Office, April 1971).

10. Gerald N. Wogan, Stephen S. Hecht, James S. Felton, Allan H. Conney, and Lawrence A. Loeb, "Environmental and Chemical Carcinogenesis," *Seminars in Cancer Biology* 14, no. 6 (December 2004): 473–486.

11. Michel Morange, "From the Regulatory Vision of Cancer to the Oncogene Paradigm, 1975–1985," *Journal of the History of Biology* 30, no. 1 (1997): 1–29.

12. Ilana Löwy and Jean-Paul Gaudillière, "Localizing the Global: Testing for Hereditary Risks of Breast Cancer," *Science, Technology, & Human Values* 33, no. 3 (2008): 299–325.

13. Xaq Frohlich, "The Informational Turn in Food Politics: The US FDA's Nutrition Label as Information Infrastructure," *Social Studies of Science* 47, no. 2 (2017): 145–171; Xaq Frohlich, *From Label to Table: Regulating Food in America in the Information Age* (Oakland: University of California Press, 2023). On related shifts at the FDA regarding drug testing, see Cathy Gere, chapter 7 in this volume.

14. For example, Jamie Peck, *Constructions of Neoliberal Reason* (New York: Oxford University Press, 2010); David Harvey, *A Brief History of Neoliberalism* (Oxford: Oxford University Press, 2005); Michael Lind, *The New Class War: Saving Democracy from the Managerial Elite* (New York: Portfolio/Penguin, 2020); and, for a reflection on this trend, Harold James, "Neoliberalism and Its Interlocutors," *Capitalism* 1, no. 2 (Spring 2020): 484–518.

15. Daniel Rodgers, "The Uses and Abuses of 'Neoliberalism,' " *Dissent* 65, no. 1 (2018): 78–87; Michel Foucault, *The Birth of Biopolitics: Lectures at the Collège de France, 1978–1979*, ed. Michel Senellart, trans. Graham Burchell (New York: Picador, 2010); and, for a critical elaboration of Foucault on neoliberalism, Wendy Brown, *Undoing the Demos: Neoliberalism's Stealth Revolution* (Cambridge, MA: MIT Press, 2015).

16. To be sure, pollution before World War II was not negligible: François Jarrige and Thomas Le Roux, *The Contamination of the Earth: A History of Pollutions in the Industrial Age*, trans. Janice Egan and Michael Egan (Cambridge, MA: MIT Press, 2020). For reflections on the accumulating environmental problems related to chemicals, see Soraya Boudia, Angela N. H. Creager, Scott Frickel, Emmanuel Henry, Nathalie Jas, Carsten Reinhardt, and Jody A. Roberts, *Residues: Thinking through Chemical Environments* (New Brunswick, NJ: Rutgers University Press, 2022).

17. Christopher C. Sellers, *Hazards of the Job: From Industrial Disease to Environmental Health Science* (Chapel Hill: University of North Carolina Press, 1997).

18. Rachel Carson, *Silent Spring* (Boston: Houghton Mifflin, 1962), 6.

19. Ralph H. Lutts, "Chemical Fallout: Rachel Carson's *Silent Spring*, Radioactive Fallout, and the Environmental Movement," *Environmental Review* 9, no. 3 (Autumn 1985): 211–225.

20. Soraya Boudia, "From Threshold to Risk: Exposure to Low Doses of Radiation and Its Effects on Toxicants Regulation," in *Toxicants, Health and Regulation since 1945*, ed. Soraya Boudia and Nathalie Jas (New York: Routledge, 2013), 71–87.

21. A. H. Sturtevant, "Social Implications of the Genetics of Man," *Science* 120, no. 3115 (1954): 405–407; J. Christopher Jolly, "Thresholds of Uncertainty: Radiation and Responsibility in the Fallout Controversy" (PhD dissertation, Oregon State University, 2003), chap. 12; Ilana Löwy, *Preventive Strikes: Women, Precancer, and Prophylactic Surgery* (Baltimore: Johns Hopkins University Press, 2010), chap. 7; Creager, "Radiation, Cancer, and Mutation in the Atomic Age."

22. Theodor Boveri is usually credited with proposing the somatic mutation theory of cancer, although his conception of what a mutation consisted of was chromosomal and related to embryological development. By the time geneticists endorsed the idea in the 1950s, it was increasingly associated with mutation as DNA damage, often at the base change level. Theodor Boveri, *Concerning the Origin of Malignant Tumours*, trans. Henry Harris (Cold Spring Harbor, NY: Cold Spring Harbor Laboratory Press, 2008); Marja Sorsa, "Somatic Mutation

Theory," *Journal of Toxicology and Environmental Health* 6, nos. 5–6 (November 1980): 977–982.

23. Peter Armitage and Richard Doll, "The Age Distribution of Cancer and a Multi-Stage Theory of Carcinogenesis," *British Journal of Cancer* 8 (March 1954): 1–12.

24. National Research Council (US), *The Effects on Populations of Exposure to Low Levels of Ionizing Radiation* (Washington, DC: National Academies Press, 1972).

25. Joseph Rotblat, "Hazards of Low-Level Radiation—Less Agreement, More Confusion," *Bulletin of the Atomic Scientists* 37, no. 6 (June/July 1981): 31–36; Edward J. Calabrese, "Origin of the Linearity No Threshold (LNT) Dose–Response Concept," *Archives of Toxicology* 87, no. 9 (September 2013): 1621–1633. Ongoing challenges to the linear, no-threshold model (still used by US agencies) have been stoked by the nuclear industry and corporate-funded scientists. Calabrese, the most prominent critic of the linear, no-threshold model (and advocate of hormesis, the theory that low-level radiation is beneficial), is a prime example of industry-backed science. See Kristin S. Shrader-Frechette, *Tainted: How Philosophy of Science Can Expose Bad Science* (New York: Oxford University Press, 2014), chap. 3.

26. Claes Ramel, "Advantages of and Problems with Short-Term Mutagenicity Tests for the Assessment of Mutagenic and Carcinogenic Risk," *Environmental Health Perspectives* 47 (1983): 153–159.

27. Scott Frickel, *Chemical Consequences: Environmental Mutagens, Scientist Activism, and the Rise of Genetic Toxicology* (New Brunswick, NJ: Rutgers University Press, 2004), 56–62.

28. Frickel, *Chemical Consequences*.

29. Maureen Harris, "Mutagenicity of Chemicals and Drugs," *Science* 171, no. 3966 (1971): 51.

30. Bruce N. Ames, Frank D. Lee, and William E. Durston, "An Improved Bacterial Test System for the Detection and Classification of Mutagens and Carcinogens," *Proceedings of the National Academy of Sciences of the United States of America* 70, no. 3 (1973): 782–786.

31. Bruce N. Ames, William E. Durston, Edith Yamasaki, and Frank D. Lee, "Carcinogens Are Mutagens: A Simple Test System Combining Liver Homogenates for Activation and Bacteria for Detection," *Proceedings of the National Academy of Sciences, USA* 70, no. 8 (1973): 2281–2285. Liver homogenate from human autopsies was also tried, although the rat (not human-derived) extract became the standard addition.

32. On industry's embrace of the test, Gina Bari Kolata, "Chemical Carcinogens: Industry Adopts Controversial 'Quick' Tests," *Science* 192, no. 4245 (1976): 1215–1217.

33. Bruce N. Ames, "Carcinogens Are Mutagens: Their Detection and Classification," *Environmental Health Perspectives* 6 (1973): 115–118.

34. John Cairns, *Cancer: Science and Society* (San Francisco: W. H. Freeman, 1978), 63.

35. Ramel, "Advantages of and Problems with Short-Term Mutagenicity Tests"; Leslie A. Beyer, Barbara D. Beck, and Thomas A. Lewandowski, "Historical Perspective on the Use of Animal Bioassays to Predict Carcinogenicity: Evolution in Design and Recognition of Utility," *Critical Reviews in Toxicology* 41, no. 4 (April 2011): 321–338.

36. Arlene Blum and Bruce N. Ames, "Flame-Retardant Additives as Possible Cancer Hazards," *Science* 195, no. 4273 (1977): 17–23.

37. National Toxicology Program (US), "Bioassay of Tris (2,3-Dibromopropyl) Phosphate for Possible Carcinogenicity," National Cancer Institute Carcinogenesis Technical Report Series No. 76, 1978, CAS No. 126-72-7.

38. Consumer Product Safety Commission, "Hazardous Substances and Articles; Administration and Enforcement Regulations, Children's Wearing Apparel Containing TRIS; Interpretation as Banned Hazardous Substance," *Federal Register* 42 (1977): 18850–18854. The agency cited Ames's work in its decision.

39. Joyce McCann, Edmund Choi, Edith Yamasaki, and Bruce N. Ames, "Detection of Carcinogens as Mutagens in the Salmonella/Microsome Test: Assay of 300 Chemicals," *Proceedings of the National Academy of Sciences of the United States of America* 72, no. 12 (1975): 5135–5139; Joyce McCann and Bruce N. Ames, "Detection of Carcinogens as Mutagens in the Salmonella/Microsome Test: Assay of 300 Chemicals: Discussion," *Proceedings of the National Academy of Sciences of the United States of America* 73, no. 3 (1976): 950–954; Errol Zeiger, "History and Rationale of Genetic Toxicity Testing: An Impersonal, and Sometimes Personal, View," *Environmental and Molecular Mutagenesis* 44, no. 5 (2004): 363–371, on 364. Further validation studies showed the predictive value of the Ames test for carcinogens to be somewhat lower.

40. Library of Congress Environment and Natural Resources Policy Division and John E. Blodgett, *Legislative History of the Toxic Substances Control Act: Together with a Section-by-Section Index* (Washington, DC: US Government Printing Office, 1976), 562.

41. Angela N. H. Creager, "To Test or Not to Test: Tools, Rules, and Corporate Data in US Chemicals Regulation," *Science, Technology, & Human Values* 46, no. 5 (2021): 975–997.

42. Robin Wolfe Scheffler, "Managing the Future: The Special Virus Leukemia Program and the Acceleration of Biomedical Research," *Studies in History and Philosophy of Science Part C: Studies in History and Philosophy of Biological and Biomedical Sciences* 48 (2014): 231–249; Doogab Yi, "Governing, Financing, and Planning Cancer Virus Research: The Emergence of Organized Science at the U.S. National Cancer Institute in the 1950s and 1960s," *Korean Journal for the History of Science* 38, no. 2 (2016): 321–349; Robin Wolfe Scheffler, *A Contagious Cause: The American Hunt for Cancer Viruses and the Rise of Molecular Medicine* (Chicago: University of Chicago Press, 2019).

43. Henry Harris, *The Cells of the Body: A History of Somatic Cell Genetics* (Cold Spring Harbor, NY: Cold Spring Harbor Laboratory Press, 1995), 109.

44. James B. Murphy, "Discussion of Some Properties of the Causative Agent of a Chicken Tumor," *Transactions of the Association of American Physicians* 46 (1931): 187; see also Neeraja Sankaran, *A Tale of Two Viruses: Parallels in the Research Trajectories of Tumor and Bacterial Viruses* (Pittsburgh: University of Pittsburgh Press, 2021).

45. Sankaran, *A Tale of Two Viruses*, 177–178. Dulbecco was at Caltech.

46. Michel Morange, *The Black Box of Biology: A History of the Molecular Revolution*, trans. Matthew Cobb (Cambridge, MA: Harvard University Press, 2020), 213.

47. Dominique Stehelin, Ramareddy V. Guntaka, Harold E. Varmus, and J. Michael Bishop, "Purification of DNA Complementary to Nucleotide Sequences Required for Neoplastic Transformation of Fibroblasts by Avian Sarcoma Viruses," *Journal of Molecular Biology* 101, no. 3 (1976): 349–365; D. Stehelin, H. E. Varmus, J. M. Bishop, and P. K. Vogt, "DNA Related to the Transforming Gene(s) of Avian Sarcoma Viruses Is Present in Normal Avian DNA," *Nature* 260, no. 5547 (March 1976): 170–173.

48. Deborah H. Spector, Harold E. Varmus, and J. Michael Bishop, "Nucleotide Sequences Related to the Transforming Gene of Avian Sarcoma Virus Are Present in DNA of Uninfected Vertebrates," *Proceedings of the National Academy of Sciences of the United States of America* 75, no. 9 (September 1978): 4102–4106; Gregory J. Morgan, *Cancer Virus Hunters: A History of Tumor Virology* (Baltimore: Johns Hopkins University Press, 2022), 150.

49. Robert A. Weinberg, *Racing to the Beginning of the Road: The Search for the Origin of Cancer* (New York: W. H. Freeman, 1996), 110.

50. Paul Nurse, "A Long Twentieth Century of the Cell Cycle and Beyond," *Cell* 100, no. 1 (2000): 71–78.

51. Weinberg, *Racing to the Beginning of the Road*, 151.

52. Weinberg, *Racing to the Beginning of the Road*, 159.

53. Morange, *The Black Box of Biology*, 214.

54. Morange, *The Black Box of Biology*, 218.

55. J. Michael Bishop, "Enemies Within: The Genesis of Retrovirus Oncogenes," *Cell* 23, no. 1 (January 1981): 5–6.

56. I am drawing on Greg Morgan's helpful metaphor here: Morgan, *Cancer Virus Hunters*, 216. See also Bert Vogelstein, Surojit Sur, and Carol Prives, "p53: The Most Frequently Altered Gene in Human Cancers," *Nature Education* 3, no. 9 (2010): 6.

57. Arnold J. Levine, "p53, the Cellular Gatekeeper for Growth and Division," *Cell* 88, no. 3 (1997): 323–331; Morgan, *Cancer Virus Hunters*, chap. 13.

58. SV40 was widely studied among virologists after being identified by Maurice Hilleman and coworkers as a contaminant of the polio vaccine. Bernice Eddy then found it was tumorigenic in hamsters. Morgan, *Cancer Virus Hunters*, 165.

59. Daniel I. H. Linzer and Arnold J. Levine, "Characterization of a 54K Dalton Cellular SV40 Tumor Antigen Present in SV40-Transformed Cells and Uninfected Embryonal Carcinoma Cells," *Cell* 17, no. 1 (1979): 43–52.

60. Cathy A. Finlay, Philip W. Hinds, and Arnold J. Levine, "The p53 Proto-Oncogene Can Act as a Suppressor of Transformation," *Cell* 57, no. 7 (1989): 1083–1093.

61. Suzanne J. Baker, Eric R. Fearon, Janice M. Nigro, Stanley R. Hamilton, Ann C. Preisinger, J. Milburn Jessup, Peter VanTuinen, David H. Ledbetter, David F. Barker, Yusuke Nakamura, et al., "Chromosome 17 Deletions and p53 Gene Mutations in Colorectal Carcinomas," *Science* 244, no. 4901 (1989): 217–221.

62. Eric R. Fearon and Bert Vogelstein, "A Genetic Model for Colorectal Tumorigenesis," *Cell* 61, no. 5 (1990): 759–767.

63. Morgan, *Cancer Virus Hunters*, 224.

64. Alfred G. Knudson Jr., "Mutation and Cancer: Statistical Study of Retinoblastoma," *Proceedings of the National Academy of Sciences* 68, no. 4 (1971): 820–823; Anya Plutynski, *Explaining Cancer: Finding Order in Disorder* (New York: Oxford University Press, 2018), chap. 1.

65. Harold Varmus, "An Historical Overview of Oncogenes," in *Oncogenes and the Molecular Origins of Cancer*, ed. Robert A. Weinberg (Cold Spring Harbor, NY: Cold Spring Harbor Laboratory Press, 1989), 26.

66. Huei-Jen Su Huang, Jiing-Kuan Yee, Jin-Yuh Shew, Phang-Lang Chen, Robert Bookstein, Theodore Friedmann, Eva Y.-H. P. Lee, and Wen-Hwa Lee, "Suppression of the Neoplastic Phenotype by Replacement of the RB Gene in Human Cancer Cells," *Science* 242, no. 4885 (1988): 1563–1566.

67. Raul Necochea, "From Cancer Families to HNPCC: Henry Lynch and the Transformations of Hereditary Cancer, 1975–1999," in *Cancer in the Twentieth Century*, ed. David Cantor (Baltimore: Johns Hopkins University Press, 2008), 269.

68. Necochea, "From Cancer Families to HNPCC," 272.

69. Necochea, "From Cancer Families to HNPCC," 274–277.

70. Bert Vogelstein and Kenneth W. Kinzler, *The Genetic Basis of Human Cancer* (New York: McGraw-Hill, Health Professions Division, 1998), 565.

71. Robert A. Aronowitz, "The Dilemma of Genetic Testing: The 'Breast Cancer Gene' and the Physician's Role, an Ethics Case Study," *ACP Observer* 18, no. 3 (March 1998): 1.

72. Beth Newman, Melissa A. Austin, Ming Lee, and Mary-Claire King, "Inheritance of Human Breast Cancer: Evidence for Autosomal Dominant Transmission in High-Risk Families," *Proceedings of the National Academy of Sciences of the United States of America* 85, no. 9 (May 1988): 3044–3048.

73. Jeff M. Hall, Ming K. Lee, Beth Newman, Jan E. Morrow, Lee A. Anderson, Bing Huey, and Mary-Claire King, "Linkage of Early-Onset Familial Breast Cancer to Chromosome 17q21," *Science* 250, no. 4988 (1990): 1684–1689.

74. Kevin Davies and Michael White, *Breakthrough: The Race to Find the Breast Cancer Gene* (London: Macmillan, 1995), 134–140, 178–182.

75. Yoshio Miki, Jeff Swensen, Donna Shattuck-Eidens, P. Andrew Futreal, Keith Harshman, Sean Tavtigian, Qingyun Liu, Charles Cochran, L. Michelle Bennett, Wei Ding, et al., "A Strong Candidate for the Breast and Ovarian Cancer Susceptibility Gene *BRCA1*," *Science* 266, no. 5182 (1994): 66–71; P. Andrew Futreal, Qingyun Liu, Donna Shattuck-Eidens, Charles Cochran, Keith Harshman, Sean Tavtigian, L. Michelle Bennett, Astrid Haugen-Strano, Jeff Swensen, Yoshio Miki, et al., "*BRCA1* Mutations in Primary Breast and Ovarian Carcinomas," *Science* 266, no. 5182 (1994): 120–122.

76. Myriad's patents were challenged legally all the way to the US Supreme Court: Jorge L. Contreras, *The Genome Defense: Inside the Epic Legal Battle to Determine Who Owns Your DNA* (Chapel Hill, NC: Algonquin Books of Chapel Hill, 2021).

77. Fangjian Guo, Matthew Scholl, Erika L. Fuchs, Abbey B. Berenson, and Yong-Fang Kuo, "BRCA Testing and Testing Results among Women 18–65 Years Old," *Preventive Medicine Reports* 26 (April 2022): 101738; Shobita Parthasarathy, *Building Genetic Medicine: Breast Cancer, Technology, and the Comparative Politics of Health Care* (Cambridge, MA: MIT Press, 2007).

78. Office of the Commissioner, FDA, "FDA Authorizes, with Special Controls, Direct-to-Consumer Test That Reports Three Mutations in the BRCA Breast Cancer Genes," FDA, March 24, 2020, https://www.fda.gov/news-events/press-announcements/fda-authorizes -special-controls-direct-consumer-test-reports-three-mutations-brca-breast-cancer; Heather Murphy, "Don't Count on 23andMe to Detect Most Breast Cancer Risks, Study Warns," *New York Times*, April 6, 2019, https://www.nytimes.com/2019/04/16/health/23andme-brca-gene -testing.html.

79. Löwy, *Preventive Strikes*, chap. 7; Parthasarathy, *Building Genetic Medicine*; Löwy and Gaudillière, "Localizing the Global."

80. Richard Doll and Richard Peto, "The Causes of Cancer: Quantitative Estimates of Avoidable Risks of Cancer in the United States Today," *JNCI: Journal of the National Cancer Institute* 66, no. 6 (June 1981): 1192–1308.

81. Doll and Peto, "The Cause of Cancer," 1251.

82. Martin Walker, "Sir Richard Doll: A Questionable Pillar of the Cancer Establishment," *The Ecologist* 28, no. 2 (March/April 1998): 82–92. In the 1980s, Doll worked as a legal consultant for the Chemical Manufacturers Association, the largest trade group for the chemical industry, to try to overturn epidemiological evidence that the monomer vinyl chloride (the ingredient for the ubiquitous plastic polyvinyl chloride, or PVC) was carcinogenic. Lennart Hardell, Martin J. Walker, Bo Walhjalt, Lee S. Friedman, and Elihu D. Richter, "Secret Ties to Industry and Conflicting Interests in Cancer Research," *American Journal of Industrial Medicine* 50, no. 3 (2007): 227–233

83. Cathy Gere, *Pain, Pleasure, and the Greater Good: From the Panopticon to the Skinner Box and Beyond* (Chicago: University of Chicago Press, 2017), esp. 205–210; Cathy Gere, chapter 7 in this volume.

84. Brown, *Undoing the Demos*, esp. chap. 5; on economization: Koray Çalışkan and Michel Callon, "Economization, Part 1: Shifting Attention from the Economy towards Processes of Economization," *Economy and Society* 38, no. 3 (2009): 369–398.

85. Brown, *Undoing the Demos*, 132–133.

86. Robert A. Aronowitz, *Risky Medicine: Our Quest to Cure Fear and Uncertainty* (Chicago: University of Chicago Press, 2015).

CHAPTER SEVEN

"Drugs into Bodies"

AIDS Activism and the Constitutional Limits of Biocapital

Cathy Gere

Introduction: Biocapital

On Tuesday, October 14, 1980, against the backdrop of the steep ascent of Deng Xiaoping, Margaret Thatcher, and Ronald Reagan, the bioengineering company Genentech issued a million shares of stock in its initial public offering (IPO). The opening price was $35 a share. Within an hour it leapt to $88, closing at $71.25. The *Wall Street Journal* called it "one of the most spectacular market debuts in recent history." The *LA Times* described it as "a frenzy."[1]

As evidenced by its multiple appearances in this volume, that 1980 scene is *the* acknowledged origin point of greedy science. In chapter 6 of this volume, "Neoliberal Mutations," for example, Angela Creager invokes it as a myth that obscures as much as it illuminates. For the purposes of the present chapter, the event's significance lies exactly in those mythical dimensions: in the spooky, performative, world-altering way it conjured up a new form of scientific soothsaying. After that fateful Tuesday, so the story goes, future-casting about biotech became its own form of economic value.

This mode of gambling on pharmaceutical possibility has been dubbed "biocapital" by the anthropologist of science Kaushik Sunder Rajan. According to Sunder Rajan, the essence of biocapital is its *promissory* nature. Beginning with Genentech's IPO, pharmaceutical hype acquired enormous value in its own right, capable of raising venture capital and driving up share prices far in advance of any proven progress in clinical outcomes.

Of course, nature is hard to subdue. Sometimes the gamble pays off; more often, it doesn't. As Kaushik Sunder Rajan describes the process, "a vision of that future has to be sold, even if it is a vision that will never be realized."[2] This fake-it-till-you-make-it tendency recently reached its apogee in the celebrity trial of Elizabeth Holmes, founder of the fraudulent blood-testing company Theranos. Holmes is a truly degenerate heir to Genentech's legacy—consider the nondisclosure agreements that everyone had to sign, the bullying and excommunication of anyone who tried to point out that the devices did not work, the Potemkin laboratory they threw together to impress Joe Biden, Holmes's pseudo-messianic insistence that her useless little machines were "the best thing humans have ever built," and so on.[3] Is it possible we have reached *peak greed*?

Kaushik Sunder Rajan focuses his analysis of promissory science on genetic technologies, analyzing how that stream of new therapeutic possibilities was developed with and through 1980s casino capitalism. His emphasis is on the "upstream" terrain of biotech start-ups, small companies developing promising therapeutic molecules for licensing to big pharmaceutical multinationals "downstream." Biocapital, in his formulation, unfolds in these high-risk, high-reward wellsprings of novelty and innovation, funded by venture capital, fueled by hype, exuberantly exceeding the traditional norms of scientific method.

This chapter focuses on the way that promissory biocapital also emerged "downstream," in the domain of traditional drug testing by big pharma. Just as the Cold War was whimpering to its end, AIDS activists' campaigns for access to experimental drugs produced a decidedly more market-friendly legal regime for the pharmaceutical industry. In a ferocious drive to get "drugs into bodies," as their slogan went, the AIDS Coalition to Unleash Power (ACT UP) forced through a series of reforms of pharmaceutical licensing that made it easier, cheaper, and faster to get new therapies onto the market.

Although not as eye-catching as Genentech's recombinant DNA technology, there was a form of promissory future-casting at work in these regulatory changes. At the activists' urging, the US Food and Drug Administration (FDA) started to allow companies to obtain market licenses on the basis of "surrogate endpoints," markers of disease that show improvement in laboratory tests, regardless of the ultimate effects on morbidity or mortality. Rather than going through lengthy and expensive phase 3 clinical trials, the industry could begin to sell a new drug based on promising lab results rather than clinical outcomes. The last phase of testing would then be done, in theory, on patients who had been prescribed the drugs by their physicians, rather than on trial participants.

There was every reason to make this change in the context of the AIDS epidemic. In this case, the relevant biomarker might be viral load or T-cell count, eminently plausible surrogates for disease progression. For people with AIDS, taking experimental drugs was a rational course of action in the face of the prospect of an untimely death. In the short term, this loosening of regulatory and evidentiary standards helped save the lives of a generation of people with AIDS who got early access to the first round of treatments. In the medium to long term, they also saved the pharmaceutical companies a great deal of money.[4]

The use of surrogate endpoints to fast-track drug development is a classic example of a neoliberal policy. By loosening regulation and hastening drugs to market, it encouraged and rewarded entrepreneurialism within a capitalist incentive structure. ACT UP was more *anti-* than pro-capitalist, however, and so its emboldening of neoliberal reforms was both an unintended consequence and a bitter irony, a case study in the tendency of left libertarianism and market fundamentalism to align in practical and consequential ways. In the parlance of this volume and its predecessor, the outcome of AIDS activism is a story of greedy appropriating groovy, that familiar fable of capitalism's ever-renewing ability to assimilate, digest, and find nourishment in even the fiercest forms of countercultural rebellion.

Of course, the role of grassroots patient activist groups in encouraging medical commerce must be put into the context of other changes at the time tending in the same direction. Think of the technology-transfer initiatives of the eighties and nineties, set in motion by the Bayh-Dole Act examined in the Introduction to this book. Then there was the ascendance of an ideology that moved the risk calculus to individual patients, as in the neoliberal "responsibilization" analyzed in the chapters by Agar (chapter 4) and Creager (chapter 6). And because of Cyrus Mody's contribution to this volume (chapter 1), going forward, we must now account for the powerful effect of energy-industry funding for bioscience in the period under consideration. ACT UP's complicity with the triumph of the for-profit pharmaceutical sector was part of an irresistible wave, bearing everyone along, free-market enthusiasts and naysayers alike.

But once the wave crashed to shore, the limitations of greedy science became apparent. After first exploring how AIDS activists unwittingly helped the pharmaceutical companies deregulate their industry, this chapter moves onto subsequent developments that saw "drugs into bodies" campaigners pull the tide of history in the other direction. Once the first round of effective antiretroviral therapies reached the market, a cadre of AIDS activists in South Africa inspired and trained by ACT UP successfully challenged the new patent regime that kept these drugs unaffordable by low-income countries. Section 27 of the postapartheid South African Constitution enshrines a "right to health care." In 2001, this provision was invoked by activists to deliver a stinging defeat to neoliberal enclosure on a global scale.

In the United States, AIDS activists aligned with the private sector to get access to treatments; in South Africa, campaigners invoked the state's obligation to protect its citizens from capitalist profiteering. Taken together, the two cases illustrate how biocapital emerged alongside its partner "bioconstitutionalism," a form of democratic governance pursued through courts and legislatures, in which the "right to health care," especially in postcolonial states, turned out to be actionable against the pharmaceutical bottom line. Sunder Rajan, discussing the overturning of a global patent regime in India, describes bioconstitutionalism thus: "*A spirit of constitutionalism that . . . has emerged as a democratic counterweight to the logics of multinational pharmaceutical capital.*"[5]

Although biocapital and bioconstitutionalism have a distinct right–left polarity, pitting the market against the state, they are, in truth, the two faces of a single coin of vital politics. Greedy science was encouraged and facilitated at some moments, only to be audited, circumscribed, and curtailed at others. Driven along by the supremely practical goal of saving human life, the twists and turns of the story reveal how the incentive structures of neoliberal drug licensing constitute but one element in the complex terrain of new-millennial medicine.

ACT UP New York

In the winter of 1987, I joined ACT UP, recruited to the organization by an Irish American contractor called Gerri Wells. I was a Lower East Side squatter doing occasional, under-the-table carpentry jobs for Gerri when she nudged me toward ACT UP's Monday night meetings at the Gay and Lesbian Center on West Thirteenth Street. I remember those gatherings as a heady mixture of anger and mourning, electric with sexual tension, darkened by existential dread.

My boss Gerri was a founding member of ACT UP's women's caucus, which had formed partly in response to an article published in *Cosmopolitan* magazine saying that women were not at risk of contracting AIDS from vaginal intercourse with HIV-positive men. As with so many of ACT UP's publicity drives, the campaign against *Cosmo* was almost too successful and we ended up on a local cable talk show, having a shouting match with the doctor who wrote the piece. This was followed by a series of television appearances debating his arguments. In fairly short order, *Cosmo* ran an article on safer sex for heterosexual women.[6] I did not stay with the organization much beyond that one campaign, but it left an indelible impression of the urgency, fury, and effectiveness of life-and-death politics.

In the summer of 2020, I made a pilgrimage to the Thirteenth Street LGBTQ Center, hoping to refresh my memory of ACT UP. COVID-19 had driven away the tourists and the super-rich, turning the streets into an uncanny simulacrum of the 1980s, but the center itself was completely unrecognizable as the grungy place it had once been. There was a soaring atrium, in the middle of which a modernist chandelier cascaded down from a skylight. A bearded youth at the sculptural front desk kindly unlocked the room where the Monday night meetings had taken place, but the space had been divided into two glossy sunlit cubes. All along the hallways were occasional gaps in the smooth white plaster where scruffy murals and graffiti by Keith Haring and other queer downtown artists had been preserved with museological piety. I failed to wring memories from the visit, but the way the center's bland interior walls were broken up by glimpses of the gritty ferment of the ACT UP years can stand as a fitting symbol of my argument.

For sociologist of science Steve Epstein, ACT UP's historical significance lies in the way it engendered a newly demotic species of scientific medicine, based in the lived experience of gay patients and their communities. Epstein describes how an unruly group of queer activists with little to no technical training ended up understanding HIV as well as the certified experts. By plowing through medical journals—in public libraries in those days of print—compiling glossaries and making summaries for their comrades in the movement, ACT UP's treatment activists mastered the technical literatures of virology, pharmacology, and clinical medicine. At first, the infectious disease people—including Anthony Fauci—did not know what to make of these scientific interlopers, but they eventually gave them a place at the table. Through this unprecedented level of grass-roots advocacy, ACT UP succeeded in leveling the epistemic playing field between sufferers and scientists.[7]

With its denunciations of the authoritarianism of the medical establishment, ACT UP extended the medical and antipsychiatry activism of the 1960s and 1970s through the Reagan, Bush, and Clinton years. It also *reversed* some of the hard-won gains of the earlier movement by taking antipaternalism to its logical extreme. In keeping with its anti-authoritarian tendencies, ACT UP mounted a highly effective critique of pharmaceutical industry regulation. As a result, protections that had been fought for by previous generations were rolled back in the name of access. By turning the autonomous citizen's right to *refuse* treatment into the empowered consumer's right to *choose* it, the activists performed an unwitting service to the pharmaceutical industry's bottom line.

Sins of Omission

ACT UP did not set out to facilitate pharmaceutical profiteering. To better appreciate the irony, we can travel back to 1972, when a group of neoliberal economists at the University of Chicago started to argue for the comprehensive deregulation of the whole pharmaceutical sector. At the time, these academic economists failed, and they looked on in dismay as the regulations only got more and more stringent. In the late 1980s, however, a group of activists with none of the same ideological goals, but driven by the stakes of bare survival, succeeded in carrying out a version of their program.

The burdensome drug regulations so loathed by neoliberals had their roots in a pharmaceutical catastrophe. Thalidomide was a molecule developed in 1954 by a German chemical company as an antiseizure medication. Finding that it did not work to control epilepsy but instead had a slight soporific effect, Chemie Grünenthal marketed thalidomide as a safe sedative, on the grounds that it was impossible to overdose on it. It was licensed in forty-six countries under a variety of brand names. When reports of severe side effects began to filter back, the company suppressed them, doubling down on the safety and beneficence of the drug. In the late 1950s, Grünenthal circulated a pamphlet recommending the pills as a treatment for morning sickness in pregnant women. It turned out that as few as one or two tablets taken during pregnancy was enough to cause complete deformation of whatever organ or limb happened to be developing in the fetus at the time.

Because of a knowledgeable scientist at the FDA, the drug did not receive a license in the United States. On the face of it, this should have boosted the agency's credibility. But millions of tablets had nonetheless been distributed as free samples in the United States without the patients being informed that it was an experimental substance. As a result, the Kefauver–Harris Drug Amendments to the Federal Food, Drug, and Cosmetics Act were signed into law in 1962, greatly tightening and extending the hitherto lax procedures for drugs to reach the market.

After thalidomide, American medicine's days of reckoning kept coming. In 1963, a group of doctors at the Jewish Chronic Diseases Hospital resigned in protest at being asked to inject cancer cells into terminally ill patients without their consent. In 1965, anesthesiologist Henry Beecher published a summary of twenty-two cases of American research conducted in violation of the principle of informed consent that lay at the heart of the 1948 judgment in the Nazi Doctors' Trial. When 1972 saw the revelation of the four-decade-long "Tuskegee Study of Untreated Syphilis in the Male Negro," the racial dimension of the scandal made the family resemblance between American and Nazi medical research seem inarguable.[8]

But even as comparisons with fascist atrocities were becoming a journalistic commonplace in the aftermath of the Tuskegee exposé, quiet preparations were under way for ideological battle. In December 1972, a group of free-market fundamentalists gathered at the University of Chicago to argue for the *dismantling* of the FDA, in cheerful defiance of the prevailing mood of liberal handwringing. The precise target of their zeal was the 1962 Kefauver–Harris Amendments. Afterward, they published a volume of proceedings, which they characterized as "controversial before it ever went to press." In it, they argued for complete deregulation of the drug approval process, up to and including the abolition of the FDA.[9]

At the heart of the economists' case was a counterfactual claim. Without regulation, they argued, drug companies would be incentivized to produce therapeutic molecules at a much faster rate. One professor of economics developed pages of mathematical formulae in an attempt to assign a dollar value to the benefit of suppressing harmful drugs versus the harm of suppressing beneficial ones. He came to the conclusion that the cost of delaying a useful medication by two years would amount to ten to one hundred times the cost of allowing a toxic drug onto the market.[10] In place of the FDA, he argued, a deregulated market would protect patients by allowing them to try drugs and then to abandon them if they did not work, bringing the pharmaceutical companies into medical line with the simple pressure of consumer demand.[11]

Milton Friedman subsequently popularized this work in an article in *Newsweek*, concluding that the "1962 amendments to the Food, Drug, and Cosmetic Act should be repealed. They are doing vastly more harm than good." He ended the piece with the "even more shocking conclusion that the FDA itself should be abolished."[12] In his suggestion that the market will solve every problem it creates even unto *thalidomide*, we see neoliberal ideology at its most raw and provocative.

Friedman admitted the "present climate of opinion" made his suggestion to abolish the FDA akin to an attack on "motherhood or even apple pie." He was right. Only a few short weeks after the publication of his article, Edward Kennedy presided over the first of a series of congressional hearings on the problem of human experimentation, prompted by the revelation of the Tuskegee Syphilis Study. The National Research Act of 1974, a piece of legislation arising directly from the hearings, mandated the informed consent of all research subjects, its first draft taking its language verbatim from the judgment in the Nazi Doctors' Trial. As a result of the ensuing legislative reforms, the huge prison research complex was deemed illegal, cutting the pharmaceutical industry's access to thousands of cheap human subjects for clinical trials.

Freedom for the Chicago economists may have been synonymous with deregulation, but for most activists, journalists, and lawmakers at the time, "patient autonomy" was secured by ever-*tighter* rules governing the practice of medical research on humans, centering on the principle of informed consent, and requiring rigorous proof of safety and efficacy. By the 1980s, the process of getting a drug to market had grown into a years-long process, costing millions of dollars and taking up to eight years.[13]

A Holocaust of Inaction

If the Chicago Boys' denunciations of the FDA's sins of omission gained no traction in 1972, a decade and a half later, they acquired tremendous salience in the context of the AIDS epidemic. In one of his speeches, the founder of ACT UP, rabble-rousing gay playwright Larry Kramer, argued that "a holocaust does not require a Hitler to be effective. . . . Holocausts can occur, and probably most often do occur, because of *inaction*. This inaction can be unintentional or deliberate."[14]

Kramer's reworking of Tuskegee-era rhetoric comparing American to Nazi medicine bespeaks a generational inversion of values. For the medical activists rocked by the revelation of the Tuskegee study, the problem resided in a research establishment that trampled the human rights of vulnerable experimental subjects in the rush to develop treatments. In the context of the AIDS epidemic, anything that got in the way of drug development was a crime against humanity, including the very regulatory edifice that had been put in place to protect ordinary citizens from research abuses.

In line with Milton Friedman's fondest hopes, the FDA subsequently became the target of ACT UP's most relentless campaigns. In the words of writer and activist Sarah Schulman, whose 2021 volume *Let the Record Show* is a thoughtful chronicle of the movement, "access to new treatments was a right, and government obstructions were not to be obeyed."[15] ACT UP's first demonstration was on Wall Street, where they rallied to demand "immediate release by the Federal Food and Drug Administration of drugs that might help save our lives."[16]

For Schulman, ACT UP "was a movement that addressed corporate power and greed as strongly as government indifference."[17] And indeed, a few months later, a group of ACT UP members disguised as Wall Street workers infiltrated the New York Stock Exchange and climbed up to an interior balcony above the trading floor, where they dropped a banner inscribed with "Sell Wellcome" and threw fistfuls of fake 100-dollar bills into the crowd below, printed with the message "FUCK YOUR PROFITEERING. People are dying while you play business."[18]

The Stock Exchange action was a tribute to the Yippee founder Abbie Hoffman's anti-capitalist prank of 1967, in which he threw real dollar bills down at the traders, turning the floor into instant chaos as the breadheads scrambled to grab the cash. But in truth, after the first gestures of antagonism had been weathered, private industry had every reason to support ACT UP's agenda. As one ACT UP member interviewed by Schulman acknowledges, "The drug company [Hoffman-La Roche] agreed, everything was slow, so they had an ally in us."[19] One of the FDA commissioners at the time later observed, "So you had this synergy between these ultra-liberal AIDS activists and these almost right-wing conservatives who wanted regulation reduced."[20]

Surrogate Endpoints

Of all the reforms urged by ACT UP, perhaps none has been more helpful to the pharmaceutical bottom line than the lowering of standards of efficacy through the use of "surrogate endpoints" in the place of clinical outcomes.[21] The first drug to be approved this way was the antiretroviral ddC, developed by Hoffman-La Roche. In 1992, in response to AIDS activist demands, the FDA announced that ddC had been approved under "Accelerated Approval" licensing procedures that had been reduced from between two and eight years to a mere eight months. Clinical trials evaluated the drug using surrogate markers of HIV disease—the T-cell count or the concentration of an antigen associated with HIV—rather than waiting to see if it made any difference in disease burden or mortality. In vitro blood samples replaced in vivo outcomes, making for a cheaper, faster, and more flexible pathway to FDA approval. Surrogate endpoints are a form of promissory science embedded into the landscape of pharmaceutical regulation.

Peter Staley, organizer of the Stock Exchange action (a closeted Wall Street bond trader before an HIV diagnosis turned him into a fearlessly creative activist), later wrote about the way the force of necessity drove ACT UP's political realism. Discussing the members who thought the organization should hold out for the goal of socialized medicine, Staley observed, "The folks with AIDS would remind us 'We don't have time to overthrow capitalism,' and that was that. The ticking clock for half our community forced us to adopt hard-nosed pragmatism as one of our guiding forces."[22]

Hard-nosed pragmatism resulted in easier access to experimental drugs, but this new world was no free-market utopia; it was a bitter struggle for some measure of agency over the prospect of an untimely and painful death. Here is the testimony of one member of ACT UP about the death of a friend, poignantly illustrating the way that people with AIDS traversed the rocky terrain of the pharmaceutical promise: "[My friend] picked the [road] where you are taking interferon, which turned out in the long run to be not the one to take, but, you know, who knew, right? Other people took the road that was AZT, and maybe they lived or maybe they didn't, but the interferon way just didn't work." The same activist remarked on a similar gamble, this one involving a desperate assessment of a surrogate-endpoint test. After describing how a group of friends had pooled their money to buy an experimental drug, he ruefully recalled, "It didn't work. But we thought it did, because it worked in a test tube. You had to order it from Japan, and so we each had contributed, like, a bunch of money. I know it was like hundreds of dollars to get it, but it didn't work."[23] This is casino capitalism played with the currency of bare survival.

The greedy science that triumphed as a result of AIDS activism was always going to be more complex and contradictory than the frictionless markets imagined by the Chicago School. The fact that queer anti-capitalists turned out to be the most effective drivers of the neoliberal agenda was not the only political irony of the story. Most tellingly, it was President Reaganomics himself, in his capacity as a champion of family values, who created the research vacuum in the first place. Pandering to the homophobes in his party, Reagan prevented his surgeon general from addressing the epidemic for his first four years in office. Journalists received instructions in advance of press conferences that the surgeon general would not answer questions about AIDS. The Reagan administration quietly sponsored some research but otherwise remained silent on the disease, allowing fear, stigmatization, and prejudice to proliferate.[24] A public health crisis exacerbated by one dimension of Reaganism—the reassertion of the heteronormative family—was to be ameliorated by another—the conviction that the amoral workings of the market would most effectively serve patient-consumers.

Corruption and Remission

In the 1970s, the neoliberals argued that individual patients would happily pick and choose among new drugs and simply discard the ones that did not work for them. As this agenda came to fruition in the context of the epidemic, however, vociferous demands for access on the part of people with AIDS was coupled with quieter insistence on more government-regulated research. In the mid-1990s, the Treatment Action Group (TAG), a committee within ACT UP that had broken away and formed its own organization, mounted an increasingly fierce critique of the way that the FDA's Accelerated Approval was corrupting research. The drug companies were supposed to conduct postmarketing trials but had little incentive to do so. Once the genie of licensing was out of the regulatory bottle, it proved hard to stuff back in. Most concerning, the long-term trials that had been conducted were showing that surrogate endpoints were poor predictors of actual survival rates.[25] Accordingly, when Roche filed their next accelerated approval request, the activists wrote to the FDA to object: "We feel that such an approval would penalize people with AIDS/HIV by setting an inappropriately low standard of evidential requirements."[26]

The wider activist community was shocked by this challenge to the gospel of access. Roche took advantage of the divisions within the movement, faxing copies of TAG's letter to scores of AIDS organizations, claiming that a breakaway faction was trying to undermine the achievements for which everyone had worked so hard. TAG stuck to its position, urging the FDA to insist that the drug companies collaborate with one another to enroll people with AIDS in what they called "large simple trials," in which low barriers to entry would be compensated for statistically by the sheer weight of numbers of participants. To their credit, FDA scientists promoted TAG's recommendations, and the drug companies made some adjustments to their trial protocols, even to the extent of pooling results with their market rivals.

In the end, however, the debate about the shoddiness of accelerated-approval protocols was blunted by jubilation about the results of the new round of trials. The Roche drug that TAG had objected to was a protease inhibitor, representing the third molecular strategy that had been attempted to stop the virus from multiplying. None of the drugs worked alone or in double combination. But in the large trials of the mid-1990s, it quickly emerged that taking antiretrovirals in triple combination seemed to fully control HIV in patients. By multiplying the mechanisms of action, these combinations outwitted the virus's notorious ability to mutate and become resistant. After a few months on the regimens, most patients had undetectable viral loads. With the advent of triple-combination treatments, AIDS in high-income countries went from a death sentence to a life sentence, from a terminal diagnosis to a chronic manageable condition.

The good news about combination therapy, announced in 1996, came on the heels of a period of extraordinary political optimism. In 1989, the Berlin Wall fell; in 1990, Nelson Mandela was released from prison after twenty-seven years; in 1991, the Soviet Union collapsed; in 1994, majority rule was achieved in South Africa. Everything and anything seemed possible, as if some of the most intractable political pathologies afflicting the world had gone into unexpected remission. The effectiveness of the "triple cocktail" against HIV was like the physiological analog of events on the world stage. And as with events on the world stage, the bad news quickly followed, as the Lazarus-like recovery of people in rich countries was mirrored by almost unchecked devastation in sub-Saharan Africa, where the virus was spreading with the speed and ferocity of wildfire, and the new medications were unaffordable.

Four factors converged in the mid-1990s to make South Africa the crucible of new-millennial biopolitics. The AIDS epidemic hit southern Africa with full force, during the Republic's transition to democracy, at the same time as combination therapy proved effective against HIV, just as the World Trade Organization was crafting a global intellectual property regime for the post-Soviet world. Starting in 1998, organizations affiliated with ACT UP, working in collaboration across the Northern and Southern Hemispheres, engaged in an ongoing campaign to get "drugs into bodies" using, among many other tactics, the powerful instrument of constitutional law.

Bioconstitutionalism in South Africa

The Constitutional Court of South Africa is a courthouse like no other: constructed on the site of a notorious prison in Hillbrow, one of Johannesburg's inner-city neighborhoods, it comprises a layered series of modestly scaled modern buildings, symbolizing the transition from white supremacist state to multicultural democracy. During business hours, anyone can walk into the building through its mighty timber door, divided into panels carved with the rights enshrined in the South African Constitution. The rights are carved in English, plus others of the eleven official languages of the Republic, plus sculpted hands performing sign language, an aspirational symbol of the postapartheid nation.

Section 27 of the postapartheid Bill of Rights declares that "everyone has a right to healthcare services, including reproductive healthcare." This wording allowed South African AIDS activists to insist on drug access as a legal right, an argument they pursued successfully, first against the global intellectual property laws that kept antiretroviral drugs unaffordable and then against the South African government's own AIDS denialism.

The regulatory regime through which the pharmaceutical industry kept prices high across the world was the 1994 "Trade-Related aspects of Intellectual Property RightS" (TRIPS) agreement of the World Trade Organization (WTO). TRIPS required all WTO members to adopt a minimum standard of intellectual property protection in their domestic laws, including twenty-year patent terms for medicines. It effectively required all nations to adopt the patent standards of industrialized countries. In some low-income nations, TRIPS introduced American-style patents into domestic law for the first time.

At the same time, American intellectual property was undergoing its own changes, most notably the 1980 court case *Diamond v. Chakrabarty* that allowed the patenting of life forms, as mentioned in the Introduction. In chapter 8, this volume, Robin Wolfe Scheffler describes patent races among Cambridge biotech companies and shows how these legal developments fueled a high-octane style of scientific entrepreneurialism on the domestic front. But as soon as the products of the US patent regime were marketed globally, the same spirit of enclosure threatened to exacerbate health disparities between nations. South Africa was particularly hard hit by the epidemic, enduring hundreds of thousands of AIDS deaths every year, much of it carried by heterosexual transmission, with half the sufferers being women. TRIPS threatened to put the new antiretrovirals out of reach of these patients.

The Republic of South Africa promptly hit back. In response to the epidemic, Nelson Mandela's government drafted "The Medicines and Related Substances Bill" of 1997, providing for several methods to improve access to HIV drugs. Specifically, the bill enshrined the authority of the Ministry of Health to issue compulsory licenses. Compulsory licensing compels the holder of a patent, in this case the multinational pharmaceutical giants, to grant a license to local manufacturers to produce or import generic versions for cheap domestic sale. The law would allow South Africa to manufacture its own drugs, or procure them from India and Brazil, and distribute them to its own citizens at affordable prices.[27]

In 1998, gay activists in South Africa formed the Treatment Action Campaign (TAC), inspired by ACT UP: "The model that was adopted came from the United States, where AIDS activists, led by people with HIV, had pioneered the idea of 'treatment literacy' among people with HIV." In 1999, a group from ACT UP traveled to South Africa to provide training to the first cadre. TAC volunteers who were trained and passed an examination were given a small bursary for a year and then assigned to clinics, hospitals, and community organizations, where they would conduct further training among patients and advocate for the right to treatment.[28] This organizing strategy resulted in a mobilization of hundreds of HIV-positive women in villages and townships, transforming TAC into a truly grassroots movement. Back in the United States, ACT UP staged a protest accusing presidential candidate Al Gore of killing South African citizens with his support of the drug companies' patent regime.[29]

In 2001, the Pharmaceutical Manufacturers Association, representing the biggest companies, took the South African government to court to block the passing of the Medicines Act. The case was heard in the Pretoria High Court. A turning point came when Judge Bernard Ngoepe allowed testimony from TAC, against strenuous opposition from the pharmaceutical companies. The activists invoked Section 27 of the Constitution: "The ability of many persons with HIV to purchase and take medicines that treat HIV and its attendant illnesses, has a profound and inseparable bearing on the constitutional rights to human dignity and life, and access to health care," their affidavit stated. "In this respect people with HIV are directly dependent on the State's ability to fulfill its constitutional duty to bring about the progressive realization of their rights to health care services."[30]

Judge Ngoepe ordered the Pharmaceutical Manufacturers Association to make their pricing policies transparent, a move that threw the industry into disarray.[31] The US government, once the industry's staunchest ally in the case, finally backed off, declaring that it would not seek sanctions against low-income countries ravaged by AIDS even if American patent laws were broken. So long as the countries abided by *international* treaties governing intellectual property, *national* sovereignty could be asserted in issues affecting public health. In April 2001, the case against the South African government was dismissed, amid a tsunami of bad publicity for the drug companies.[32]

All forty drug makers that had sued South Africa in 1998 conceded that the South African law complied with international trade agreements—those parts of TRIPS that allowed a degree of national sovereignty. The WTO's Fourth Ministerial Conference was in November 2001 in Doha, Qatar. At the end of the meeting, the WTO (in the face of opposition from the United States, Japan, Switzerland, and other countries with powerful pharmaceutical lobbies) promulgated the "Doha Declaration on Access to Medicine and TRIPS," the fourth provision of which admits that "the TRIPS Agreement does not and should not prevent members from taking measures to protect public health. Accordingly . . . we affirm that the Agreement can and should be interpreted and implemented . . . to promote access to medicines for all." Inasmuch as bioconstitutionalism provides a "counterweight," in Sunder Rajan's words, to the "logics of multinational pharmaceutical capital," the most important constitution in question is that of the Republic of South Africa.

Of course, the story does not rest with this neat alignment between AIDS activists and the African National Congress. Even as TAC was helping the government win the lawsuit brought against it by the pharmaceutical industry, it was grappling with intensifying AIDS denialism on the part of the recently elected president Thabo Mbeki. The very day of the legal victory, standing in front of the courthouse, Mbeki's health minister gave a speech equivocating on the government's commitment to antiretroviral treatment. In private, Mbeki was accusing the Treatment Action Campaign of being funded by the pharmaceutical industry.[33] Under his presidency, the African National Congress (ANC) mobilized postcolonial suspicion of globalized corporate greed into an ever-intensifying rejection of the science around AIDS, denying the viral cause of the disease, promoting ineffective homegrown treatments, and refusing the products of the pharmaceutical industry, all in the name of African empowerment.

The TAC leadership, many of whom had been activists in the struggle against apartheid, felt keenly their disappointment in their government.

At first, it seemed as if antiretroviral access would be a politically straight-forward case of speaking truth to corporate power. But when Nelson Mandela was replaced by Thabo Mbeki, the activists learned the hard way that "governmental neglect of public health, even by democratic pro-poor governments such as the African National Congress (ANC) in South Africa, can be as much of a barrier to the right to health as profiteering by pharmaceutical companies."[34]

During this period, Nelson Mandela was conspicuously supportive of the Treatment Action Campaign, donning one of their "HIV Positive" T-shirts and helping them to win a human rights award.[35] Mbeki remained intransigent. A few short months after the Pharmaceutical Manufacturers Association case was dismissed, TAC turned around and sued the South African government, demanding the government provide antiretrovirals to prevent parent-to-child transmission in accordance with the Constitution.[36]

The High Court ruled in TAC's favor. The government appealed the decision in the Constitutional Court. The lead counsel for TAC gave evidence about the lethal consequences of the government's denial of antiretroviral medications to women giving birth. On July 5, 2002, the Court announced to the world that it had denied the government's appeal. Edwin Cameron, the constitutional court judge who is one of the very few public figures in South Africa to be open about his own HIV-positive status, has described this judgment as "a pivotal victory for the rule of law and for constitutionalism itself."[37]

In 2003, TAC organized a demonstration for the opening of parliament, consisting of "a magnificent cross-section of South African society," which marched under the banner of "Dying for Treatment." In November that year, the ANC committed to making antiretroviral medication available to every district in the country. Jubilation was followed by disillusion, as the government dragged its feet.[38] Even as the battle against denialism seemed to be won, the hard grind of pushing for implementation got under way.

A Luta Continua

One of the artworks in the foyer of the Constitutional Court in Johannesburg includes a red neon sign declaring "A Luta Continua," "the struggle continues" in Portuguese, adopted as a rallying cry during Mozambique's fight for independence. As a never-ending struggle for bodily autonomy in the face of a terrifying disease, AIDS activism has always defied the political polarity of left versus right. By giving rise to the "surrogate endpoint" as a measure of efficacy, it emboldened neoliberal deregulation; by forcing through the Doha admission that public health is more important than industry profit, it pushed back against globalization. Galvanized by ACT UP, the FDA got out of the way of Wall Street and let the drugs flow into bodies. Shamed by TAC, the pharmaceutical industry got out of the way of the South African Medicines Act, to let drugs flow into bodies. The story of AIDS activism reveals biocapital and bioconstitutionalism as the two faces of a single coin of vital politics.[39]

In both settings, the victory was short-lived and partial. In the United States, the pharmaceutical companies took advantage of easier access to do shoddy science. In South Africa, the ANC appealed to anticolonial rhetoric to deny the basic mechanisms of viral transmission and treatment. AIDS activists kept up the pressure, using lawsuits, demonstrations, media, civil disobedience, workshops, conferences, and education. Pushing forward relentlessly, using every available instrument of civil society, they created many of the economic, ethical, judicial, legal, and regulatory forms of biovalue that still shape our world today.

In the United States, the deregulatory legacy shapes drug licensing to this day. Despite the fact that drugs approved using surrogate endpoints so often fail, the practice continues apace. In June 2021, for example, pharmaceutical giant Biogen received a license for their Alzheimer's drug Aduhelm, "using the Accelerated Approval pathway, under which the FDA approves a drug . . . when the drug is shown to have an effect on a surrogate endpoint."[40] Aducanumab is the latest in a long line of drug candidates targeting amyloid-beta plaques, the proteins that accumulate in the brains of patients with Alzheimer's disease. Because every drug of this type has so far failed to improve cognition, it is now unclear whether amyloid-beta is even a promising direction in Alzheimer's research. Biogen is supposed run a "postmarketing" trial to confirm the efficacy of the drug, but it has up to nine years to complete the research, during which time the company could in theory make a great deal of money.[41] Its share price went through the roof as soon as the license was announced.

The FDA explained that "ultimately, the decision on whether aducanumab will be used for treatment will be made by patients, their families and caregivers, and health care professionals," an almost complete abdication of its responsibility to protect consumers.[42] On the face of it, this looks as if Milton Freidman's hopes for a consumer-driven pharmaceutical sector have come to pass. But Biogen quickly ran up against formidable roadblocks from other quarters.[43] The fatal blow was dealt in June 2022, when Medicare announced that it would only cover the drug for patients enrolled in clinical trials, a victory for the protective role of the state and a measure of its market power as a large-scale purchaser of pharmaceutical products.[44] In December 2022, a congressional investigation into the licensing process condemned the FDA's "lapses in protocol" and characterized the licensing procedure as "rife with irregularities."[45] For all the private sector was encouraged and empowered in the 1980s, American drug development continues to be one of the most highly regulated industries in the world.

On the bioconstitutionalist front, in South Africa, the Treatment Access Campaign continues its work, now under the name of Section 27, a nongovernmental organization with 182 branches and provincial offices, and a wider health agenda than AIDS. The strategy remains to seek relief in South Africa's courts, using constitutional arguments, backed up by "marches, media, legal education and social mobilization," without which a legal judgment "may deliver little more than pieces of paper."[46] The range of campaigns is broad, but the pragmatism is just as deep as ever. The education campaign, for example, focuses on sanitary school toilets, nutritious school meals, and asbestos abatement in school buildings. Their pro bono legal work includes representing the families of 140 mental health care recipients who were abandoned by the Gauteng provincial government and subsequently died. In one campaign that has resonance with recent developments in the United States, Section 27 lawyers have waged a fierce war against COVID-19 misinformation, including arguing in court against the use of the Trump-touted antiparasite drug ivermectin to treat the disease.[47] As a force for accountability, responsibility, and solid evidence, TAC's health activism may be one of the best things ever to happen to a postcolonial state.

Conclusion: Corporate Drag

As the winter of 1988 turned to spring and then summer, I would sometimes keep going west after the Monday night ACT UP meetings, walking through steamy Greenwich Village to the Chelsea Piers, a weather-beaten cruising ground of ruined warehouses and splintering docks, surrounded by the dark glitter of the Hudson River, where music from multiple boom boxes played deep into the night. I was a voyeur, all but invisible, mesmerized by the crowds of gay men and transwomen who gathered there to dance and have sex and get high. In 1991, when *Paris Is Burning*, the documentary about Harlem's drag ball culture, came out, I recognized some of the performers from those nights at the piers.

Paris Is Burning holds up a mirror to the greed-is-good decade, its African American stars donning corporate drag for "executive realness" competitions, dressing up as the heroines of the oil-baron soap opera *Dynasty*, perfecting the rural attire and proud posture of the horsey super-rich. Somewhere between homage and satire, these elaborate productions were a way for members of a doubly and triply marginalized community to imagine themselves into a slice of the good life, providing an outlet for talents unrecognized and unacknowledged by the white straight mainstream, a place to nourish dreams of fame and fortune. Like the ACT UP stock exchange demonstration, "executive realness" competitions lay somewhere between protest and application for admission, a subversive, satirical, furious, hilarious bid from the dispossessed for a place at the table.

Corporate drag was in the air in the go-go eighties. In December 1987, to play my tiny part in a protest involving stopping subway trains and blocking the Brooklyn Bridge—I was the look-out for a comrade who pulled the emergency cord at Wall Street Station—I bought a skirt suit and earrings at a thrift store and borrowed some makeup. It was easy to blend into the crowd, easy to feel like one of the commuters, the screeching subway functioning as the great leveler. Organized by a group of Black lawyer-activists against police violence and impunity, the action was reported admiringly by the *New York Times* as a series of "protest tourniquets" applied with "surgical precision," allowing around 500 demonstrators to disrupt the rush-hour commute of tens of thousands of New Yorkers.[48]

Battle lines that seemed so stark to me at the time—between ACT UP and pharmaceutical executives, Black activists and Wall Street workers, Harlem drag artists and the celebrity culture they parodied—now seem as much continuum as conflict. Between the municipal divestment that made parts of New York City look like they had been fire-bombed and the limbic capitalism that was rising from the ruins, there arose a species of freedom in which anarchism and libertarianism clashed and then conspired. Greedy pharma was but one aspect of the world that resulted from AIDS activism, channeled and constrained by the same existential politics of the deathbed that had facilitated its rise in the first place. For anyone engaged in vital politics today, including the harrowing threat of climate change and biodiversity loss, it should come as both comfort and challenge to know that we can place constitutional limits on capitalism in the name of our own survival, as long as there are enough people willing to take the fight to the court and the streets.

NOTES

1. Laura Fraser, "Genentech Goes Public." *Genentech*, April 28, 2016, https://www.gene .com/stories/genentech-goes-public.

2. Kaushik Sunder Rajan, *Biocapital: The Constitution of Postgenomic Life* (Durham, NC: Duke University Press, 2006), 115–116.

3. John Carreyrou, *Bad Blood: Secrets and Lies in a Silicon Valley Start-Up* (New York: Knopf, 2018), 137.

4. Sydney Lupkin, "Drugmakers Are Slow to Prove Medicines That Got a Fast Track to Market Really Work," *NPR Shots—Health News*, July 22, 2022, https://www.npr.org/sections /health-shots/2022/07/22/1110830985/drugmakers-are-slow-to-prove-medicines-that-got-a -fast-track-to-market-really-wo.

5. Kaushik Sunder Rajan, *Pharmocracy: Value, Politics, and Knowledge in Global Biomedicine* (Durham, NC: Duke University Press, 2017), 156. Emphasis in original.

6. See the documentary by ACT UP leader Maria Maggenti, *Doctors, Liars and Women: AIDS Activists Say No to Cosmo* (1988), directed by Jean Carlomusto and Maria Maggenti.

7. Steven Epstein, *Impure Science: AIDS Activism and the Politics of Knowledge* (Berkeley: University of California Press, 1996).

8. Cathy Gere, *Pain, Pleasure, and the Greater Good: From the Panopticon to the Skinner Box and Beyond* (Chicago: University of Chicago Press, 2017), chap. 1.

9. Richard L. Landau, ed., *Regulating New Drugs* (Chicago: University of Chicago, Center for Policy Study, 1973).

10. The symmetry between positive and negative outcomes in Peltzman's analysis is one of the shibboleths of neoliberal economics, which would come under attack at the end of the 1970s in the form of the discovery of the almost-universal response that a negative result outweighs a positive one with the same dollar value, published in 1979 by Daniel Kahneman and Amos Tversky. There may be no better example than Peltzman's claim that the "cost" of a few thousand thalidomide births is completely offset by the "benefit" of a widely available tranquilizer (Peltzman's example of the kind of drug kept off the market by the newly stringent regulations).

11. Edward Nik-Khah, "Neoliberal Pharmaceutical Science and the Chicago School of Economics," *Social Studies of Science* 44, no. 4 (2014): 489–517.

12. Milton Friedman, "Frustrating Drug Advancement," *Newsweek*, January 8, 1973, 49, in *The Collected Works of Milton Friedman*, ed. Robert Leeson and Charles G. Palm (Stanford, CA: Hoover Institution Press, 2017).

13. Gere, *Pain, Pleasure, and the Greater Good*.

14. Epstein, *Impure Science*, 221.

15. Sarah Schulman, *Let the Record Show: A Political History of ACT UP New York* (New York: Farrar, Straus and Giroux, 2021), 42.

16. ACT UP, "Flyer of the First ACT UP Action, March 24, 1987, Wall Street, New York City," http://www.actupny.org/documents/1stFlyer.html.

17. Schulman, *Let the Record Show*, 102.

18. Peter Staley, *Never Silent: ACT UP and My Life in Activism* (Chicago: Chicago Review Press, 2022), 12.

19. Schulman, *Let the Record Show*, 71.

20. Lucas Lichert, *Conservatism, Consumer Choice, and the Food and Drug Administration during the Reagan Era: A Prescription for Scandal* (Lanham, MD: Lexington Books, 2014), 145.

21. Damayanthi Divineni and James M. Gallo, "Zalcitabine: Clinical Pharmacokinetics and Efficacy," *Clinical Pharmacokinetics* 28 (1995): 351–360.

22. Staley, *Never Silent*, 144.

23. Schulman, *Let the Record Show*, 95.

24. "AIDS, the Surgeon General, and the Politics of Public Health," in *The C. Everett Koop Papers*, National Library of Medicine, https://profiles.nlm.nih.gov/spotlight/qq/feature/aids.

25. Epstein, *Impure Science*, 299–315.

26. Mark Harrington, "Access versus Answers," Treatment Action Group (1996), https://www.treatmentactiongroup.org/publication/access-versus-answers/.

27. Lest this should seem straightforward, see Cori Hayden, *The Spectacular Generic: Pharmaceuticals and the Simipolitical in Mexico* (Durham, NC: Duke University Press, 2022), for all the confounding complexities of the market for generic drugs.

28. Mark Heywood, "South Africa's Treatment Action Campaign: Combining Law and Social Mobilization to Realize the Right to Health," *Journal of Human Rights Practice* 1, no. 1 (March 2009): 14–36.

29. "AIDS Activists Badger Gore Again," *Washington Post*, June 18, 1999.

30. Pat Sidley, "Drug Companies Sue South African Government over Generics," *British Medical Journal* 322 (2001): 447, https://www.ncbi.nlm.nih.gov/pmc/articles/PMC1119675/.

31. Chris McGreal and Sarah Boseley, "Pretoria Pressures Drug Giants," *The Guardian*, March 6, 2001.

32. Rachel Swarn, "Drug Makers Drop South Africa Suit Over AIDS Medicine," *New York Times*, April 20, 2001.

33. Nathan Geffen, *Debunking Delusions: Inside the Story of the Treatment Action Campaign* (Aukland: Jacana Media, 2010), 56.

34. Heywood, "South Africa's Treatment Action Campaign," 16.

35. Geffen, *Debunking Delusions*, 62.

36. Heywood, "South Africa's Treatment Action Campaign," 22.

37. Edwin Cameron, *Justice: A Personal Account* (Cape Town: Tafelberg, 2014), 161.

38. Geffen, *Debunking Delusions*, 62–64.

39. Legally speaking, given the constitutional imperative in the United States "to promote the Progress of Science and the useful Arts by securing for limited Times to Authors and Inventors the exclusive Right to their respective Writings and Discoveries," biocapital is in fact another version of constitutionalism.

40. Patrizia Cavazzoni, "FDA's Decision to Approve New Treatment for Alzheimer's Disease," News & Events for Human Drugs, US FDA, June 7, 2021, https://www.fda.gov/drugs/news-events-human-drugs/fdas-decision-approve-new-treatment-alzheimers-disease.

41. Cavazzoni, "FDA's Decision."

42. Jamie Talan, "Dementia Experts on Why the FDA Approval of Aducanumab for Alzheimer's Gets Mixed Grades," *Neurology Today*, July 8, 2021, https://journals.lww.com

/neurotodayonline/fulltext/2021/07080/dementia_experts_on_why_the_fda_approval_of
.3.aspx.

43. See the August 2021 report by the Institute for Clinical and Economic Review, a non-profit that evaluates drug pricing practices, concluding that "the evidence is insufficient to conclude that the clinical benefits of aducanumab outweigh its harms or, indeed, that it reduces progression of AD." Grace A. Lin, Melaine D. Whittington, Patricia G. Synnott, Avery McKenna, Jon Campbell, Steven D. Pearson, and David M. Rind, "Aducanumab for Alzheimer's Disease: Effectiveness and Value; Final Evidence Report and Meeting Summary," *Institute for Clinical and Economic Review*, August 5, 2021.

44. "Medicare Officially Limits Coverage of Aduhelm to Patients in Clinical Trials," *New York Times*, April 7, 2022.

45. "Congressional Inquiry into Alzheimer's Drug Faults Its Maker and FDA," *New York Times*, December 29, 2022.

46. Heywood, "South Africa's Treatment Action Campaign," 22.

47. "2021: Year in Review," *Section 27*, December 2021, 11, https://section27.org.za/wp-content/uploads/2021/12/0170_Section27_Year-In-Review_20211215_V2.pdf.

48. "Protest against Racism Disrupts New York Rush Hour," *New York Times*, December 22, 1987.

Part III / Regions

CHAPTER EIGHT

Far from Footloose

The Greedy Localization of Biotechnology in Cambridge

Robin Wolfe Scheffler

Introduction

Silicon Valley. Route 128. The Research Triangle. In the generation after World War II, these regions were icons of America's turn to a "high-technology" economy, their nature closely studied by groups around the world seeking to construct their own engines of technoscientific economic development, such as Peter Westwick's oil-flush Texans in chapter 10 of this volume. Although these regions promised to define the future of the economy, their appearance was of a piece with the Industrial Revolution's concentration of manufacturing in specialized districts.[1] Yet if earlier urban industrial districts were rooted in the needs of labor, the geography of high technology reflected the desires of leisure. Companies in high-technology fields such as computing took advantage of improvements in transportation and communications infrastructure to set themselves in suburban settings that satisfied the lifestyle and recreation preferences of their "knowledge workers." As the 1980s dawned, prognosticators asserted that flows of information in the new knowledge economy would "decouple" work from cities and drive an "exodus" of businesses from major cities to the sub-urbs, if not from metropolitan centers entirely.[2]

Just as commentators issued these predictions, greedy science produced a different geographic distribution of technoscientific development. The spatial behavior of biotechnology, a quintessentially "greedy" scientific in-dustry of the 1980s, diverged from its high-technology predecessors in two respects. First, rather than following flows of knowledge around the globe, the industry clustered in a small subset of regions with vibrant bi-ology communities, particularly the San Francisco Bay Area, San Diego, and the Greater Boston Area (GBA). Moreover, within these regions, the industry gathered in industrial neighborhoods rather than the suburbs. The biotechnology and biopharmaceutical cluster in the GBA, now the densest and largest in the world, highlighted these divergences from high technology.[3]

While boosters of biotechnology in the GBA were quick to laud the natural advantages that firms drew from proximity to the region's many world-class biomedical research institutions, the fact that the epicenter of the cluster lay in gritty industrial Cambridge, whose city council had considered banning a core technology of the new biotechnology boom, recombinant DNA, in the late 1970s, suggests that forces unanticipated by prophets of the information age shaped the geography of greedy science. Following scientists, entrepreneurs, investors, corporations, and govern-ments as they embraced Cambridge as a necessary site for their new in-dustry illuminates the blend of speculation and rationality that animated these developments.

Although scientific knowledge enjoys a reputation as a placeless, mobile "view from nowhere," the processes undergirding its production have long rested upon local, communal, tacit, and unwritten practices. Like their predecessors, scientists in the emerging field of molecular biotechnology communicated through face-to-face interactions that conveyed a wealth of vital, unwritten, "tacit" information. Those who knew how to make ex-periments work were not jet-setting laboratory directors or scientific stars but the technicians and postdoctoral fellows who remained behind. While famous scientists might burnish the reputation of a firm when they joined advisory boards, these firms also needed scientific workers to produce new knowledge.[4] Faced with the prospect of moving these workers from their existing academic communities, founts of cutting-edge molecular biology techniques, to new locations, or shifting capital and resources to create new enterprises near universities, entrepreneurs and their backers preferred the latter.

Even as the importance of tacit knowledge transfers restrained the global ambitions of early biotechnology companies, the uneven urban development of biotechnology within the GBA reflected the fact that the geography of biotechnology was entwined with another notoriously greedy industry of the of the 1980s: real estate (also discussed by Hallam Stevens in chapter 9 of this volume). Even compared to other speculative technoscientific investments, investors struggled to define the value of early biotechnology companies. In the absence of alternatives, they were willing to value a company based on narratives of its future potential, stories that drew deeply on the cachet of specific scientific locations. When firms sought proximity to universities, the physical spaces they occupied were created by developers seeking to capitalize on the value of these neighborhoods for speculation in biotechnology. These different varieties of greed shaped the industry's geography (figure 8.1).

This chapter explores the processes that drove the concentration of the biotechnology industry in the GBA during the 1980s. I start with a discussion of the spatial consequences of understanding biotechnology as a greedy science rather than a commercial science. I then explore how the greedy nature of biotechnology guided it to locations in the GBA through three episodes. The first of these, a groundbreaking affiliation between Massachusetts General Hospital in Boston and West German pharmaceutical firm Hoechst AG, shows the rapidity with which multinational corporations recognized the value to their biotechnology efforts of establishing proximity to academic communities rather than recruiting scientists to distant corporate laboratories. Biotechnology firms funded by speculative venture capital reached the same conclusion, as I examine my second case: Biogen's long, reluctant journey to placing its main operations in Cambridge—the first company to do so. Finally, I look at the development of an urban science park, University Park, by MIT as emblematic of efforts of real estate investors and new biotechnology companies to profit from specific locations—a feature of greedy science that has become especially striking in the following decades.

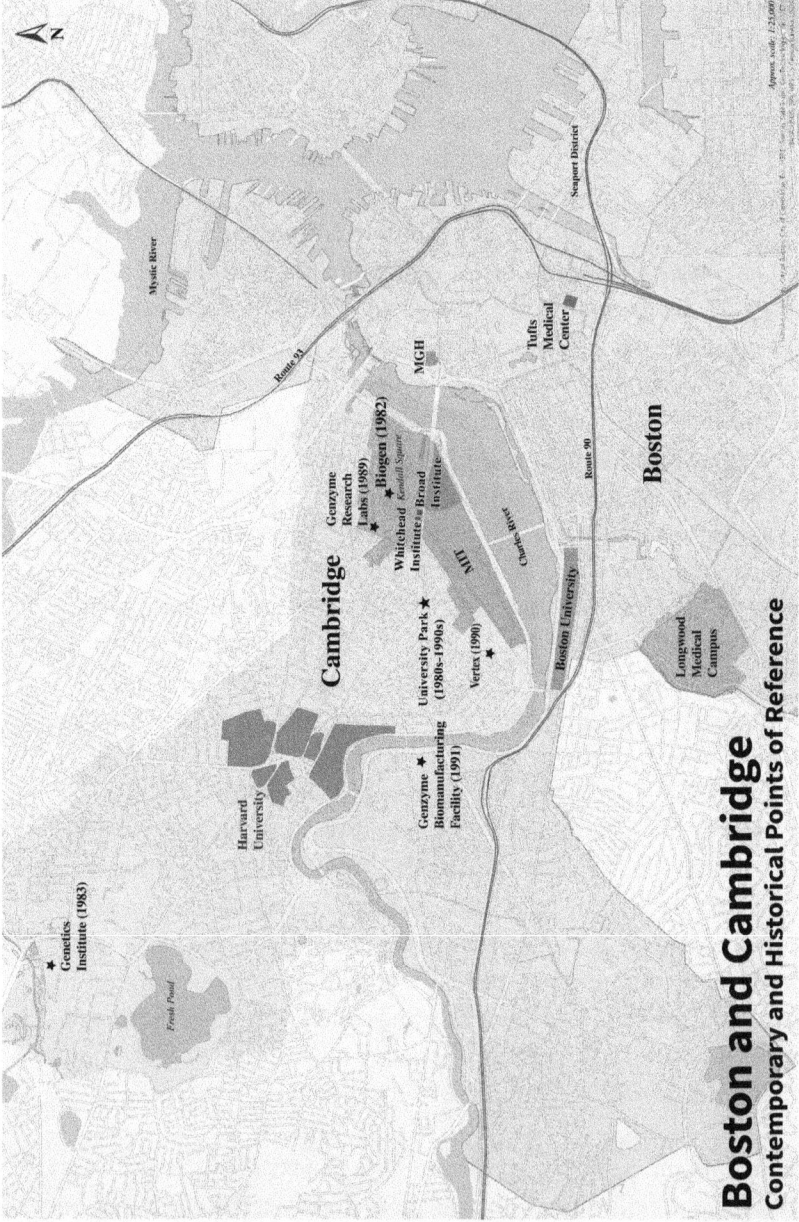

Figure 8.1 A map of the Cambridge and Boston area showing some of the institutional anchors of the local biotechnology economy, the locations of some early biotechnology companies, and other locations mentioned in this chapter. Based on locational data compiled by author and other sources. Map created by Rustam Khan

Biotechnology as Greedy Science

While industries ranging from agriculture to pharmaceuticals had drawn on and created biological knowledge for centuries, the modern biotechnology industry stood out for its new blend of greed and science. The industry emerged in the late 1970s, supported by new techniques for manipulating life (especially recombinant DNA), the growth of venture capital investing thanks to loosening regulations, the expansion of intellectual property in living things by the US Supreme Court (in *Diamond v. Chakrabarty*), and the passage of the Patent and Trademark Law and Amendments Act of 1980 (the "Bayh-Dole" Act). The initial public offering of South San Francisco–based Genentech in 1980 showcased the greedy returns that investors in biotechnology might enjoy—on the first day of public trading, its scientist-founders saw their initial $500 investment transformed into stock worth $70 million.[5] This example inspired imitators. "Our goal is not to make a 10 per cent return on our research effort," the CEO of Cambridge-based Biogen blustered on behalf of his as-of-then unprofitable company. "Our goal is quite different. We view our research as an investment on which we want to make a ten- or a hundredfold return."[6]

Despite the promise of such riches, individual biotechnology start-ups faced precarious circumstances. As Cathy Gere describes in chapter 7 of this volume, the 1980s were a moment when the embrace of market-driven drug development as a solution to public health problems ignited ongoing questions regarding how to judge the value and efficacy of pharmaceuticals writ large. In the eyes of many investment analysts, biotechnology companies had "nebulous" value because of the tenuous relationship between their research and the potential for a product that might, at best, be marketable a decade or more in the future. Conventional accounting methods used by stock analysts counted research expenditures solely as a loss on balance sheets, to be offset by the revenues generated by the sale of products, which most biotechnology firms did not have. By this standard, even a firm arriving at exciting scientific discoveries was a poor investment prospect.[7]

In the face of these challenges, biotechnology firms and their investors placed greater emphasis on research, not less. Scientific research might or might not result in marketable products, but it was vital to a firm's management and early backers as a means of raising equity. Companies exhibited a "quantum" pattern of growth as they raced to secure more investment by filing patents or announcing findings that satisfied stock analyst "checkpoints."[8] Critically, while firms might get credit from investors through affiliating themselves with scientific "stars" on their advisory boards, they relied on a larger, unsung body of staff scientists whose knowledge of molecular techniques produced research products, such as patents or publications, that cemented a firm's reputation with investment analysts.[9] In this setting, the line between good science and good marketing was thin—a mixture of research, prestige, and hype.[10]

While scientific productivity played a central role in the financial prospects of new firms, they found that money alone was a poor means of attracting skilled scientists. Molecular biologists who came of professional age in the late 1970s were the product of the rapid expansion of federal support for research in the life sciences after World War II. Ironically, this support, which had been predicated on the hope that advances in esoteric corners of biology would improve human health, underwrote a community of molecular biologists who lauded "pure" scientific knowledge and devalued its applications.[11] Academic molecular biology had few links to large commercial agricultural, chemical, or pharmaceutical companies, and its scientists tended to frown upon the secrecy associated with corporate research and development.[12]

Biotechnology firms learned that they could better lure scientists by offering to slake their thirst for credit and recognition—greed of a distinct sort present in molecular biology during the 1960s and 1970s. Impresarios of the new field, such as James Watson, enshrined competition for scientific credit, especially racing to results, as a central element of what it meant to be a first-rate scientist. For many biologists, the scandal of Watson's *The Double Helix* (1968) was not his derogatory portrayal of Rosalind Franklin but the fact that he discussed divining the structure of DNA as a race to be won for personal glory. By the 1970s, Watson had company from many others in the field of molecular biology who embraced both a commitment to scientific knowledge and a desire for professional accolades.[13] Competition for scientific recognition provided a critical bridge between the social worlds of academic molecular biology and biotechnology. Firms were quick to incorporate policies that allowed scientists to feel as if they were still a part of a larger intellectual community involved in "state of the art research," such as encouraging publication in prominent journals, sponsoring travel to academic conferences, or purchasing expensive equipment.[14]

Geographically speaking, this bridge did not reach far. Academic communities were not only wellsprings of (federally sponsored) scientific ideas but also anchors for the prestige economy of molecular biology. The vice president of the biotechnology start-up DNA Science reflected that as he worked to recruit researchers, "most scientists and many of the exceptionally good ones want to remain on campus," a feature of the fact that "things have progressed so rapidly in [recombinant DNA] that scientists haven't had the opportunity to make the transition totally to industry." The crux of the issue was that "expertise lies on campus and any single, small genetic engineering company sometimes cannot pull together all the various forces that are necessary to make it successful."[15] Oil companies interested in biotechnology, as Cyrus Mody writes in chapter 1 of this volume, also recognized this feature of the new field, investing in university laboratories or academically adjacent start-ups rather than setting up their own in-house research operations.

Investors found it much easier to move capital to these existing centers of molecular biology instead of moving molecular biologists—no amount of money could quickly re-create the blend of prestige and community found around universities.[16] An early biotechnology investor from the pharmaceutical industry explained, "You had postdocs coming out of places like Stanford and Harvard and MIT, and to the extent they're willing to work in private industry, they want to work for [Nobel Prize winners at Stanford and MIT]. . . . So they're going to stay where those people are . . . so the venture capitalists set the companies up on the doorsteps of those labs."[17]

Stationary Communities, Mobile Companies

Not all doorsteps were equally accommodating. Many factors, including institutional policies, safety concerns, or municipal zoning, limited the locations available to the biotechnology industry.[18] By virtue of its patchwork terrain of cities and biomedical institutions, each with its own institutional ambitions, the GBA offered the industry many possible points of entry. Although it shared few of the venture capital sensibilities of American biotechnology firms, the West German pharmaceutical firm Hoechst

AG's 1981 decision to provide $50 million over ten years to establish a new department of molecular biology at Massachusetts General Hospital (MGH) of Boston suggests the greedy logic that led investors to move capital to scientific communities and the academic greediness that facilitated these new connections.[19]

Like most pharmaceutical companies, Hoechst had an extensive in-house research program, but when it came to the new field of molecular biology, it decided to outsource its science. It originally sought to establish an institute at the University of California, San Francisco (UCSF), a hotbed of biotechnology research, under biochemistry professor Howard M. Goodman. Goodman had been a central participant in the use of recombinant DNA to clone genes that were the focus of early biotechnology ventures, such as insulin, human growth hormone, and the antigen of hepatitis B. Goodman, oral histories suggest, felt unjustly overshadowed by his UCSF biochemistry colleague and Genentech cofounder Herbert Boyer. After his work on insulin brought him into contact with Hoechst, he saw an opportunity. In 1980, the year Genentech went public, Goodman proposed that Hoechst sponsor a new center for molecular biology at UCSF under his supervision, employing up to 100 people. Hoechst, seeking, in the words of its representative, a "window into science," accepted his proposal.[20]

Creating such a center at UCSF, however, required reaching agreement with the sprawling University of California system. These negotiations took longer than Goodman or Hoechst were interested in waiting to start. Harvard Medical School, meanwhile, had hired a close colleague of Goodman's from the National Institutes of Health, Phillip Leder, to chair its Genetics Department with an eye toward expanding its research in molecular genetics. MGH saw an opportunity to recruit Goodman and gain resources to expand its own research in molecular biology. While MGH operated in close conjunction with the Medical School, it had an independent director and board of trustees. MGH could quickly approve an agreement without seeking approval from a complex bureaucracy, but Goodman and members of his new department—and Hoechst, by extension—would enjoy access to Harvard's prestigious biological sciences community.[21]

The sponsorship of an entire department by a non-US corporation at a prominent American biomedical institution drew widespread scrutiny. MGH's leadership at first refrained from publicizing the arrangement.

However, when its existence was revealed during congressional hearings on the commercialization of biomedical research in 1981, legislators called in the MGH's director of research, Ronald Lamont-Havers, to account for its details. Lamont-Havers professed little concern for the nature of Hoechst's sponsorship. He reminded the committee that similar fears in the 1950s that federal funding might undermine the integrity of biomedical research had come to naught. At a moment when the Reagan administration had proposed across-the-board cuts to research spending, the MGH was entitled to protect its institutional interests and those of its scientists: "The jeopardy of their careers and the people who work for . . . has put a great deal of strain on them." Rather than marking corporate intrusion, the "pressure for us to get involved with industry came out of our scientific community."[22]

While debates over decorum and the disbursement of intellectual property shadowed the Hoechst–MGH partnership, Hoechst saw the greatest immediate value of the arrangement in the access that its researchers would gain to the penumbra of ideas and techniques around MGH in the GBA. Given that German universities had less experience in molecular biology, the new department allowed Hoechst scientists to gain experience in this field. This could not be done at a distance; Lamont-Havers insisted that scientists "have to interact . . . they cannot be isolated." He underscored that MGH offered "access to new knowledge" rather than products—"what they're getting is information."[23] Supporting American molecular biology was a useful means of technology transfer through people, a testament to the tacit dimensions of work in this new field. Moreover, while Hoechst had sought a window into molecular biology through its sponsorship, MGH made sure that the new department acted as a door, adding terms to consulting agreements that faculty members in the department signed with local businesses to ensure Hoechst had access their discussions as well.[24]

Biogen's Pedestrian Choice

Hoechst's decision highlighted the ways in which the geography of greedy science started to diverge from commercial science, but these differences crystalized when the biotechnology start-up Biogen chose Cambridge as its research headquarters—the first of many to do so. Biogen initially promised to fulfill predictions that the knowledge economy would usher in mobile and global scientific entrepreneurship. Born of a meeting in Geneva in 1978, Biogen's scientist-founders envisioned it as an international company from the outset. Legally incorporated in Luxembourg and then the Netherlands Antilles, to avoid taxes on its prospective revenues, Biogen adopted a decentralized model of operations, distributing its research throughout the laboratories of the famous members of its Scientific Advisory Board (SAB) in Western Europe and the eastern United States rather than establishing a standalone research laboratory. Like MGH, members of the SAB were lured by the promise of using venture capital as means of "putting work in their labs" in the face of declining state funding. Early investors liked this model as well—professors could take advantage of existing academic infrastructure to expand into new research areas "relatively cheap[ly]" by hiring postdoctoral fellows.[25]

Investors, however, soon became wary of the lack of control that this diffuse model implied and advocated for a dedicated research headquarters. Harvard chemistry professor Walter Gilbert, SAB member and future CEO, explained, "We are a Swiss company owned by a Dutch company owned by a Netherlands Antilles holding company." "You can do R and D in university laboratories. There are large numbers of people. It is easy to move and do things. But there is no pace and focus. So we set up a lab to get focus and hopefully revenue."[26] Given the presumed mobility of science, Biogen's leadership favored placing its main research operations in Europe—and eventually settled on Geneva. Working in Switzerland placed Biogen close to a promising set of markets as well as to the Zurich Institute of Molecular Biology, where Charles Weissman, another member of the SAB, had just found a means of producing recombinant alpha-interferon, its first prospect for a marketable product.[27]

Even as Biogen headed for Europe, many of its investors urged it to stay connected to American centers of molecular biology. Ray Schafer, representing the venture capital arm of INCO, a Canadian nickel mining company backing Biogen, thought that the "Cambridge area" would be an ideal spot for such a branch. "In addition to proceeding with our main research laboratory in Europe people in Biogen have suggested that we consider establishing . . . a daughter lab in the US. . . . Special relationships with MIT and Harvard as institutions or professors might be possible. Staffing might be easier since we could attract scientists not wanting to leave the US for Europe. We could establish a physical presence in an attractive, highly technological area before someone else does."[28]

Yet Cambridge itself seemed inauspicious for this purpose. In 1976, the city had drawn national attention for placing a municipal moratorium on research using recombinant DNA, a core technique of the biotechnology industry. Gilbert was one of the Cambridge-based scientists whose research was shifted by the uncertainty these regulations fostered—his laboratory had traveled to Britain's Porton Down chemical and biological warfare laboratories in 1978 for a series of gene cloning experiments. Although Cambridge appeared to have moved toward more predictable and lenient regulations, it remained possible that new decisions would complicate the future of Biogen's research—laboratories could not easily be moved once built.[29] Nonetheless, two members of Biogen's SAB, Philip Sharp (MIT) and Gilbert (Harvard), were adamant that a location in Cambridge was "of key importance for the performance of the lab." Biogen's vice president of finance and administration examined sites in Cambridge and several neighboring communities as fallbacks if Cambridge decided not to welcome a biotechnology firm.[30]

Biogen came to see the pedestrian proximity to MIT and Harvard offered by a site in Cambridge as essential. Meeting in the Frantel Hotel on Martinique in January 1980, Schafer reported to Biogen's Board of Supervisors that there was "large amount of space available" in Cambridge, much of it "within walking distance" of MIT. "Almost any size facility was available and could be refurbished within 3 months." Looking further afield, he noted that there was no laboratory space available in Boston itself but that there were sites in "Summervile [Sommerville]," "Maldon [Malden]," and other towns along the nearby suburban Route 128 corridor that hosted many technology companies. These sites suffered, however, from being "some distance from a university campus."[31]

Engineers explained that the primary building they identified, 241 Binney Street, "satisfies several stated objectives." "The space is mostly open, permitting a flexible and efficient layout and is abundant with windows, thereby providing natural light and ventilation to all laboratories and offices." Moreover, "it is within a few blocks of MIT" (figure 8.1).[32] Amid uncertainty over the future of biotechnology, Sharp, later Biogen's research director, recalled that he favored keeping Biogen in Cambridge because it reassured scientists that they could rejoin their academic communities should the company fail.[33]

Biogen adopted many of the customs of academic research, encouraging rapid publication of results and the interaction of its employees with their academic peers. Even though such encounters might divulge some "commercially useful information," Biogen maintained that its "policy in this regard is of significant importance in obtaining the services of skilled and resourceful research scientists in the field of biotechnology." These academic relationships were part of the value Biogen promised to investors as it continued to raise money. "Access to universities where biotechnological research is being done have been essential to Biogen's success to date . . . and continue to be a distinguishing characteristic of Biogen and an important factor in its future commercial success."[34]

The emphasis on remaining within walking distance of academic molecular biology communities marks a moment when the movement of knowledge, which in the rueful view of economic geographers often leaves "no paper trail," left many traces.[35] Joe Rosa, hired in 1981 from a postdoctoral fellowship at Yale, recalled that the academic conversations cultivated by Biogen and all-company scientific meetings produced intense discussions of research that represented "science as it should be." He would take "busman's holidays" to Harvard and MIT to attend seminars of interest, although he was not willing to travel "across the river" to MGH or the Longwood Medical Campus.[36] These habits made the informal exchange of information with local universities substantially easier than the reception larger pharmaceutical firms received. Another Biogen employee reflected that he would have never worked in an environment where he would have had to forgo publications, accusing some large pharmaceutical companies at the time of acting as "parasite[s]" for taking knowledge without providing anything in exchange.[37]

Biogen expanded its Cambridge operations dramatically in 1982. This growth highlighted the strains of attempting to work across different locations. The pace of research slowed, and Biogen worried that companies in a single location, such as Genentech in South San Francisco, were able to arrive at, and patent, valuable methods and findings just before Biogen's scientists. Biogen acknowledged that its international structure and divided "physical locations ma[de] it difficult to create the critical mass required for research and marketing." To create a "one company atmosphere," Biogen planned more "face to face interactions through frequent group meetings, small group discussions, and exchange programs" as well as a new computer data system.[38]

Gilbert became the CEO of the company in 1982, promising to focus its research. However, the split nature of the company drew criticism from all sides. The president of Biogen reported that scientists in Geneva expressed "blatant concern" that Gilbert's leadership augured the closure of the company's European laboratories and that they wanted to be "told and retold that headquarters is in Geneva and that Biogen is an international company."[39]

In 1984, Biogen's challenges bringing products to market and declining financial prospects precipitated Gilbert's resignation, but if anything, these developments accelerated its consolidation in Cambridge.[40] The SAB met in February 1985 to consider the company's future in the face of its severe operating losses. Maintaining two separate laboratories was costly and unwieldy. Representatives of corporate stockholders questioned why they continued to pay for postdoctoral fellows in academic laboratories when their expense did not advance the production of marketable products.[41]

During this moment of crisis, Biogen made the decision to locate most of its operations in Cambridge. Scientific knowledge did not travel easily, and only Cambridge allowed coordination of researchers with biomanufacturing facilities for the fermentation, cell culture, purification, and testing work that would bring molecular discoveries to clinical trials. "We need a [manufacturing] facility near research such that we can exert control without excessive traveling," one manager urged. The lack of production faculties and appropriate coordination forced Biogen into the "wasteful" use of PhD-level researchers to manufacture recombinant proteins of sufficient quality for clinical trials, a process that brought "everything else . . . to [a] stop." Biogen faced the prospect of needing to carry out this process not once but nearly a dozen times as more trials loomed, undermining its effort to produce further patentable discoveries.[42]

Biogen's next CEO, the businessman James Vincent, recognized the need for localization and sold Biogen's Geneva laboratories to the British pharmaceutical firm Glaxo in 1987, committing the company to Cambridge.[43] Although its business fortunes proved turbulent, the work that Biogen did to brave Cambridge's recombinant DNA licensing process soon convinced other firms that the city's regulations were worth navigating for the sake of a desirable location—anchors of the early GBA biotechnology industry such as Genzyme and the Genetics Institute soon followed Biogen to Cambridge.[44]

Local Branding, Global Value: University Park

If given a choice between moving capital and moving people, greedy science preferred to move money. Yet while economic geographers often speak of the concentration of capital and industry as if it happened of its own accord, creating infrastructure for research where none had previously existed offered biotechnology a particular challenge. Venture capital investors were unwilling to spend money on the construction of laboratories for biotechnology research—this money was irretrievably fixed in bricks and mortar rather than mobile intellectual property or other assets that might partially recoup their investment in the likely event that a start-up failed. For want of space for their research activities, much of the $1.2 billion in venture capital raised by GBA biotechnology firms in the early 1990s sat in banks.[45] Creating the physical spaces to instantiate the economic logic of clustering depended on investment from another notoriously greedy industry of the 1980s: real estate. These developers sought to profit from biotechnology's location not through its products or equity but through its rents. As Hallam Stevens observes in his account of the later construction of Singapore's Biopolis, these real estate developers brought with them the logistical, political, and financial acumen to fashion the spaces for technoscience to operate.

During the 1960s and 1970s, Cambridge underwent a period of wrenching deindustrialization, and this process provided both the physical and the political backdrop for the biotechnology industry. Biogen, for example, sought approval of its laboratories from the city on the promise that it would bring blue- and white-collar employment back to the city.[46] The process of deindustrialization also created opportunities for MIT. The university had moved to East Cambridge earlier in the century to be close to the vibrant manufacturing neighborhoods of Kendall Square and Cambridgeport, but with deindustrialization, its planners saw a chance to cheaply acquire land on the edges of its campus in these neighborhoods for other uses. In 1969, for example, MIT purchased the former site of Simplex Wire and Cable in Cambridgeport.[47]

This purchase had longstanding implications for the growth of biotechnology in Cambridge. MIT had not obtained the Simplex site to expand its campus. Instead, like other projects MIT carved out of industrial Cambridge, such as Technology Square, Simplex would be developed as commercial property to generate revenue for the university by capitalizing on its proximity to MIT's classrooms and laboratories.[48] Toward this end, MIT partnered with Cleveland-based developer Forest City, an early specialist in university-centered urban redevelopment, to create "University Park."[49] The developer bet that it would be able to charge premium rents by capitalizing on "on the recent historical trend of new businesses stemming from work at" MIT, Harvard, or Boston University.[50] The initial plans for University Park heavily emphasized research and development over any other uses.[51] Landscaping and paths would encourage pedestrian flows between buildings and MIT—fashioning University Park as an extension of the MIT campus. Proximity to MIT ensured informal "access to a wide variety of organizations, meetings, seminars, and other activities"—the types of encounters that fostered the exchange of tacit knowledge.[52]

Local activists took exception to this plan, portraying MIT as a greedy, grasping octopus set on destroying a working-class neighborhood for the sake of maximizing its rents—replacing homes and businesses with "luxury-, high-rise development providing neither jobs or homes" to residents of the neighborhood.[53] However, on the verge of the free-market Reagan Revolution, MIT countered that "what's going to happen is going to happen in the marketplace" rather than modifying its plans to serve the public interest articulated by neighborhood activists.[54]

However, the market that MIT's spokespeople invoked did not appear to value high-technology development at University Park. Semiconductor manufacturer Temptronic declined to consider the Simplex site on the grounds that operating in Cambridge exposed its business to capricious regulations and raised safety concerns for its employees, one of whom had been held up "at knifepoint" in the neighborhood.[55] Like the hit 1984 movie, Boston Properties warned that high-technology firms were "footloose"—ready to move to the suburbs for "workforce availability and parking."[56] The difficulties that Forest City encountered signing high-technology tenants for its research spaces threatened to undermine the financing of the project, especially as the savings and loan scandal in the middle of the 1980s produced a credit crunch for commercial real estate development.[57]

However, Forest City found a new set of enthusiastic tenants: biotechnology firms. The Boston real estate firm Meredith and Grew advised the developer that local biotechnology companies could be captured by University Park rather than watching them follow high-technology firms to the suburbs—if it could construct enough laboratory space with a "campus-style setting."[58] Biotechnology firms were excellent tenants—facing chronic shortages of laboratory space, they were eager to sign leases in buildings whose construction had yet to begin![59] Biotechnology firms accounted for 80 percent of Forest City's first phase of its development. The industry dominated the remaining million square feet of laboratories and office space planned for construction through the early 2000s, including future anchors of biotechnology in Cambridge such as Vertex, Alkermes, and Millennium Pharmaceuticals.[60] A real estate development case study observed that "although Forest City did not initially intend to develop a biotech park," it had.[61]

In the late 1980s and early 1990s, biotechnology companies and real estate investors flocked to Cambridge while largely ignoring neighboring communities that offered the same physical proximity to academic research and intensely lobbied for their business. What explained the value of a location in Cambridge? Surveying the high value of real estate at the center of global cities such as New York, Tokyo, or London at the end of the 1980s, the sociologist Saskia Sassen observed that global investors "were clearly willing to pay an extremely high premium for a central location and had no interest whatsoever in a less central location."[62] Biotechnology expressed a similar dynamic. A real estate agent reported that companies were more than happy to pay a premium for a Cambridge address because "all the tenants want to be near MIT."[63] A spokesperson for one biotechnology company explained that they chose their Cambridge location to satisfy the "overriding consideration" to be "at the hub of all that intellectual activity."[64] Another biotechnology firm set in University Park was so convinced of the intellectual fecundity of these contacts that it ran its own shuttle vans to nearby subway stops out of concern that chance meetings on public buses would cause their scientists to inadvertently reveal trade secrets.[65]

The value of a Cambridge location was especially clear when entrepreneurs and companies with no prior connection to the city sought space there. This process of locational branding reflected the interplay of the pragmatic and the speculative. When Joshua Borger, a Merck research director based in Rahway, New Jersey, founded the biotechnology company Vertex in 1989, he put its laboratories in Cambridgeport near University Park, "within courier distance" of its Harvard-based scientific advisory board. Borger hoped that the visibility that came from an address in one of the nation's acknowledged biotechnology centers would help him in meetings with investors during his initial "roadshow."[66]

Araid Pharmaceuticals, founded by a biotechnology executive from Pennsylvania, emphasized to its prospective investors that the "location of the company in Cambridge offers the advantage of close proximity to a number of the nation's outstanding hospitals, academic institutions, and research centers, including Harvard and the Massachusetts Institute of Technology." This assurance was important for a company with no products and a share price that bore "no inherent relationship to the Company's assets, book value, net worth, cash flow, or any other recognized criteria of value."[67] These assurances appeared to have succeeded—Araid completed the single most valuable biotech stock offering to date in 1991.[68]

Location also offered local biotechnology firms a resource as they sought to establish partnerships with distant pharmaceutical companies, a common strategy to gain revenue while products were in development. Pharmaceutical analysts forecast that the 1990s would be defined by two major classes of compounds: protein drugs and traditional small-molecule chemical drugs. Whereas small molecules could be synthesized using the expertise in chemistry cultivated by pharmaceutical researchers, complex proteins were produced through biotechnology. Even if they did not believe that this new class of drugs would be effective, the pharmaceutical industry sought to keep connections to small biotechnology firms as a means of ensuring their own investors that they had a "pipeline" of future projects.[69]

Real estate developers welcomed the arrival of these corporate clients because their financial stability made them more appealing construction risks than biotechnology start-ups.[70] The appeal of Cambridge to both start-ups and pharmaceutical companies produced the intense cluster of biotechnology that exists in the city—and the real estate industry to serve it. The dominant private developers in East Cambridge today are real estate investment trusts, or REITs, which are publicly traded companies that include investors on a global basis—major shareholders in Alexandrea Real Estate Equities, one of the largest lab-based REITs, include the Bank of Norway and the Canadian Government Pension Fund.[71]

Cutting Loose?

In the decades since the greedy 1980s, the GBA biotechnology cluster deepened even as other forces worked to disperse the industry, including the widespread availably of scientific talent, subsidies for the industry in other regions, and relative lack of lifestyle amenities within the GBA—as any number of complaints about traffic, housing costs, or slushy northeastern winters attest. This observation runs askance of many implicit assumptions about the geography of scientific and technical work in the United States at midcentury, formalized in the 1970s and recently revived in a postpandemic world of telework: that in a knowledge-based economy, work could take place nearly anywhere, and the geography of technoscientific industries would cut loose from cities to follow the lifestyle and leisure preferences of its workers.[72] Biotechnology highlights a paradox: the greater mobility of people, capital, and ideas ushered in by globalization deepened rather than dispersed clusters of technoscientific activity as investors and entrepreneurs sought maximum advantage from the locations of their operations.[73]

However, this clustering was not solely the product of rational calculations regarding scientific productivity—greed and speculation played a vital role. Cambridge's ascendence as a locational brand for biotechnology drew real estate investment out of proportion to the pragmatic locational advantages it offered firms, accelerating the city's transformation into a biotechnology hub. Biotechnology real estate in Cambridge now constitutes an investment in its own right—from individual scientist-entrepreneurs willing to pay thousands of dollars to rent a single bench to globally traded REITs funneling billions of dollars into single neighborhoods.[74] At the height of the first wave of the COVID-19 pandemic, investors converted sites in the GBA ranging from gyms to malls to horse racing tracks into research space—a testament to the ongoing power of greed to shape both science and cities.[75]

NOTES

I thank Michael D. Gordin and W. Patrick McCray for inviting me to take part in this volume and for their comments on earlier drafts, along with those of my fellow greedy regionalists, Hallam Stevens and Peter Westwick. Research for this chapter was supported by an award from the National Science Foundation (#1947087), the James A. and Ruth Levitan Prize in the Humanities from the Massachusetts Institute of Technology, and a Sydney Brenner Research Fellowship from the Cold Spring Harbor Laboratory and Archives.

1. On the history of these regions and other high technology districts in the United States, see (in chronological order of publication) Anna Lee Saxenian, *Regional Advantage: Culture and Competition in Silicon Valley and Route 128* (Cambridge, MA: Harvard University Press, 1994); Christophe Lécuyer, *Making Silicon Valley: Innovation and the Growth of High Tech, 1930–1970* (Cambridge, MA: MIT Press, 2006); Margaret Pugh O'Mara, *Cities of Knowledge: Cold War Science and the Search for the Next Silicon Valley* (Princeton, NJ: Princeton University Press, 2005); Thomas J. Misa, *Digital State: The Story of Minnesota's Computing Industry* (Minneapolis: University of Minnesota Press, 2013); Robyn Klingler-Vidra, *The Venture Capital State: The Silicon Valley Model in East Asia* (Ithaca, NY: Cornell University Press, 2018); Alex Sayf Cummings, *Brain Magnet: Research Triangle Park and the Idea of the Idea Economy* (New York: Columbia University Press, 2020). On the history of American industrialization and geographic specialization, see Philip Scranton, *Endless Novelty: Specialty Production and American Industrialization, 1865–1925* (Princeton, NJ: Princeton University Press, 1997).

2. Peter F. Drucker, "Information and the Future of the City," *Wall Street Journal*, April 4, 1989. Drucker's essay crystalizes anti-urbanist themes of the 1960s and 1970s and the exhalation of the suburbs and regions further afield as new locations for economic activity. For an overview of the idea of the suburban knowledge worker, see Cummings, *Brain Magnet*; Patrick Vitale, *Nuclear Suburbs: Cold War Technoscience and the Pittsburgh Renaissance* (Minneapolis: University of Minnesota Press, 2021). "Knowledge economy" has several definitions; an overview can be found in Walter W. Powell and Kaisa Snellman, "The Knowledge Economy," *Annual Review of Sociology* 30, no. 1 (2004): 199–220.

3. While density can be reckoned in many forms, the GBA scores at or near the top of most measures. "Top 10 U.S. Biopharma Clusters 2021," *GEN—Genetic Engineering and Biotechnology News* (blog), March 10, 2021, https://www.genengnews.com/a-lists/top-10-u-s-biopharma-clusters-8/; Shiri M. Breznitz and William P. Anderson, "Boston Metropolitan Area Biotechnology Cluster," *Canadian Journal of Regional Science* 28, no. 2 (Summer 2005): 249–263; Jason Owen-Smith and Walter W. Powell, "Accounting for Emergence and Novelty in Boston and Bay Area Biotechnology," in *Cluster Genesis: Technology-Based Industrial Development*, ed. Pontus Braunerhjelm and Maryann P. Feldman (New York: Oxford University Press, 2006), 61–83.

4. "View from nowhere" is Steven Shapin's notable formulation. For one such technique in early biotechnology, see Kathleen Jordan and Michael Lynch, "The Sociology of a Genetic Engineering Technique: Ritual and Rationality in the Performance of the 'Plasmid Prep,'" in *The Right Tools for the Job: At Work in Twentieth-Century Life Sciences*, ed. Adele E. Clarke and Joan H. Fujimura (Princeton, NJ: Princeton University Press, 1992). For the importance of working scientists at firms, see David B. Audretsch and Paula E. Stephan, "Company-Scientist Locational Links: The Case of Biotechnology," *American Economic Review* 86, no. 3 (1996): 641–652. On tacit knowledge and the challenges of distance to the practice of experimental science even in an era of improving communications and transportation, see H. M. Collins, "The TEA Set: Tacit Knowledge and Scientific Networks," *Science Studies* 4, no. 2 (1974): 165–185; Steven Shapin, "Here and Everywhere: Sociology of Scientific Knowledge," *Annual Review of Sociology* 21 (1995): 289–321; Martina Merz, "'Nobody Can Force You When You Are across

the Ocean'—Face to Face and E-Mail Exchanges between Theoretical Physicists," in *Making Space for Science: Territorial Themes in the Shaping of Knowledge*, ed. Crosbie Smith and Jon Agar (London: Palgrave Macmillan, 1998), 313–329.

5. Sally Smith Hughes, *Genentech: The Beginnings of Biotech* (Chicago: University of Chicago Press, 2011), 158–159. For a critical sociological observer's tour of the emerging industry, see Martin Kenney, *Biotechnology: The University-Industrial Complex* (New Haven, CT: Yale University Press, 1986).

6. John Elkington, *The Gene Factory: Inside the Genetic and Biotechnology Business Revolution* (New York: Carroll & Graf, 1985), 69–70.

7. Kenneth W. Clarkson, *Intangible Capital and Rates of Return: Effects of Research and Promotion on Profitability* (Washington, DC: American Enterprise Institute for Public Policy Research, 1977), 27–28.

8. Kelvin W. Willoughby and Edward J. Blakely, "Making Money from Microbes: Finance and the California Biotechnology Industry," UC Berkeley Center for Real Estate and Urban Economics Working Paper no. 89-166 (Institute of Business and Economic Research, 1989), 9; Jeffery L. Fox, "Biotechnology: A High-Stakes Industry in Flux," *Chemical & Engineering News Archive* 60, no. 13 (1982): 10–15.

9. Audretsch and Stephan, "Company-Scientist Locational Links"; Lynne G. Zucker and Michael R. Darby, "Star Scientists and Institutional Transformation: Patterns of Invention and Innovation in the Formation of the Biotechnology Industry," *Proceedings of the National Academy of Sciences* 93, no. 23 (1996): 12709–12716.

10. Anthropologists of science have called this a species of "venture science." Kaushik Sunder Rajan, *Biocapital: The Constitution of Postgenomic Life* (Durham, NC: Duke University Press, 2006), 114, 128–129.

11. These values were very much in display during the resistance that many molecular biologists articulated to federal investment in molecular biology with the aim of curing cancer during the 1970s. See Robin Wolfe Scheffler, *A Contagious Cause: The American Hunt for Cancer Viruses and the Rise of Molecular Medicine* (Chicago: University of Chicago Press, 2019), 166–182.

12. These questions relate to a larger discussion of the "moral economy" of molecular biology research; one point to start with is Nicolas Rasmussen, *Gene Jockeys: Life Science and the Rise of Biotech Enterprise* (Baltimore: Johns Hopkins University Press, 2014), 94–100. For a critique of the "pure" versus "applied" divide in this scholarship, see Doogab Yi, *The Recombinant University: Genetic Engineering and the Emergence of Stanford Biotechnology* (Chicago: University of Chicago Press, 2015), 12–15.

13. Sociologist of science Robert Merton noted this theme in his early review of *The Double Helix*, Robert Merton, "Making It Scientifically," *New York Times*, February 25, 1968. Steven Shapin has written about this new mode of brash self-presentation by biologists in the biotechnology industry; see Steven Shapin, *The Scientific Life: A Moral History of a Late Modern Vocation* (Chicago: University of Chicago Press, 2008), 217–229.

14. David Botstein [Genentech] to Dr. Ruth Kirschstein [director NIGMS], October 10, 1988, Department of Distinctive Collections, MIT Libraries (Cambridge, MA), AC 455, Box 5, Series 3. "Committee on Biotechnology NRC." See also Nicolas Rasmussen on the creation of a "postdoc's republic" at Genentech. Rasmussen, *Gene Jockeys*, 96.

15. *Commercialization of Academic Biomedical Research: Hearings, Day 2, Before the Subcommittee on Investigations and Oversight and the Subcomm. on Science, Research, and Technology of the Committee on Science and Technology*, 97th Cong., 117–118 (1982) (testimony of Dr. Harsanyi [VP DNA Science Inc.], formerly of Cornell Medical College and the head of an Office of Technology Assessment study on applied genetics).

16. Companies seeking access to the insights of molecular biologists could still attempt to recruit scientists, such as when the Swiss pharmaceutical company Roche established an institute of molecular biology through hiring personnel from the National Institutes of Health to work near other pharmaceutical companies in Nutley, New Jersey, but this approach could take decades to bear commercial fruit. Even when established, these research institutes could not necessarily gain advantages from the free exchange of ideas with their corporate neighbors. Carsten Timmermann, *Moonshots at Cancer: The Roche Story* (Basel: Editones Roche, 2019), 64–89.

17. Hugh A. D'Andrade, interview by Sally Hughes and Leo Slater in Madison, NJ, November 8, 1998 (Philadelphia: Science History Institute, Oral History Transcript #0172), p. 18.

18. Maryann Feldman and Nichola Lowe, "Consensus from Controversy: Cambridge's Biosafety Ordinance and the Anchoring of the Biotech Industry," *European Planning Studies* 16, no. 3 (April 2008): 395–410.

19. This was one of the largest corporate–academic partnerships in biotechnology at that time. See figures compiled on early university–corporate partnerships in Kenney, *Biotechnology*, 56.

20. Goodman was involved in this wide range of projects because of his knowledge of restriction endonucleases, enzymes that cut DNA at specific points. Barbara J. Culliton, "The Hoechst Department at Mass General," *Science* 216, no. 4551 (1982): 1200–1203. On Goodman's possible conflicts, see Herbert Heyneker, interviews by Sally Hughes in San Francisco, CA, 2002 (Berkeley: University of California Regional Oral History Office, Bancroft Library), 38–39.

21. Culliton, "The Hoechst Department at Mass General."

22. Federal funding formed two-thirds of the MGH's budget. *Commercialization of Academic Biomedical Research*, 87–96 (1982) (testimony of Ronald Lamont-Havers, MGH research director).

23. *Commercialization of Academic Biomedical Research*, 93, 96 (1982) (testimony of Ronald Lamont-Havers, MGH research director).

24. Stephen Budiansky, "Academic Consultancy: Mass. General Placates Hoechst," *Nature* 300, no. 5890 (1982): 305.

25. Daniel Adams, interview by Mark Jones in San Francisco, California, December 10, 2010, March 20, 2011, and November 17, 2011 (Philadelphia: Chemical Heritage Foundation, Oral History Transcript #0972), 27–28.

26. Joe Bower [Professor Harvard School of Business] to "The Files" re: "Lunch with Wally Gilbert," March 31, 1981, Cold Spring Harbor Laboratory Archives (Cold Spring Harbor, Y), Walter Gilbert Papers [hereafter Gilbert Papers], Series 3, Box 6, Biogen March–April 1981.

27. Daniel Adams interview, 25.

28. Memo by Ray Schafer re: Summary Biogen Laboratory Research Sites, January 5, 1978, Cold Spring Harbor Laboratory Archives (Cold Spring Harbor, NY), Charles Weissman Papers [hereafter Weissman Papers], Series 2, Box 15, Annual Board of Directors Documents, 1978.

29. Stephen S. Hall, *Invisible Frontiers: The Race to Synthesize a Human Gene* (New York: Atlantic Monthly Press, 1987), 249–265. For more on the recombinant DNA safety, see Everett Mendelsohn, "'Frankenstein at Harvard': The Public Politics of Recombinant DNA Research," in *Transformation and Tradition in the Sciences: Essays in Honor of I. Bernard Cohen*, ed. Everett Mendelsohn (Cambridge, UK: Cambridge University Press, 1984), 317–335; John Durant, "'Refrain from Using the Alphabet': How Community Outreach Catalyzed the Life Sciences at MIT," in *Becoming MIT: Moments of Decision*, ed. David Kaiser (Cambridge, MA: MIT Press, 2010), 145–163. The Cambridge episode is touched upon in Sheldon Krimsky, *Genetic Alchemy: The Social History of the Recombinant DNA Controversy* (Cambridge, MA: MIT Press, 1982);

Susan Wright, *Molecular Politics: Developing American and British Regulatory Policy for Genetic Engineering, 1972–1982* (Chicago: University of Chicago Press, 1994).

30. The vice president also noted that Biogen's US legal counsel, the Boston firm of Mintz & Levin, gave them "very good information and public relations channels, at both the local and state levels," for such a move. Although "it [was] well known that Cambridge has established [recombinant DNA guidelines] . . . that are more stringent than elsewhere in the USA," it was in many ways better that "the battle in Cambridge has been fought, whereas there is uncertainty in surrounding communities." AJ Muller (Biogen SA, Genva) to RE Crawthorne & JE Davies, "US Laboratory Current Status," September 9, 1980, Weissman Papers Series 2, Box 19, Biogen Board of Director Documents September 1980.

31. "Board of Supervisors Meeting Minutes," January 14, 1980, 10, Gilbert Papers, Series 3, Box 5, Biogen 1980. Biographical information on RH Schaefer from Phil Sharp Interview with MIT News in 2014, http://news.mit.edu/2014/how-to-build-a-biotech-renaissance-mit-in-kendall-square (accessed and archived November 13, 2018).

32. Biogen Laboratories Cambridge, "Massachusetts Architectural Feasibility Study by Huygens and Tappé (Architects and Planners, Boston)," Weissman Papers, Series 2, Box 19, Biogen Board of Director Documents, September 1980.

33. Philip Sharp, interview by the author, March 22, 2018.

34. Biogen NV Placing Memorandum 522,000 Ordinary Shares, September 28, 1981, 14, Weissman Papers, Series 2, Box 21, Biogen Board of Directors, September 1981.

35. Paul R. Krugman, *Geography and Trade* (Leuven: Leuven University Press, 1991), 53.

36. Joseph J. Rosa interview by the author, March 11, 2022. Quoted with permission.

37. Jeffery Browning interview by the author, March 4, 2022. Quoted with permission.

38. Biogen (Inc) 1983 Long Range Plan, 13, Gilbert Papers, Series 3, Box 6, Biogen 1982–1984.

39. Peter Osgood to Walter Gilbert and Robin Nicholson [managing director INCO], August 25, 1981, Gilbert Papers, Series 3, Box 6, Biogen May–September 1981.

40. Brian Dick and Mark Jones, "The Commercialization of Molecular Biology: Walter Gilbert and the Biogen Startup," *History and Technology* 33, no. 1 (2017): 140.

41. Ken Murray [University of Edinburgh] to Biogen Scientific Advisory Board, November 27, 1984, Gilbert Papers, Series 3, Box 6, Biogen 1982–1984.

42. Julian Davies to Walter Gilbert [April 1982?], Gilbert Papers, Series 3, Box 6, Biogen 1982–1984.

43. John Wilke, "Biogen Shedding Its Loss-Making Swiss Operation," *Boston Globe*, July 29, 1987.

44. Cambridge Department of Public Health Recombinant DNA Permit data compiled by the author. Urban planners have also observed the paradoxical attraction of the stability of a predictable, if more onerous, regulatory process for the expansion of biotechnology in Cambridge: Feldman and Lowe, "Consensus from Controversy."

45. Ronald Rosenberg, "Expanding Biotech Hits a Money Snag," *Boston Globe*, June 28, 1992, City edition.

46. Prepared remarks of Frederick A. Eustis (Biogen Legal Counsel) for a Cambridge Biohazards Committee public hearing on Biogen Application, May 13, 1982, Cambridge Public Health Department (Cambridge, MA), rDNA Collection, Box 1, Biogen 1980–2002.

47. On MIT's land purchasing strategy, see O. Robert Simha, *MIT Campus Planning, 1960–2000: An Annotated Chronology* (Cambridge: Massachusetts Institute of Technology, Office of the Executive Vice President, 2001), 106, 116.

48. Although other private universities invested their endowments in real estate on a national basis, MIT was unique for having its investments concentrated on the edges of its

campus. Walter Milne, "Statement on Former Simplex Property," June 1974 [statement sent to all members of Cambridge City Council], MIT Archives AC-69, Series 2 Box 43, Correspondence, News Items, University Park Development. On MIT's role in urban redevelopment in Cambridge, see also LaDale C. Winling, *Building the Ivory Tower: Universities and Metropolitan Development in the Twentieth Century* (Philadelphia: University of Pennsylvania Press, 2017), 156–166.

49. Materials on Forest City such as *A Landmark Development Comes to Cambridge: University Park at MIT* (brochure), MIT Archives AC-69, Series 2 Box 43, Correspondence, News Items, University Park Development. See also James Rater [Forest City President] to MIT Real Estate Office, stage one proposal, May 5, 1982, MIT Archives AC-0205, Box 186, Folder: "[P78-05] Simplex Development Area Proposal to MIT Forest City Enterprises 1982."

50. Notes of Cambridgeport Blue Ribbon Committee prepared by Roger Boothe, Director of Urban Design, April 1, 1986, MIT Archives AC 408, Box 8, Blue Ribbon General Information.

51. Forest City favored research and development over office space by roughly 2:1, or 1,250,000 to 600,000 square feet. Melvin Rock [Forest City] to Philip Trussell [MIT Real Estate] response to comments, November 29, 1982, MIT Archives AC-0205, Box 186, Folder: [P78-05] Forest City 1982.

52. *A Landmark Development Comes to Cambridge: University Park at MIT* (brochure), MIT Archives AC-69, Series 2 Box 43, Correspondence, News Items, University Park Development.

53. Simplex Steering Committee flyer included in MIT Archives AC-69 [MIT News Office Records], Series 2 Box 43, Correspondence, News Items, University Park Development. For MIT as Octopus, see flyer in Cambridge Historical Society Archives (Cambridge, MA), Bill Cavelini Papers, Box 2, Folder 9.

54. David McKay Wilson, "MIT Agrees to Stop Land Purchases," *Cambridge Chronicle* 134, no. 11 (1980): 1, 9.

55. Henry A. Lynden [VP, Temptronic Corp Administration] to Philip A. Trussell [MIT], January 4, 1980, MIT Archives AC-69, Series 2 Box 43, Correspondence, News Items, University Park Development.

56. Notes of Cambridgeport Blue Ribbon Committee on presentation from Boston Properties prepared by Roger Boothe (Director of Urban Design), April 28, 1986, MIT Archives AC 408, Box 8, Blue Ribbon General Information.

57. "University Park at MIT Cambridge, Massachusetts," *ULI Development Case Studies* 31, no. 10 (2001): 7.

58. Meredith & Grew Incorporated, "Forest City's Cambridge Market Overview" [1984?], MIT Archives, AC 408, Box 8, Blue Ribbon General Information.

59. Upendra Mishra, "Biotech Companies Find Space Tight in Cambridge," *Boston Business Journal*, September 3, 1993. One of Forest City's financing strategies was to only start building with lease agreements signed, as opposed to building on a speculative basis. "University Park at MIT Cambridge, Massachusetts," 7.

60. J. K. Dineen, "Slow Start at University Park," *Cambridge Chronicle* 149, no. 9 (1995): 8. For these companies, see map c. 2002 of "Biotechnology Cluster in Kendall Square" produced by the Biomedical Enterprise Program of the MIT Sloan School of Management, in author's possession.

61. Biotechnology firms then occupied 90 percent of the 700,000 square feet it had completed. "University Park at MIT Cambridge, Massachusetts," 6.

62. Saskia Sassen, *The Global City: New York, London, Tokyo*, 2nd ed. (Princeton, NJ: Princeton University Press, 2001), 186.

63. Mishra, "Biotech Companies Find Space Tight in Cambridge."

64. Amy Miller, "Biotech Lags, City Gauges Impact," *Cambridge Chronicle* 149, no. 9 (1995): 1, 8.

65. Brandon Mitchell, "Economic Development Challenges in the City of Cambridge: The Biotechnology Industry" (master's thesis, Massachusetts Institute of Technology, 2000), 60.

66. Barry Werth, *The Billion-Dollar Molecule: One Company's Quest for the Perfect Drug* (New York: Simon & Schuster, 1994), 21–22.

67. Summary of the Offering: Business: Industry Overview, 25, 29, ARAID 1992, Barry Werth Papers (private collection), ARAID.

68. Another important device behind this branding was the focus of biotechnology companies in this era on private stock placements, where the effects of branding and personal charisma could flourish unchecked by fiduciary concerns. Udyan Gupta and David Stripp, "Fledgling Biotechnology Firm Scores a Financing Coup," *Wall Street Journal*, March 27, 1992.

69. Lynn Klotz, Robert Schatz, Christopher Morris, Andrew Kerr, and Robin Rodgers, *Drug Discovery and Development Strategies: A Look at Six Leading Biotechnology Companies* (Waltham, MA: Decision Resources, 1993), 17.

70. Massachusetts Centers of Excellence (Organization), *An Assessment of the Massachusetts Biotechnology Industry*, 1990, 16–17.

71. Anonymous, "Clusterluck; The Biotechnology Industry," *The Economist*, January 16, 2016; Mark R. Trusheim, Ernst Brendt, Fiona Murray, and Scott Stern, "American Entrepreneurial Chaos or Collaborative Industrial Policy: The Emergence of the Massachusetts Biotechnology Super-Cluster," in *CONCORD-2010* (Corporate R&D: An Engine for Growth, A Challenge for European Policy, Seville: European Commission, 2010), 5–57. On Alexandria Real Estate Equities, see https://money.cnn.com/quote/shareholders/shareholders.html?symb =ARE&subView=institutional (accessed December 19, 2023).

72. Edward Gunts, "Richard Florida Outlines His Vision for a 'Post-Pandemic City,'" *The Architect's Newspaper* (blog), April 26, 2021, https://www.archpaper.com/2021/04/richard -florida-outlines-a-post-pandemic-city/. This builds on a set of ideas about cities and the lifestyle demands of the "creative class" that Florida popularized in the late 1990s and early 2000s. Mark Lorenzen and Kristina Vaarst Andersen, "Centrality and Creativity: Does Richard Florida's Creative Class Offer New Insights into Urban Hierarchy?" *Economic Geography* 85, no. 4 (2009): 363–390.

73. Krugman, *Geography and Trade*; Sassen, *The Global City*.

74. Jonathan Saltzman, "It Costs $4,600 a Month to Rent a Lab Bench Here. But the Place Has Plenty of Fans," *Boston Globe*, March 5, 2019, sec. Business.

75. Catherine Carlock, "Lab Space Goes Boom. But What Comes Next?" *Boston Globe*, June 5, 2022, sec. Business.

CHAPTER NINE

Science as Speculation

State Capitalism, Real Estate, and Singapore's
Jurong Town Corporation

Hallam Stevens

Crazy Rich Singaporeans

The 2018 movie *Crazy Rich Asians* (dir. Jon M. Chu) opens in London in 1986. Turned away by a snooty and racist hotel concierge, Eleanor Young makes a phone call to her husband in Singapore. Within minutes, the problem is solved: the Youngs have purchased the ancient London establishment in its entirety and the insufferable concierge is put in his place. The message is clear: Singaporeans are now rich beyond imagining and everyone better get on board with it.

How did Singapore and Singaporeans become "crazy" rich?

In 2003, Singapore opened the doors to Biopolis, a science park focused on the biomedical and life sciences. Biopolis, developed and managed by Jurong Town Corporation (JTC), aimed to stake Singapore's claim on the world scientific stage. Seeking to compete with other science-intensive zones, like the Greater Boston Area, while taking advantage of the 2001 George W. Bush ban on stem cell research in the United States, Biopolis aimed to attract scientific talent from around the globe and provide state-of-the-art facilities for pursuing groundbreaking biomedical research. For Singapore, this was about wealth as well much as science. Although Biopolis's backers certainly hoped to put Singapore firmly on the global scientific map, they also wanted to see a rapid return on their investment.[1] This could come in the form of spin-offs, local startups, or foreign investment. JTC's plans included ample space for commercial tenants—biotech companies and especially pharmaceutical companies who could use Biopolis as a base of operations in Asia.[2] Biopolis was a hub for Singapore's economic as well as scientific growth.

These kinds of linkages between scientific and economic goals are hardly new for Singapore. A small island nation sitting at the end of the Malay Peninsula with a land area roughly the size of New York's five boroughs, the "Singapore story"—from self-government in 1959 to the present—is a narrative of growth and intensive modernization driven by industrialization. Under British rule from 1819, Singapore thrived as an entrepôt trading post, becoming a wealthy, multiethnic society dominated by Chinese migrants to Southeast Asia. As an independent nation, Singapore pursued a successful policy of export-oriented industrialization that generated rapid economic growth in the 1970s and 1980s. During this period, the government aggressively courted multinational companies who would base factory operations in Singapore, providing employment for locals and export revenue. Lee Kwan Yew, who ruled the country from its independence until 1990, was both feared and admired for his ruthlessness in dispensing with obstacles to this economic and social development. Broad government powers to acquire and clear land, suppression of political opponents, control of the media, co-optation of trade unions, and heavily curtailed citizen activism meant that the state had significant scope to pursue its agenda.[3] By the 1980s, Singapore had become a home to global finance and multinational corporations, a tropical outpost of greedy Wall Street capitalism.

Science and technology have played a significant role in these developments. The massive expansion of transport (including road, subways, ports, and airports) and telecommunications infrastructure represented a faith in the transformative potential of engineering and technology. Likewise, the government's cultivation of high-tech manufacturing industries, including electronics and semiconductors, suggests the close relationship between economic growth and the nation's technological development.[4]

Since the 1960s, one of the most significant architects of this growth has been JTC. JTC's role as the master planner, developer, and manager of Singapore's industrial estates has meant that it has played a critical role in infrastructure development and the expansion of high-tech industries on the island. But it has also played a key role in fostering spaces for science, beginning with the Singapore Science Park in the early 1980s and continuing through to the ongoing development of the Biopolis site from the 2000s. JTC has also overseen similar developments outside Singapore, including in Thailand, Indonesia, and China in the 1980s and 1990s. In these projects, JTC has brought to bear a suite of technical, infrastructural, managerial, and financial expertise that could be sold as a (Singapore) model or package for consumption in foreign markets. Scientific and technological development could be marketed as a Singapore export.

This chapter traces a particular history of JTC, focusing on the 1980s as the key period during which they were able to formulate and consolidate this model. I argue that the development of science parks and high-tech industrial parks has been a key mechanism through which Singapore has voraciously expanded both its wealth and its soft power within Asia. Critical to Singapore's success was the close coupling between capitalism and the state. What Chua Beng Huat has called "state capitalism" involved a "string of state-owned enterprises" (of which JTC is one) that buttressed Singapore's wealth and success as well as ensuring the government's continued political legitimacy.[5] By maintaining a significant stake in key industries like petrochemicals and ports (as well as land), the Singapore government generated revenue that could then be reinvested into other emerging industries or elsewhere in the economy.

Crazy Rich Asians is a spectacle of wealth: mansions, gardens, and swimming pools are the motifs of an ultra-elite on a land-scarce island. Indeed, much of Singapore's wealth is linked—in various ways—to land. JTC's ambiguous institutional status allowed it wide scope to greedily pursue the acquisition and development of land. During the 1980s, JTC increasingly embraced its roles as developer and "accelerator" for scientific and technological endeavors. After amendments to the JTC Act in 1986, JTC's powers to operate overseas were expanded. Moving in step with the privatization of government services in other developed nations, at the end of the 1980s, JTC began to further corporatize various aspects of its business. As such, JTC's operations became increasingly similar to those of a private property developer, even competing with private developers both in Singapore and internationally.

Of course, JTC's primary aim was to attract investment in Singapore by multinational corporations. Selling Singapore to investors meant selling its real estate. One way to do this was to sell Singapore as a "cluster" or an ecosystem—to persuade potential investors that they should come because their partners and competitors were already there.[6] Several other chapters in this volume illustrate the ways in which cities (like the Greater Boston Area) or larger regions (like Texas) have harnessed such clustering effects. In Cambridge, Massachusetts, for example, the "walkability" between company and university labs in the cluster was seen to be of central importance by biotech and pharmaceutical investors.[7] Certainly, this form of clustering—emulating Silicon Valley and other high-tech regions—was one of JTC's goals.[8] Like Texas, Singapore had the money (including from its significant petrochemical industry) to fund the kinds of infrastructure development that were needed to seed scientific and industrial collaboration.[9]

However, JTC also sought to attract investors in ways that went beyond clustering. JTC was not only a developer but also a landlord. It worked to manage its estates in ways that continued to enhance their economic value for Singapore and for their tenants. JTC created a "package" that included not merely industrial or technological spaces but also mechanisms, systems, rules, and "software" for how to design, build, and operate them. This included infrastructural, legal, financial, social, and political elements that could be bundled together and replicated both inside and outside Singapore. In other words, JTC developed a technology park "cookie-cutter" model that could ultimately be exported beyond Singapore. JTC mobilized science and high-tech industrial spaces into a real estate investment product that could be packaged and sold. This model offered an alternative for attracting high-tech science and industry that went beyond mere clustering.

Singapore has been called a "landlord state" for its policies in housing its population.[10] But the role of JTC suggests the importance of Singapore's "landlordism" for its industrial and economic development too. Management of industrial space came to play a central role in making Singapore rich. Scholars have demonstrated the multifarious ways in which Silicon Valley innovation was driven by the "entrepreneurial state," making critical investments in research and development.[11] In Singapore's case, "state capitalism" played an even more dominant role, building and managing discrete scientific and technological spaces that it could replicate both internally, onto Singapore's own landscape, and externally, onto other territories. Science and technology became valuable not just for their own sakes but also as part of a package that Singapore could sell and was critical to Singapore's image and ideology of a modern, incorruptible, successful, and profitable state. Science and technology became important components in Singapore's global brand.

The Origins of JTC

In 1960, the Dutch economist Albert Winsemius was dispatched by the United Nations (UN) to Singapore as part of a UN Development Program team. Winsemius and his colleagues were asked to advise on Singapore's capacity for industrialization and make recommendations for how to develop the city's economy. His 1961 report suggested that Singapore pursue industrialization driven by foreign investment and export to foreign markets.[12] The government was quick to respond. Goh Keng Swee, Singapore's minister for finance, proposed transforming the swampy and sparsely populated area of Jurong, in Singapore's west, into a 69-square kilometer industrial zone called Jurong Industrial Estate (JIE).[13] By August 1961, the government had secured investment from a Chinese Indonesian businessman, Goh Tjoei Kok, to begin development of National Iron and Steel Mills as the flagship factory on the site.

Jurong's industrial development was to be overseen by the newly formed Economic Development Board (EDB).[14] Also the brainchild of Goh Keng Swee, EDB was endowed with $100 million to develop infrastructure and other supports for industrialization.[15] EDB's activities included attracting foreign and local investors, managing investment and lending, technical consultation on projects, and overseeing the provision and management of land for industry.[16] After achieving full independence in 1965, the Singapore government redoubled its industrialization efforts, aggressively pursuing foreign investors. On the one hand, the imminent departure of the British Army (announced in 1967)—a major employer of locals—posed a significant challenge. On the other hand, the intensifying Vietnam War provided a crucial opportunity for the island to become a site for American resupply and repair. Through strategic investment and the development of infrastructure and facilities, the government fostered new industries such as shipbuilding, ship repair, and petrochemicals.

By 1967, JIE had attracted foreign and local firms specializing in medium to heavy industries. This amounted to a total foreign investment of $178 million and the creation 6,500 jobs.[17] With JIE already growing rapidly and the need to expand beyond Jurong, the tasks associated with estate management were becoming increasingly complex. In 1968, Goh announced the government's intention to carve away various functions from the EDB: financing would now be managed by a newly formed Development Bank of Singapore (DBS), and the development and management of the industrial estates would fall to the newly created Jurong Town Corporation (JTC).[18]

Under the Jurong Town Corporation Act (1968), JTC would have a chairperson and a small group of other members, appointed by the relevant minister, acting as public servants but holding the powers of a corporation. The government would provide funds to the corporation in order to "develop and manage industrial estates and sites in Singapore or elsewhere" and to "provide facilities to enhance the operation of industries and social amenities for the advancement and well-being of persons living and working in such industrial estates and sites."[19] Moreover, the corporation was empowered to compulsorily acquire, lease, or sell land; to develop, sell, or lease housing; and to provide loans.

Inheriting EDB's assets and mission, JTC quickly set to work expanding and upgrading JIE and creating new industrial estates in other parts of the island. In Jurong itself, JTC took over or commenced the construction of wharves, railways, roads, and water and electricity supplies.[20] The development of a deep-water pier as well as cargo offloading and storage facilities formed the core of the new Jurong Port, linked by rail and road to both Singapore and the West Malaysia railway network.[21] Following a model adapted from Hong Kong, JTC also took over the construction of "standard" or "flatted" factories. These were multistory, multipurpose factory spaces that could be rented out by businesses that did not need specialized floorplans or builds. Between December 1968 and December 1969, JTC rapidly increased the size of the estate (by over 30 percent) and grew the number of operational factories from 153 to 202.[22]

Oil-refining operations around Jurong were also vital to Singapore's early independent economy. Founded in 1969, the Singapore Petroleum Company (SPC) was a joint venture between American Oil Company (Amoco), Oceanic Petroleum, and DBS.[23] SPC led Singapore's drive to become a hub for processing oil coming from Indonesia and the Malay Peninsula.[24] Through DBS, the Singapore government retained a direct stake in the petrochemical industry. As with the development of science in Texas described in this volume (see Peter Westwick, chapter 10), oil became one important source of revenue that could be redirected toward investment in other emerging industries, including high-tech ones.[25] The state capitalist model powered cycles of investment that would fuel technoscientific development.

Beyond JIE, JTC took over the management of industrial estates in Kranji and Sungei Kadut, Kallang Basin, Tiong Bahru, Redhill, Tanglin Halt, Tanjong Rhu, and Kampong Ampat. At these sites, JTC embarked on rapid programs of flatted factory construction, aiming to create jobs for tens of thousands of workers. Within JIE, JTC assumed responsibility for housing workers, building and managing thousands of worker dormitories and apartments on the estate. During the 1970s, JTC expanded this para-industrial development, constructing housing on and adjacent to many of their industrial sites. By 1978, JTC housed over 93,000 workers in their facilities.[26] The development of recreational facilities (aimed at workers and their families) also became an important part of JTC's mission. This included commercial premises (for shops, supermarkets, restaurants, and department stores) and market and food (hawker) centers. The development of Jurong Park (completed 1976) became a flagship JTC project, expanding the corporation's work to include landscaping and the development of sports facilities.[27]

Through the 1970s, JTC increasingly focused its efforts on the development of high-tech industries. This was consonant with the government's aim to move Singapore's economy from low-value manufacturing toward higher-value add goods and services. Of particular importance in this respect was the influx of multinational microelectronics firms into JTC's factories and estates in the late 1960s and early 1970s. Efforts by EDB officers abroad and in Singapore succeeded in convincing several large US and Japanese firms to establish assembly operations in the city state.[28] Texas

Instruments moved into a factory in the Kallang Basin in 1969; in 1970, Hewlett Packard rented a flatted factory space in Redhill; General Electric followed, leasing several factory spaces in Kallang through the early seventies.[29] By the early 1980s, Singapore had transformed itself into a hub for the electronics and electrical industries. The establishment of several Seagate plants (the first in JTC's Kallang Bahru Industrial Estate in 1982) made Singapore into the premier global site for hard-drive manufacturing. By 1980, the focus of JTC's work became planning for, cultivating, and managing the growth of these high-tech industries.

From its inception in 1968, JTC played a dual role. On the one hand, it was an arm of the government, developing and promoting state interests in industrial development. On the other, it was acting as a property developer, acquiring, developing, and managing real estate for multinational clients. Already in these early years, too, JTC's remit had quickly expanded beyond merely building and operating factories—housing, recreation and leisure, and commercial spaces were already part of its portfolio. High-tech industrial spaces were beginning to be packaged together with these infrastructural and social elements.

Singapore Science Park

In 1979, JTC Chairman I-Fang Teng announced a "masterplan" for the further industrial and technological development of Singapore during the 1980s. The masterplan aimed to shift Singapore toward high-tech and high-value industries, particularly by upgrading the quality of industrial estates. "More aesthetically pleasing small industrial parks housing modern multi-story factories" would be a priority.[30] These would be served by improved transportation, infrastructure, and services and include more emphasis on planning and building "clean and green" spaces for industry.[31] The further decentralization of industrial estates, the deployment of more sophisticated town planning, and facilities for "skill and technology-oriented industries" were also important priorities.[32]

One of the centerpieces of the masterplan was JTC's vision for a science park. Tang labeled it the "most important" aspect of their plan.[33] Explicitly following the examples of Stanford and the Massachusetts Institute of Technology, the science park would be located at Kent Ridge, adjacent to the National University of Singapore (NUS). Like MIT's science park, situated adjacent to the Charles River, Singapore's science park was planned to be closely integrated with research and development activities at the university. Like the newly planned industrial estates, the science park would be situated within a "green, lush, and secluded setting."[34]

Tang promised that 70 percent of those employed at the park would be scientists, engineers, economists, and accountants and that multinational and Singapore companies would be able to make use of the "research services and facilities" provided at the park.[35] The initial plan for the park called for 115 hectares (285 acres) of space that would be shared between university researchers, industry, and government research institutes. Space was set aside for the Singapore Institute of Standards and Industrial Research (SISIR), the National Productivity Board (NPB), and the Applied Research Corporation. The Kent Ridge site was proximate not only to NUS but also to other educational institutions (including Singapore Polytechnic) and to JTC flatted factory complexes at Ayer Rajah, Telok Blangah, Tanglin Halt, and Redhill.[36]

The design of spaces for research at Singapore Science Park mirrored JTC's industrial developments. The park included significant room for "standard type premises" that would act, like flatted factories, as "starter units" that could be rented to companies or other research operations that wanted to establish operations quickly and with minimal investment.[37] Like JTC's industrial sites, the science park included more than workspaces. As well as developing transportation and telecommunications infrastructure for the site, JTC's plans afforded the construction of a specialized administration center, a recreation center, "middle-income" housing, restaurants, parks, a ferry terminal, and factories.[38] A computer center, science and engineering libraries, and convention and seminar facilities would be shared with NUS.[39]

The financial structure of the Science Park also resembled JTC's other facilities. JTC operated a tender system through which construction work within the park was offered to contractors. In April 1982, the first contract was won by Evan Lim, a local construction and civil engineering company involved in several other JTC, the Housing and Development Board, and Public Utilities Board projects.[40] Land in the park remained in the possession of JTC. Prospective tenants had two options. Those wishing to construct their own laboratory or research facilities could lease "prepared and serviced sites" (empty plots with provision for transport, utilities, and telecommunications) for $11.60 per square meter per year, with a lease of thirty years. Companies or organizations interested in occupying JTC's "standard" R&D buildings (various sizes available) could do so by leasing space for $205 per square meter per year.[41] Like JTC's industrial tenants elsewhere, the corporation acted as the landlord for the site as well as the developer.

By 1982, as ground was broken for construction in the park, JTC established an "admissions committee" to screen potential tenants. Consisting of representatives from NUS, EDB, JTC, SISIR, and the Science Council, the committee sought to grant tenancies to companies that would make the most significant contributions to "high technology production."[42] Of particular interest were companies specializing in advanced electronics development, product testing and analysis, medical and scientific instrument development, biotechnology, process control and automation, optical and opto-electronic applications, and chemicals and plastics research.[43] In March, the park confirmed its first industrial tenant—Det Norske Veritas—a Norwegian company specializing in shipping. Veritas would set up a $10 million laboratory and data-processing complex for marine insurance and the oil and petroleum industries. Among other projects, Det Norske began a large-scale study of environmental conditions in Southeast Asian waters and the South China Sea, surveying physical, chemical, and biological conditions in the nearby oceans. Such a study would contribute to corrosion control and the safe design of ships and oil rigs for the region.[44]

Alongside Det Norske, the newly formed National Computer Board (NCB) would take up residence in a specially designed 10,000-square foot building that would also house two private computing software and education companies—Singapore Computer Systems Pte Ltd. and Systems Education Centre Pte Ltd. These would form the core of a "Software Technology Centre" within the Science Park.[45] Such arrangements exemplified the way in which JTC hoped to foster the growth of private enterprise by seeding the park with government agencies and institutes.

By early 1984, however, although park construction continued apace, there were worries that demand might not be as strong as expected. Vincent Yip—head of the EDB's advanced technology and research department—was called on to defend the Science Park against rumors that it was not attracting tenants and that the park was "not progressing well."[46] EDB was responsible for promoting the park overseas and selling space to multinational companies that might wish to lease space. Despite their efforts, the park had attracted only six tenants, occupying only 30 percent of the space available in phase I (and only 6 percent of the planned total space). Yip argued that the park was a "long term project and not a piece of real estate like shopping complexes."[47] EDB would intensify its efforts.

After opening its doors in 1984, the Science Park's fortunes improved, and its popularity grew quickly. New tenants included Scientech-Intraco (robotics), Robot Leasing and Consulting (robotics), Plantek International (plant tissue culture), Tata-Elxsi (software and hardware), Mentor Graphics (software), Austek Microsystems (microprocessors), Seagate (hard drives), Robin Electronic Investment (R&D in microprocessor application), and Everbloom Mushrooms (culturing shiitake mushrooms for medical use).[48] Alongside the existing tenants, this represented a strong mix of well-established industries (like petrochemicals and shipping) and aspirational industries for Singapore (like software and biotech). There was a rapid uptake of occupancy after 1984, with the number of private-sector tenants growing from nine to thirty-four.[49] Phase I of the park project had involved just 29 hectares of the 125-hectare site. Indicating their confidence in further growth, in early 1985, JTC began clearing land for the second (22-hectare) phase and acquired land for the fourth (and final) phase.[50]

Both spatially and conceptually, the science park was aimed directly toward supporting Singapore's economic development. The park was integral to the state's broader plan to upskill workers, attract multinational companies to the island, and move into high-tech industries that would bring higher value for Singapore workers and companies. Like the Stanford Industrial Park and MIT's Science Park (on which it was modeled), the park aimed to draw research workers—and Singapore's educational institutions—more closely into serving the ends of industrial development. But the development of the Singapore Science Park can also be seen as a more self-contained space—as a place that brought together laboratories, support services, housing, recreational facilities, and commercial spaces all into one zone, controlled and managed by JTC. Unlike the Stamford and MIT spaces, the Science Park could be sold to potential tenants as a complete package for living, working, and playing.[51]

From Science to Business

JTC's operations were incredibly lucrative. The success of the Science Park and other estates led to a surplus of $248 million for the fiscal year of

1988–1989 (an increase of 60 percent from the previous year). In 1989, JTC held over $2.2 billion in fixed assets.[52] This was in part because their cultivation of science and high-tech industries had expanded beyond the narrow mission of managing industrial spaces. JTC increasingly operated as a real estate developer and landlord, creating and managing highly curated zones for specific forms of living and working.

From the early 1980s, JTC began several other projects that expanded its holdings beyond their industrial estates. In the early 1980s, JTC's ambitions in residential real estate expanded beyond housing factory workers. In 1982, JTC announced "The Lakepoint," a condominium complex near Jurong Lake that included two hundred maisonettes, forty townhouses, forty-four apartments, and twenty penthouses.[53] These luxury residences would be aimed at "executives"—especially those from multinational companies—who came to live in Singapore. Companies sited on JTC's industrial estates would be able to purchase two residences for their employees. Since part of JTC's role was to attract and retain multinational companies on the island, providing desirable accommodation for high-level (and often foreign) employees was perceived as a critical component of this work. If executives or managers could not live comfortably, JTC's projects would be compromised. In early 1983, JTC reported strong demand for the development. This "overwhelming response" led to plans for three more developments, beside and across the road from The Lakepoint.[54] The majority of applications had come from "professionals, executives, and businessmen" working in Jurong. Although JTC stressed that Jurong-based employees would continue to receive priority, the development was open to "others irrespective of citizenship, marital status, and income."[55] JTC was now competing with private developers on the real estate market (and offering substantially cheaper prices).

Housing was just one kind of real estate required by high-tech businesses in Singapore. In April 1988, several officers of the corporation—together with officials from EDB and the Planning Department—embarked on a study trip to visit business parks in the United States, the United Kingdom, France, and Japan. The object of this trip was to develop an understanding of how business parks operated and to consider adapting them to Singapore's needs.[56] Although an expansive campus-style park might not be suitable for land-scarce Singapore, the *Business Times* argued that buildings and facilities suitable for high-tech industries did not exist in Singapore and that a business park would solve this problem.[57] JTC—the editors suggested—with "significant land banks at its disposal," should set up a such a park.[58]

The notion of a business park was commensurate with JTC's promotion of science and high-tech industries. These "mixed-use environments" would include attractive landscapes and "aesthetically designed" buildings to support high-tech industries.[59] This would, JTC's Annual Report argued, "harmonise very well with the economic objective of evolving Singapore into a global city with total business capabilities."[60] More specifically, JTC reasoned that multinational businesses wanted to integrate several of their business operations (such as design, development, manufacturing, and marketing) in a single location. This required appropriate buildings and work environments.[61] In other words, business parks would allow Singapore to maximize the investments that overseas companies (particularly high-tech companies) would make in the island.

Following the study trip, JTC established a working group to consider four possible sites for development as a business park: Jurong East, Loyang, Changi North, and Changi South. Quickly, they selected a zone in Jurong East adjacent to public transport and expressways as the most suitable site. The International Business Park encompassed 100 hectares (later reduced to 40 hectares) and would include a 17-hectare "International Technology Centre" and the 6.4-hectare "International Merchandise Mart" (a wholesale and warehouse center) slated to open in mid-1990.[62] JTC's investment in the first phase would be over $100 million.[63] As with the Science Park, JTC retained ownership of the land and acted as landlord; multinationals could now rent flatted factories for their manufacturing and ready-made offices for their corporate operations, all from JTC.[64]

Aside from its business park, JTC was continuing to develop a range of industrial estates across Singapore. Many of these specifically aimed to attract high-tech industrial tenants. Ang Mo Kio Industrial Estate, for example, had been established in 1979 as an "industrial parkland" that would be integrated into a residential town. The $13.3 million project aimed to create a "garden industrial estate" that would serve as a template for future projects.[65] Through the 1980s, Ang Mo Kio estate became a particularly important site for high-tech manufacturing, especially microelectronics and semiconductor-related industries. In 1983, the Italian electronics firm SGS-Ates announced plans for a $60 million factory in the estate.[66] This was followed by Motorola's establishment of a plant and a research and development center, as well as a $10 million investment by General Motors (making integrated circuits for car radios).[67] In a further effort to promote such high-tech zones, JTC instituted a policy allowing nonmanufacturing firms to acquire space for their operations. The "JTC Accommodation for Selected Service Industries" allowed software, engineering consultancy, and technical and laboratory testing companies to be allocated space in flatted factories in Ang Mo Kio and four other selected industrial estates.[68]

After a temporary dip in demand for factories in the mid-1980s, by 1987 demand for JTC space once again accelerated. Miniscribe Peripherals, Maxtor Singapore, Apple Computers, and Micropolis were among the new arrivals in Ang Mo Kio. In the 1988–1989 financial year, Ang Mo Kio estate housed eighty-seven firms and employed over 13,000 people, mostly in the electronics industry.[69] High levels of occupancy across the estates (including 96 percent in Ang Mo Kio itself) encouraged JTC to go on another building spree, adding flatted factories and more space for lease on the Ang Mo Kio estate and elsewhere.[70] In 1987, five new blocks of factories were planned for Ang Mo Kio, Tai Seng, Ayer Rajah, and Kallang Basin.[71] A global uptick in demand for electronics led to a local boom in land prices, rents, and construction. As a landowner, developer, and landlord, JTC was the direct beneficiary of this growth.

In the second half of the 1980s, Singapore's state capitalism began to adopt new forms. In line with Reaganite and Thatcherite ideologies, the Singapore government began a campaign of privatization of government services. Unlike in the United States or the United Kingdom, however, such privatization usually made provision for the retention of significant government ownership and control, usually through the state's sovereign wealth fund, Temasek Holdings.[72] In 1989, JTC created two subsidiary companies—Technology Parks Pte Ltd. and Singapore Leisure Industries Pte Ltd. Technology Parks, commencing operations in April 1990, would manage Singapore Science Park and the Ang Mo Kio Industrial Estate. JTC's "leisure facilities"—the Chinese and Japanese Gardens, Jurong Park, the Jurong Country Club, and the Raffles Country Club—would be taken over by Singapore Leisure Industries. This "corporatization" of some of JTC's business areas would "pave the way for eventual privatization of JTC's assets and activities" and allow the subsidiaries to be more responsive to market demands and able to enter joint-venture projects more readily.[73] The corporatization of key parts of JTC's operations only deepened the complex entanglements between the state and the private sector. Although this move seemed to divide JTC operations between "technology" and "leisure," in fact both companies were able to pursue projects of any kind, including by collaborating with one another. In practice, by allowing the subsidiaries to act more like private developers, the move had the effect of deepening connections between technoscientific developments, on the one hand, and real estate, economic, and business development, on the other. The new corporate structure only reinforced the extent to which technoscientific spaces could be configured within a model of neatly packaged real estate development expertise. This expertise could now also be exported and deployed overseas.

The Singapore Model: JTC in the World

By the late 1980s, there was a huge diversity of scientific research and development work being undertaken on JTC's properties. In addition to electronics, this included both fundamental research on biotechnology, radar, aviation, optics, computer science, and robotics, as well as more applied work on shipping, food production, logistics, petroleum products, plastics, and chemical products.[74] The Singapore government certainly hoped that its investments in science and technology would pay off in terms of new inventions, a more highly skilled workforce, and national prestige. However, JTC's mission was far narrower: to develop land, buildings, and infrastructure for research and industrial usage. Success meant selling or renting land or building to science and tech-oriented occupants, attracting foreign investment and expertise to the island. However, their work also provided the corporation with significant experience as developers of such high-tech parks. This generated a further opportunity: for JTC to deploy its expertise overseas, building and managing parks offshore. This idea was first mooted in the mid-1970s, with Singapore looking toward its near neighbor, Indonesia. Singapore lies within a small island chain, known as the Riau Archipelago, stretching southward from the tip of the Malay Peninsula. The Indonesian islands of Batam and Bintan lie within 50 kilometers of Singapore (figure 9.1).

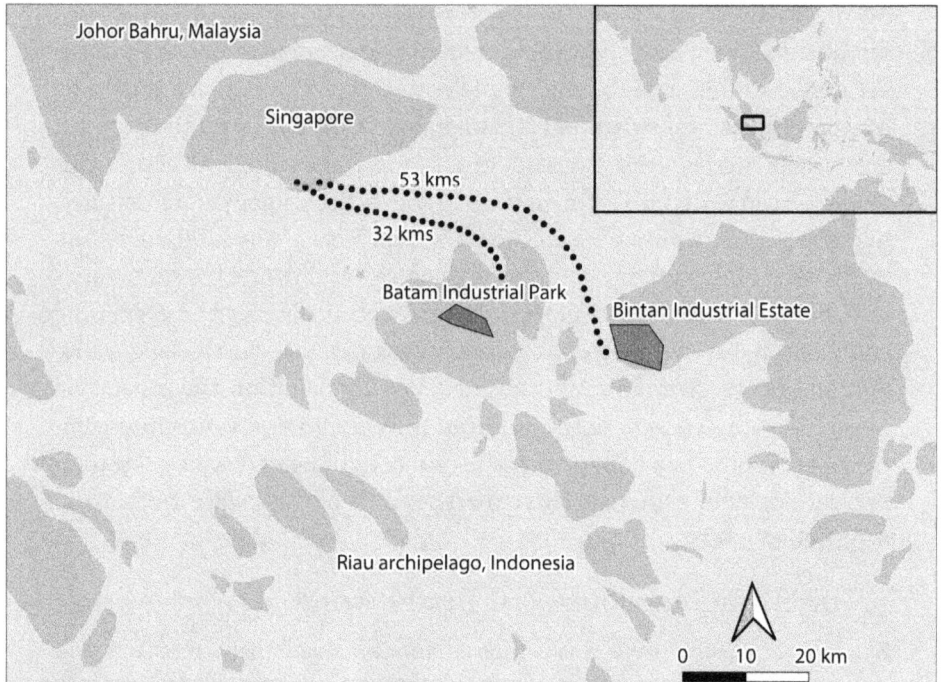

Figure 9.1 Singapore and the Riau Archipelago (Indonesia) showing ferry connections between Singapore, Batam, and Bintan Islands. Image by author

In October 1980, the governments of Indonesia and Singapore signed a formal agreement for cooperation in the development of Batam. This agreement established policies for trade, communication, banking, taxation, and immigration between Singapore and Batam. But most important, it created the framework within which Singapore would take the lead in "industrializing" Batam. "Batam has potential for development as it has a strategic geographical location," Singapore Minister for Trade and Industry Goh Chok Tong reminded the audience at the signing event. "It is, moreover, not cluttered by industries, settlements, or other unplanned physical structures. It is better than what Singapore's Jurong Industrial Estate was before its development in the Sixties."[75] For Indonesia, the industrialization of Batam would provide foreign investment and jobs for locals. For Singapore, it provided access to cheap, "uncluttered" land and cheap labor.[76]

JTC's formal involvement in Batam, however, took almost a decade to gestate. In January 1990, a deal was signed between Jurong Environmental Engineering (a wholly owned subsidiary of JTC), the Singapore Technologies Industrial Corporation, and the Indonesian company P. T. Herwido Rintis. The $400 million deal aimed toward construction of a 500-hectare industrial park to serve electronics, electrical, and light industries.[77] The Indonesian partner, Herwido Rintis, was part of the "Salim Group," established by Liem Sioe Liong (also known as Sudono Salim). With close personal ties to Suharto, Liong had expanded from food-related industries (including pig and crocodile farms on Batam) into property development in the 1980s.[78]

In February 1990, Lee Hsein Loong, Singapore's minister for trade and industry (and Lee Kuan Yew's son), traveled to Batam for the official opening of what became known as the BatamIndo Industrial Park.[79] BatamIndo, like its counterparts in Singapore, would provide not just factory space but also logistics and port facilities, housing for workers and executives, and recreational facilities. High-quality infrastructure was particularly important too. Given the lack of electrical and telecommunication facilities in Batam, in March 1990, the EDB proposed laying an undersea cable to directly pipe 50 megawatts of power as well as communications links from Singapore.[80] Park administrators also offered business services such as dealing with license applications, auditing and taxes, and recruitment of workers.[81]

Uptake of space in BatamIndo was rapid. In May 1990, the Japanese Sumitomo Electric Industries had signed up for the first tenancy.[82] By later the same month, around twenty multinationals signed letters of intent, including Philips, Western Digital, Sony, Toshiba, and Thomson Electronics.[83] BatamIndo offered a unique arrangement that coupled cheap and plentiful labor supply to many of the advantages of doing business in Singapore. Tax and free trade arrangements meant that goods manufactured in Batam could be shipped from Singapore, taking advantage of port and logistics facilities there. A shift in foreign labor regulations also meant that workers could also be trained in Singapore without counting toward the government caps and levies on foreign workers.[84] The park served as seamless-as-possible extension of Singapore onto foreign territory. JTC and its partners had developed a zone for high-tech industry that was beyond

Singapore's national borders yet an extension of Singapore's favorable business and technological environment.

Plans were soon developed not only to expand BatamIndo by a further 1,000 hectares but also to establish a second park on the neighboring island of Bintan. By December 1990, planning for a large industrial estate for textiles, woodworking, and food processing was in progress.[85] These developments were closely linked with the planned expansion of tourist and luxury residential facilities on both Batam and Bintan. In parallel with BatamIndo's development, the Singaporean company Tuan Sing Holdings had invested $1 billion in developing "Waterfront City" in the northwest of Batam. "Caribbean of the East—a heady mix of Florida-style resort living, Bali beach casual chic, Queensland Gold Coast excitement, ancient Japanese charm, and Indonesian cultural heritage,"[86] Waterfront City included two marinas, a golf course, resort hotels, and a theme park. An advertising feature noted that the development was intended to "complement the $400 million Batam Industrial Park," presumably by catering to the white-collar Singaporeans and foreign workers who would be relocated to Batam. In Bintan too—after a consortium of Singapore companies constructed a passenger ferry terminal—the Salim Group began development of a large swathe of the north of the island into tourist resorts. Club Med and Banyan Tree were early investors.[87]

The developments on Batam and Bintan allowed tech-oriented multinationals to continue to flourish in Singapore, even as wages rose. The sites leveraged JTC's existing experience in developing sites for high-tech industries, particularly electronics, to create "special zones" that were neither entirely inside nor entirely outside Singapore. In achieving this, Singapore continued to benefit from the presence of these multinationals—which maintained research and development and corporate operations in the city state—while offshoring factories (which could be pollutive) and cheap labor (making way for higher-wage employment on shore). The close links between BatamIndo and real estate development in Indonesia—via both the Salim Group and Tuan Sing—again suggest the ways in which science and technology were mobilized as part of a package of development that included industry, infrastructure, housing, recreation, and other services. Technoscientific development and real estate development went hand in hand overseas, as they had in Singapore. As well as generating continued investment from multinationals, JTC's sites now also served to strengthen Singapore's relationship with the Indonesian government, now serving Singapore's foreign policy—as well as economic—goals.

Suzhou

JTC's ventures in the Riau Archipelago were a warm-up for further overseas activities. In 1988, JTC began negotiating with Thai authorities to construct a 400-hectare light-industrial estate north of Bangkok. Yeo Seng Teck, JTC's chairman, had watched closely as several multinationals had relocated to Thailand in search of cheaper labor—JTC could follow and support them.[88] JTC would "facilitate their investments, ensuring the same standard of infrastructural facilities and factory buildings they are accustomed to in JTC estates, and provide the link to Singapore."[89] In addition, JTC would support Singaporean companies that wished to expand into Thailand, especially those already working closely with multinational companies. "JTC has acquired know-how and expertise in planning, developing, and running factory estates and is prepared to export this service internationally," Yeo explained.[90]

As well as its near neighbors in Southeast Asia, JTC had long had a close eye on China. In 1982, as China began to open its economy, Singapore made a major investment into the development of the Chiwan oil port near Shenzhen. Among the group of investors was Jurong Marine Base (a JTC subsidiary responsible for port development).[91] As China's economy expanded, Chinese businessmen approached JTC for help with other projects. In 1984, for example, Wang Guang Ying requested JTC's assistance with a new industrial estate, to be located in the Zhuhai Special Economic Zone.[92] In the early 1990s, the JTC subsidiary Jurong Environmental Engineering also invested in the Yuan Hong Industrial Park in Fuqing (Fujian Province).[93]

JTC's largest overseas project, however, took root in Suzhou, 90 kilometers to the west of Shanghai. In 1992, China's architect of reform and opening, Deng Xiaoping, suggested that China should "learn from the experience" of Singapore.[94] This meant particularly learning from Singapore's experiences in industrial and economic development. Like the parks in Indonesia, the Suzhou development would allow Singaporean and multinational companies to take advantage of cheap labor. Singapore would benefit from their continued relationship with these companies (often retaining headquarters or administrative offices in Singapore) while also profiting from the development itself and by acting as a broker with locals.

Younger Singaporeans would also develop closer ties with China and a greater familiarity with doing business there.[95]

Chan Soo Sen, the CEO of China-Singapore Suzhou Industrial Park Development Company, put the proposition for companies quite bluntly: "Look, all of you want to come to China because that is where the market is. But all of you are also scared of China because of the cultural differences, the perceived lack of transparency, the lack of investment guarantee and that kind of thing. If I have a Singapore in China, so you come to this industrial park, you can invest in a Singaporean environment. You have a launching pad into China, a piece of Singapore in China."[96]

On September 16, 1993, Singapore incorporated the Singapore-Suzhou Township Development Pte Ltd. A range of Singapore companies (including DBS Land, EDB Capital, Keppel Corporation, and JTC International) contributed 65 percent of the US$300 million in capital while Chinese investments made up the remainder.[97] As with BatamIndo, Suzhou targeted high-tech industries. Eventually, park tenants included manufacturers of semiconductors (e.g., AMD), medical devices (e.g., Beckton Dickinson), pharmaceuticals (e.g., Lilly), and electronics (Samsung).[98]

But the 70-square kilometer site was also intended to be a "township," housing 600,000 people.[99] In particular, China was searching for a suitable model under which to foster its rapid development; some worried that the successes of the Shenzhen Special Economic Zone—China's first experiment in reform and opening—had "been achieved at the expense of social order"; Singapore perhaps offered a more suitable balance between "economic growth and social cohesion."[100] As such, the park project was supported by government and corporate agreements that would foster transfer of such knowledge. In February 1994, the Singapore and Chinese governments pledged their support for the project, creating a Joint Steering Committee, chaired by Singapore Deputy Prime Minister Lee Hsein Loong and Chinese Vice Premier Li Lanqing.[101] This was backed up by a second agreement between JTC and the Suzhou Municipal Government that outlined how Singapore experience in economic and public administration would be deployed in Suzhou. This included "software transfer exercises" through which JTC and other Singapore agencies would facilitate transmission of knowledge and "philosophies" about town planning, industrial and estate planning and development, environmental planning, control and management, and building control, design, and construction.[102]

Suzhou was a social project coupled to an industrial-technical one. While aiming to develop the Suzhou site commercially and industrially, JTC now offered a model for how to plan and develop an "international modern industrial township of world standard."[103] Chan Soo Sen likened the creation of this prosperous and cohesive society to an "industrial condo": a walled-off space that remained separate from the areas around it—a self-contained model for living and working. And this model, Lee Kuan Yew imagined, would be replicable. Within five years, "we will see the beginning of Jurong Town Corporation taking shape in Suzhou, and in ten years we should be able to repeat the project anywhere else in China or Vietnam."[104]

The Suzhou project did not fare well.[105] Competition with both a Chinese-led project (Suzhou New District Industrial Park) and another Singaporean project in the same area (Wuxi Industrial Park) led to significant losses. By 1999, the Suzhou park consortium had lost $90 million and Singapore transferred its majority stake and management control to the Chinese partners.[106] Nevertheless, Singapore and JTC did continue to expand internationally—building other parks in Vietnam, India, and the Philippines in the 1990s.[107]

The expansion of JTC operations into Southeast Asia and China demonstrates the ways in which science and technology had become—for JTC and Singapore—exportable products. Packaged with real estate, infrastructure development, and social "software," research and development had become a critical part of a "Singapore model" that could be marketed and sold overseas. This was in part a consequence of the tight coupling between JTC (as a set of private corporations) and the Singapore state. Singapore's "state capitalism" meant that state interests in economic and industrial development, and even foreign policy, could be executed through the creation and sale of a real estate product.

Zoning Science and Technology

At the beginning of the 1980s, JTC's efforts had been primarily directed toward the development of infrastructure for research, development, and high-tech industry. By the early 1990s, JTC's mission had morphed and expanded into the development of real estate and the accompanying "software" for managing land and people. This software was deployed not only within Singapore but also, increasingly, overseas. Putting this another way, what began as an effort to create industrial clusters and to boost investment by multinational corporations was transformed into a mission to developed fully fledged spaces for science and technology, replete with infrastructure, housing, recreational facilities, and particular sociopolitical programming.

We can trace this corporatization of science and technology through a kind of phylogeny of JTC and its affiliated companies. Figure 9.2 suggests the ways in which JTC's activities increasingly came to include real estate development and operations, especially overseas. Ascendas Real Estate Investment Trust and Surbana Jurong, in particular, have emerged as international property development corporations. But this is not merely a story of increasing corporatization. Rather, JTC's history demonstrates how science and technology became "zoning technologies" for Singapore.

Aihwa Ong and Carolyn Cartier, among others, have developed the notion of the "zone" as a liminal space, often existing across borders, having their own rules around labor, taxation, land use, and so on. Zones, including industrial zones and business parks, are powerful enablers of global capital, they argue.[108] Building on this, Easterling argues that globalization is being driven by an "infrastructural spatial matrix"—a kind of zone

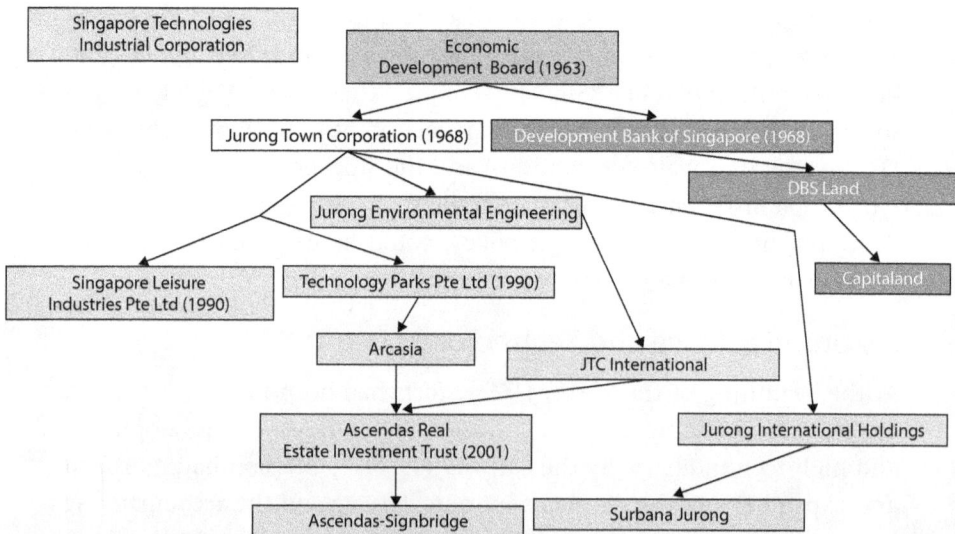

Figure 9.2 JTC Group, antecedents and descendants

enacted through plans, blueprints, and policies. These are often developed and implemented by nonstate actors and become catalysts for repeated patterns of spatial development across the globe. By organizing space in particular ways, zones often project state and corporate power across borders.[109] Understanding JTC's operations in terms of "zones" can shed light on not only their imbrication with global capital but also their ability to project Singapore's economic and political power to Asia and beyond. JTC's science and industrial parks became "zones" that could be replicated and deployed so as to order space and people in ways that were suitable for capturing investment. "Research and development" was a critical mechanism through which Singapore created and maintained special zones, operating with specialized infrastructure, financial rules, and social arrangements. These zones could be blueprinted and extended overseas, operating both as quasi-territories of Singapore and as lucrative investment and development opportunities.

Figure 9.2 also suggests the complicated relationship between the JTC (and its cousins) and the state. JTC began its life as a quasi-governmental body. Although increasingly "corporatized," many of JTC's daughter organizations remained at least partially (if not majority) owned by Temasek Holdings.[110] Under "state capitalism," JTC can be viewed as a state-owned enterprise and its operations as closely aligned with state interests. But for JTC at least, state interests have been tightly bound to increasing its own profitability. Land development, multinational investment, property management, and JTC's other activities have had both public and private benefits, realizing revenue for Singapore corporations but also bringing further investment and jobs to Singapore at large. Greed was good for the state, too.

Recent scholarship in the history of science has given more attention to the role and development of science and technology in nondemocratic contexts. Rather than assuming a special relationship between science and democracy, scholars including Benjamin Peters, Tiago Saraiva, and Sigrid Schmalzer, among others, have shown ways in which science and technology has variously aligned with communist and fascist projects.[111] Peters, for instance, demonstrates how electronic network-builders in the Soviet Union often came to operate in overtly capitalist modes while Saraiva describes the alignment of agricultural scientific projects with the autarkic desires of fascist states.

Singapore's "state capitalist" regime came to align science and the state in still different and unexpected ways. Through JTC, science and technology became spatially linked to the state's economic, (geo)political, and social goals. The packaging of space into cookie-cutter science and business parks mobilized research and development in the service of state-driven, greedy, hypercapitalist expansion. These spaces were not merely industry clusters like in Silicon Valley or the Greater Boston Area but highly curated and self-contained spaces. Peter Westwick notes how, in Texas, the shift toward a science and technology-driven economy was accompanied by a rightward political turn in the state.[112] Although the Singapore context is very different, there too, the planning, operation, and staffing of science and business parks complemented and reinforced the state's top-down models of governance and its rigid social controls. Carefully planned and regimented zones for science and high-tech industry were perfect microcosms (or "living labs") for the nation as a whole.

Singapore(ans) became rich, in part, by creating these "model" techno-scientific zones that attracted overseas investment, first by high-tech multinational manufacturers and later by financial and other service industries. Historians have long understood Singapore as a technocracy—a state run by engineer-bureaucrats and within which technological progress and success (including infrastructure development) have played a central role in the state's success. But this history of JTC illuminates other ways in which science and technology played a role in generating Singapore's wealth, through the building of special zones that could be used to avariciously expand Singapore's economy and effective territory.

NOTES

1. Singapore's scientific ambitions were signaled by talk about the possibility of winning a Nobel Prize: "Will S'pore Ever Land a Nobel Prize?" *Straits Times*, October 13, 2007. Singapore also made Nobel Prize winner Sydney Brenner an "honorary citizen" in 2003, and in 2009 appointed Bertil Anderson, former chair of the Nobel Committee for Chemistry, as the president of one of Singapore's universities.

2. Catherine Waldby, "Singapore Biopolis: Bare Life in the City-State," *East Asian Science, Technology & Society* 3, nos. 2–3 (2009): 367–383; Gregory Clancey, "Intelligent Island to Biopolis: Smart Mind, Sick Bodies, and Millennial Turns in Singapore," *Science, Technology & Society* 17, no. 1 (2012): 13–35.

3. Lee Kuan Yew, *From Third World to First: The Singapore Story, 1965–2000* (New York: Harper, 2000).

4. For a comprehensive account of Singapore's industrialization, see Garry Rodan, *The Political Economy of Singapore's Industrialization: National State and International Capital* (New York: Palgrave Macmillan, 1989).

5. Chua Beng Huat, *Liberalism Disavowed: Communitarianism and State Capitalism in Singapore* (Singapore: NUS Press, 2017), chap. 5.

6. On industrial clustering, see, for instance, AnnaLee Saxenian, *Regional Advantage: Cultural and Competition in Silicon Valley and Route 128* (Cambridge, MA: Harvard University Press, 1994).

7. See Robin Wolfe Scheffler, chapter 8 in this volume.

8. On clustering in Singapore, see Poh Kam Wong and Annette Singh, "The National System of Innovation in Singapore," in *Small Economy Innovation Systems: Comparing Globalization, Change, and Policy in Europe and Asia*, ed. Charles Edquist and Lief Hommen (Cheltenham, UK: Elgar, 2005), 71–112; Poh Kam Wong, Yuen-Ping Ho, and Annette Singh, "Toward an 'Entrepreneurial University' Model to Support Knowledge-Based Economic Development: The Case of the National University of Singapore," *World Development* 35, no. 6 (2007): 941–958; Poh Kam Wong, Yuen-Ping Ho, and Annette Singh, "Industrial Cluster Development and Innovation in Singapore," in *From Agglomeration to Innovation: Upgrading Industrial Clusters in Emerging Economies*, ed. Masatsugu Tsuji and Akifumi Kuchiki (Basingstoke, UK: Palgrave Macmillan, 2009), 50–116.

9. See Peter Westwick, chapter 10 in this volume. Unlike Texas, Singapore had no oil reserves of its own. However, by the late 1970s it had transformed itself into a regional hub for petroleum refining and storage. On the relationship between the oil industry and entrepreneurialism, see also Cyrus C. M. Mody, chapter 1 in this volume.

10. Gregory Clancey, "Hygiene in a Landlord State: Health, Cleanliness, and Chewing Gum in Late Twentieth Century Singapore," *Science, Technology and Society* 23, no. 2 (2018): 214–233.

11. Mariana Mazzucato, *The Entrepreneurial State: Debunking Public vs. Private Sector Myths* (London: Penguin, 2013). See also Margaret O'Mara, *Cities of Knowledge: Cold War Science and the Search for the Next Silicon Valley* (Princeton, NJ: Princeton University Press, 2004).

12. Loh Kah Seng, "Imaginaries of Jurong Industrial Estate, Singapore," *International Institute for Asian Studies: The Newsletter* 81 (Autumn 2018): unpaginated.

13. Seng, "Imaginaries."

14. Legislative Assembly of Singapore, *First Reading of the Economic Development Board Bill*, vol. 14 of *Debates: Official Report*, col. 1427; Legislative Assembly of Singapore, *Second Reading of the Economic Development Board Bill*, vol. 14 of *Debates: Official Report*, col. 1516.

15. Edgar H. Schein, *Strategic Pragmatism: The Culture of Singapore's Economic Development Board* (Cambridge, MA: MIT Press, 1996), 32.

16. Augustine H. H. Tan, "Official Efforts to Attract FDI: The Case of Singapore's Economic Development Board (EDB)," 1999 EWC/KDI Conference on Industrial Globalization in the 21st Century: Impact and Consequences for East Asia and Korea, August 2–3, 1999 (August 27, 1999), 1–2.

17. Tommy Koh, Timothy Auger, and Jimmy Yap, eds., *Singapore: The Encyclopedia* (Singapore: Editions Didier Millet in association with the National Heritage Board, 2006), 271.

18. "Goh Tells of Govt Plans to Speed Up Economic Growth," *Straits Times*, April 17, 1968, 8.

19. "Jurong Town Corporation Act," Part III, Section 12, Singapore Statutes Online, https://sso.agc.gov.sg/Act/JTCA1968/Historical/19870330?DocDate=19910125&ValidDate=19870330&Timeline=On&ProvIds=P1III-#pr11-.

20. *Jurong Town Corporation* (Singapore: City Press, 1968), 5–6, https://eresources.nlb.gov
.sg/printheritage/detail/011b18bb-3bd0-495c-9cdd-c35ab1bd1d8d.aspx.

21. *Jurong Town Corporation*, 5.

22. *Annual Report* (Singapore: Jurong Town Corporation, 1969), 5.

23. Lee Chiang Huat, ed., *Singapore Petroleum Company Limited 40 Years On: Looking Back* (Singapore: Singapore Petroleum Company, 2009), book 1, https://issuu.com/splash/docs
/spcbookchpt1_2_3_lores.

24. The importance of Singapore dramatically increased after the oil crises of 1973 and 1979; by the mid-1970s, Singapore was the third largest oil refining center in the world. Hing Ai Yun and Lee Kiat Jun, "Evolution of the Petrochemical Industry in Singapore," *Journal of the Asia Pacific Economy* 14, no. 2 (2009): 116–122.

25. See Peter Westwick, chapter 10 in this volume.

26. *JTC Annual Report 1977–78* (Singapore: Jurong Town Corporation, 1978), 27.

27. *JTC Annual Report 1977–78*, 30–31.

28. See Chin Bock Chan and Buck Song Koh, *Heart Work* (Singapore: Singapore Economic Development Board, 2002).

29. Hallam Stevens, "Fairchildren and Factory Girls: Gender, Family, and Space in the Singapore Electronics Industry," forthcoming.

30. *JTC Annual Report 1979–80* (Singapore: Jurong Town Corporation, 1980), 5.

31. Paul Jansen, "JTC Unveils its Master Plan for the Eighties," *Straits Times*, November 17, 1980, 1.

32. *JTC Annual Report 1979–80*, 6–7.

33. Soh Tiang Keng, "JTC Outlines Master Plan," *Straits Times*, September 18, 1979, 1.

34. *JTC Annual Report 1979–80*, 20.

35. Keng, "JTC Outlines Master Plan," 1.

36. *JTC Annual Report 1979–80*, 20.

37. *JTC Annual Report 1979–80*, 20.

38. "Science Park to Open Its Doors," 3; *JTC Annual Report 1982–83* (Singapore: Jurong Town Corporation, 1983), 21.

39. "What JTC Envisions," 3.

40. "Building of Science Park Begins on August 19," *Business Times*, April 8, 1982, 1.

41. *JTC Annual Report 1982–83*, 21.

42. Soh Tiang Keng, "Science Park to Open Its Doors in Second Quarter of Next Year," *Business Times*, March 4, 1982, 3.

43. "What JTC Envisions for Science Park," *Business Times*, February 11, 1982, 3.

44. "Improving the Safety of Oil Rig Workers," *Business Times*, February 17, 1984, 1.

45. *JTC Annual Report 1982–83*, 21.

46. "EDB Assumes Full Responsibility for Science Park," *Business Times*, January 17, 1984, 1.

47. "EDB Assumes Full Responsibility," 1.

48. *JTC Annual Report 1983–84* (Singapore: Jurong Town Corporation, 1984), 25; *JTC Annual Report 1984–85* (Singapore: Jurong Town Corporation, 1985), 25.

49. Jeffrey Tsang, "Number of Science Park Tenants Swells to 39," *Business Times*, January 1, 1988, 2.

50. "JTC Takes Over Site of Army Base for Science Park," *Straits Times*, January 21, 1985, 8.

51. This can partially explain why NUS and other educational institutions did not come to play a significant role in the Singapore Science Park. To some extent, the Park was designed to stand independently and apart from its local environment.

52. *JTC Annual Report 1988–89* (Singapore: Jurong Town Corporation, 1989), 52. Figures here and elsewhere are provided in Singapore dollars. In 1990, the Singapore dollar was worth approximately USD 0.55.

53. "Companies Can Buy Apartments in JTC Condominium," *Straits Times*, November 24, 1982, 9.

54. Ronnie Wai, "JTC to Build More Condominiums Near Lakepoint," *Straits Times*, February 8, 1983, 9.

55. Wai, "JTC to Build More Condominiums Near Lakepoint," 9.

56. *JTC Annual Report 1988–89*, 18.

57. "The Singapore Context," *Business Times*, October 24, 1989, 53.

58. "The Singapore Context," 53.

59. *JTC Annual Report 1988–89*, 18.

60. *JTC Annual Report 1988–89*, 18.

61. *JTC Annual Report 1989–90* (Singapore: Jurong Town Corporation, 1990), 34.

62. *JTC Annual Report 1988–89*, 19.

63. "Speech by Mr. Mah Bow Tan," Singapore Government Press Release, March 6, 1989, https://www.nas.gov.sg/archivesonline/data/pdfdoc/mbt19890306s.pdf.

64. *JTC Annual Report 1988–89*, 19.

65. Ronnie Lim, "JTC Sets Up Factories in Ang Mo Kio," *Business Times*, June 18, 1979, 1.

66. Amy Balan, "Joint Venture to Build Design Centres," *Business Times*, October 31, 1983, 1.

67. "Motorola to Set Up R and D Centre," *Straits Times*, November 12, 1983, 44; Richard Seah, "GM Singapore to Invest $10m in Factory Here," *Business Times*, January 21, 1983, 3.

68. *JTC Annual Report 1982–83*, 18.

69. *JTC Annual Report 1988–89*, 23.

70. Boey Kit Yin, "JTC Returns to Building after a Lapse of Almost Four Years," *Business Times*, May 15, 1987, 1.

71. Lee Han Shin, "On Common Ground," *Business Times*, November 10, 1987, 30.

72. For example, even after the Telecoms Authority of Singapore was privatized into SingTel and floated on the Singapore stock exchange, Temasek retained 89 percent ownership: Alvin Chua, "Singapore Telecommunications (SingTel)," *SingaporeInfopedia: A Singapore Government Agency Website* (2011), https://www.nlb.gov.sg/main/article-detail?cmsuuid=dd78b9f6 -29f9-4b7b-bd2a-eb6bbdfe1f8c.

73. *JTC Annual Report 1989–90*, 27–28.

74. For these examples, see *JTC Annual Report 1989–90*, 12.

75. "Speech by Mr. Goh Chok Tong, Minister for Trade and Industry, at the Signing of the Batam Agreement," National Archives of Singapore, October 31, 1980.

76. Chan Kok Pun and Kee Poir Mok, "Abundance of Land and Labour," *Business Times*, June 18, 1980, 7.

77. Gerry de Silva, "$400m Batam Deal Signed," *Straits Times*, January 12, 1990, 1.

78. Richard Borsuk and Nancy Chng, *Liem Sioe Liong's Salim Group: The Business Pillar of Suharto's Indonesia* (Cambridge, UK: Cambridge University Press, 2015).

79. "Speech by Bg Lee Hsein Loong, Minister for Trade and Industry," National Archives of Singapore, February 28, 1990.

80. Walton Morais, "Singapore-Batam Cable Link under Consideration," *Business Times*, March 1, 1990, 2.

81. Gerry de Silva, "Batam Park to Get Workers for Investors," *Straits Times*, March 5, 1990, 32.

82. Mary Kwang, "MNCs help Batam-Johor-Singapore Growth Triangle to Take Shape," *Straits Times Overseas Edition*, May 5, 1990, 14.

83. Claire Leow, "Batam Developer Looks to Nearby Island," *Business Times*, May 30, 1990, 1.

84. Gerry de Silva, "Government to Ease Rules for Training of Foreign Workers Here," *Straits Times*, May 30, 1990, 40.

85. Mary Kwang, "Plans Underway for Bintan Island to Build Industrial Estate," *Straits Times*, December 12, 1990, 40.

86. "A New City Is Born," *Business Times*, June 23, 1990, 8.

87. Borsuk and Chng, *Liem Sioe Liong's Salim Group*, chap. 11.

88. "JTC May Develop Estate in Thailand," *Straits Times*, December 29, 1988, 1.

89. "JTC Marketing Building Expertise Abroad," *Business Times*, December 29, 1988, 20.

90. "JTC May Develop Estate in Thailand," 1.

91. Boey Kit Yin and Loh Hui Yin, "Singapore in Pact to Help China Set Up Oil Supply Base," *Business Times*, December 14, 1983, 1.

92. Loh Hui Yin, "Chinese Industry Seeks JTC Aid," *Business Times*, March 22, 1984, 2.

93. Kooi Cho Teng, "JTC Plans to Take Part in More Regional Projects," *Business Times*, January 20, 1993, 2.

94. Nicholas Kristof, "The World: China Sees Singapore as Model for Progress," *New York Times*, August 9, 1992.

95. "Suzhou Project: How It All Began—SM Lee," *Straits Times*, May 20, 1994, 28.

96. Chan Soo Sen, oral history interview, reel 20, accession number 003223, National Archives of Singapore, May 5, 2008.

97. Chan Soo Sen, oral history interview.

98. "Singapore-Suzhou Township," Singapore-Suzhou Township Development Pte Ltd.

99. "Suzhou Project," 28.

100. Yong Pow Ang, "SM Lee to Sign Landmark Pact on Software Transfer to China," *Business Times*, February 24, 1994, 20.

101. Walton Marais, "PM Goh, China Vice-Premier Discuss Suzhou Project," *Business Times*, May 20, 1994, 16.

102. This language is used in "Suzhou Project Will Spur China On, Says Vice-Premier Li," *Straits Times*, May 20, 1994, 3.

103. Douglas Wong, "S'pore, Jiangsu Sign MOU on Suzhou Township," *Straits Times*, October 27, 1993, 1.

104. "Suzhou Project: How It All Began—SM Lee," 28.

105. For a longer history of the Suzhou project, see Alexius A. Pereira, *State Collaboration and Development Strategies in China: The Case of the China-Singapore Suzhou Industrial Park, 1992–2002* (New York: Routledge, 2003); John Wong and Liang Fook Lye, *Suzhou Industrial Park: Achievements, Challenges and Prospects* (Singapore: World Scientific Publishing, 2020).

106. Michael Richardson, "Singapore Industrial Park Flounders: A Deal Sours in China," *New York Times*, October 1, 1999.

107. *JTC Annual Report 1998–99* (Singapore: Jurong Town Corporation, 1999), 20.

108. Carolyn Cartier, "'Zone Fever,' the Arable Land Debate, and Real Estate Speculation: China's Evolving Land Use Regime and Its Geographical Contradictions," *Journal of Contemporary China* 10, no. 28 (2001): 445–469; Aihwa Ong, "The Chinese Axis: Zoning Technologies and Variegated Sovereignty," *Journal of East Asian Studies* 4, no. 1 (January–April 2004): 69–96. See also Andrew Barry, "Technological Zones," *European Journal of Social Theory* 9, no. 2 (2006): 239–253.

109. Keller Easterling, *Extrastatecraft: The Power of Infrastructure Space* (New York: Verso Books, 2016).

110. For example, CapitaLand (which owns Ascendas REIT, a descendant of Technology Parks) is 52 percent owned by Temasek; see Aradhana Aravindan and Anshuman Daga, "CapitaLand to Split into Two, Leaving Property Investment Business as Listed Company," Reuters, March 22, 2021, https://www.reuters.com/article/us-capitaland-reits-idUSKBN-2BE0GI. Likewise, Surbana Jurong (a descendant of Jurong International Holdings) is wholly owned by Temasek. "Our History," Surbana Jurong (2021), https://surbanajurong.com/our-history/.

111. Benjamin Peters, *How Not to Network a Nation: The Uneasy History of the Soviet Internet* (Cambridge, MA: MIT Press, 2016); Tiago Saraiva, *Fascist Pigs: Technoscientific Organisms and the History of Fascism* (Cambridge, MA: MIT Press, 2018); Sigrid Schmalzer, *Red Revolution, Green Revolution: Scientific Farming in Socialist China* (Chicago: University of Chicago Press, 2016).

112. Peter Westwick, chapter 10 in this volume.

CHAPTER TEN

Science, Texas Style

How the Lone Star State Embraced Science in a Big Way

Peter Westwick

In 1980 Texas was known for many things: barbecue, oil, *Dallas*, country music, cowboys, and Cowboys. In the 1980s it became known for something else: science and technology. One of the decade's most high-profile science projects, the Superconducting Super Collider, landed in Texas, a seemingly surprising development that was in fact part of a larger pattern, a conscious shift from an economy based on extractive industries to one based on advanced science and technology.

This chapter examines how the Lone Star State put itself on the map of American science in the 1980s and what that meant for both Texas and science. One might explain Texas's burgeoning sci/tech presence as a simple matter of scale: Texas had a big presence in science and technology just because it was big. Texas, however, was always big, yet it did not always have a major presence in science and technology. It acquired that presence in the 1980s, as a result of a series of explicit efforts.

Texas was not unique in its pursuit of a high-tech economy in the 1980s. Boosters in many other states were pursuing research parks and university–industry collaborations, and universities across the country embraced entrepreneurial enthusiasm. Texas was exceptional, however, in several ways. Most notably, in the early 1980s it was flush with cash from its oil industry. Oil money not only underwrote the ambitious expansion of science and engineering institutions but also translated into political power, exercised through elite networks connecting business, universities, and politics. The resultant political influence at the state and federal levels was evident in a series of episodes that vaulted Texas to the leading ranks of American science and technology.

The case of Texas in the 1980s will shed light on a basic question: why do certain places become epicenters of science and technology? (See chapter 8 by Robin Wolfe Scheffler and chapter 9 by Hallam Stevens in this volume.) What factors matter? Geography and natural resources, or institutions, ideology, political systems, laws, labor, culture? Bookshelves groan with the proliferating literature seeking to explain the origin of Silicon Valley and other high-tech regions.[1] Texas seems to confirm Loren Graham's argument, based on the example of science in Russia and the Soviet Union, that money is crucial. In Texas, oil fortunes provided money—and the political influence that ensued.[2]

In 1986 the University of Texas at Austin acquired a Cray supercomputer. It named the machine "Santa Rita #2." Santa Rita #1 was the famous gusher of 1923 on university land that opened the Permian Basin oil boom. The namesake represented the notion that advanced science and technology would provide a similar economic stimulus to the Lone Star State. And it did.

Precursors: Oil, Electronics, Aerospace

There were greedy Texans as well as tech-y Texans long before the '80s. Oil had long been a high-tech enterprise, deploying science and engineering to find oil, drill it, and refine it. Much of this activity centered on Houston: Howard Hughes Sr. began manufacturing his revolutionary drill bits there in 1909, and Schlumberger, a French firm specializing in well-surveying electronics, opened Houston offices starting in the 1930s. In the 1930s Shell Oil created a research lab in southwest Houston; geophysicist M. King Hubbert, a native Texan, worked there from 1943 to 1964, including several years as its director (when he formulated his famous prediction of "peak oil"). Shell opened another expansive lab, the Westhollow Research Center, nearby in 1975.[3]

The lucrative field of oilfield services expanded the tech map beyond Houston and spun off another important industry in the process. In 1930 a firm called Geophysical Service (GSI) set up shop in Dallas to pursue oilfield prospecting with the reflection seismograph. During World War II the firm shifted to electronic instruments, which soon surpassed the oil exploration business, and in 1951 the owners created a new electronics company called Texas Instruments (TI).[4]

TI was just one of several electronics firms to emerge in Texas, including Collins Radio, which expanded from Iowa to Dallas in the 1950s, and Electronic Data Systems, founded in Dallas in 1962 by a former IBM salesman named Ross Perot. In 1963 another former IBM salesman, Sam Wyly, founded University Computing Company in office space provided by SMU; early customers for its computer time-sharing services included TI and Sun Oil.[5]

Meanwhile in Austin in 1955, four physics and engineering professors at the University of Texas formed a consulting firm that later became known as Tracor. The company prospered especially from defense electronics and established an early major research and development (R&D) presence in the state capital; it was the first local *Fortune* 500 firm.[6] An Austin economic development board subsequently lured manufacturing plants for Xerox, TI, and IBM in the 1960s, followed by Westinghouse and Motorola in the 1970s.[7]

Texas also had a role in aerospace. It had no aircraft industry to speak of until World War II, when the dispersal of aircraft production away from the coasts led two California firms, Consolidated and North American, to open large branch plants near Dallas–Fort Worth. By 1950 Texas ranked second among US states in aircraft employment, with 14 percent of the total (behind California's 40 percent), all of it in Dallas–Fort Worth. The plants changed hands but kept building military planes: Chance Vought in 1948 moved from Connecticut into North American's Dallas plant; General Dynamics in 1953 took over Consolidated and its Dallas plant. Bell Aircraft, originally in Buffalo, in 1951 created a helicopter division near Dallas, which built the Huey, among other aircraft.[8] All of this activity kept Texas in the top three states for Department of Defense (DOD) prime contracts from the 1960s to the 1980s.[9]

While Dallas put the "aero" in aerospace, Houston provided the "space." The Space Race had a Texas twang. President Kennedy delivered his "We choose to go the moon" speech at Rice University in September 1962, two months after Houston had welcomed the Mercury Seven astronauts with a July Fourth barbecue featuring three tons of meat. Astronauts eventually returned the favor: the first word spoken from the moon was "Houston."[10]

Why Then, Why There? Motives, Models, Money

Astronauts were talking to Houston because NASA in 1961 had selected it as the site of its Manned Spacecraft Center, thanks in part to a thousand acres donated to Rice University by Humble Oil—an early indicator of the power of oil money and the willingness of Texans to deploy it on behalf of science and technology.

Despite the earlier sci/tech presence, or rather because of it, some Texans sought to strengthen the state's science and engineering community. There are three main explanations for this development: motivation, models, and money.

Around 1960 the founders of Texas Instruments (Cecil Green, J. Erik Jonsson, and Eugene McDermott) realized they were at a competitive disadvantage compared to firms in California or Massachusetts, where close ties to universities provided a technical labor supply and research base. They concluded that the Texas economy was overly dependent on "cattle, cotton, and oil" (as Green's biographer put it) and needed to diversify, especially into science- and technology-based industries.[11]

That was the motivation. Competing high-tech regions in other states then provided the models, especially Silicon Valley, which acquired its nickname, and broader notice as a high-tech hotbed, starting in the 1970s. After Apple's initial public offering in December 1980, other parts of the country embraced the idea, as a federal report at the time put it, of "making Silicon Valley happen here."[12] Re-creating Silicon Valley in Texas meant strengthening science and technology in local research universities and connecting academic research to industry through the creation of research parks. Stanford had started the research park trend in 1951 with the Stanford Industrial Park, followed by Research Triangle in North Carolina in 1959 and a bandwagon full of imitators in the 1960s.[13]

Those were the models. The money came primarily, although not exclusively, from oil. The price of oil tripled from 1972 to 1979, from $12 to $36 per barrel, then rose further to nearly $40 by 1981. As the 1980s began it was heady times in Texas; the state government enjoyed a billion-dollar budget surplus, and a common saying at the time was "85 by 85"—that is, oil would reach $85 per barrel by 1985. Texas provided the earliest image of the Greedy Eighties, as rich oilmen flaunted their limos, Lear jets, and lavish parties. The *Dallas* TV show, which premiered in 1978 and was the most-watched program in the United States in 1980 and 1981, captured the tone, and captivated the nation, through the story of a fictional oil family riven by greed.[14]

As Cyrus C. M. Mody shows in chapter 1 of this volume, the oil industry fed an entrepreneurial turn in American science at that time, and as Mody notes, the '80s saw huge shifts in the oil economy. Oil prices began to decline after 1981, as energy conservation cut into demand and new sources—Alaska, the North Sea, Mexico—boosted supplies. Signs of a bust emerged in 1983, when major producers, including the Organization of the Petroleum Exporting Countries, cut prices; prices then cratered in 1985 to about $10 a barrel, a fourth of the price in 1981. The oil bust sped the collapse of Texas banks and family fortunes, and it dealt a major blow to the state economy in general.[15]

So was Texas's push into science and technology driven by surplus or scarcity? The answer seems to be surplus, at first. The oil boom of the late 1970s provided the money at a time when the embrace of science and technology was already under way. The bust then provided further motivation to find new economic engines, especially in high-tech industries. But before the bust began, the models and the money had already come together in the early '80s, through several key individuals and institutions.

Entrepreneurial Institutions

The spike in oil prices was a boon to the University of Texas (UT). The state government in the nineteenth century had ceded to the university two million acres of West Texas that the government deemed too worthless even to survey—land that Texas had appropriated from indigenous peoples (primarily Jumano, Mescalero Apache, and Comanche).[16] In 1923 an exploratory well there, Santa Rita #1, struck oil, and those two million acres across the Permian Basin sprouted oil wells like weeds. The resulting gusher of royalties eventually gave UT one of the largest university endowments in the nation; by 1983 the UT system had $15 million per month in oil and gas royalties rolling into its endowment.[17] It was probably not a coincidence that the president of UT-Austin at the time, Peter Flawn, was a geologist and former director of the UT Bureau of Economic Geology, whose research included rock formations in Texas.

A straw in the wind: in 1982 Steven Weinberg, who had won the Nobel Prize in physics three years earlier, left Harvard for a position at UT-Austin as the Jack S. Josey-Welch Foundation Regents Chair in Science. Jack Josey, a UT grad and former UT regent, was a Texas oilman. The Welch Foundation, which focused on science, came from the fortune of Robert A. Welch, a Texas oilman.

At Rice University, president Norman Hackerman, an electrochemist, had received one of the first grants from the Welch Foundation in the 1950s and continued a fruitful association with the foundation, eventually (starting in 1982) chairing its scientific advisory board. As Rice president from 1970 to 1985, Hackerman quadrupled both the endowment and the number of endowed chairs and brought in several large unrestricted donations; in particular he boosted Rice's scientific research.[18]

Texas A&M meanwhile had its Texas Engineering Experiment Station (TEES), created in 1914 to support the local business community through R&D. TEES had played a small role until 1980, when it acquired a new director, W. Arthur Porter, who had previously worked at TI on semiconductor fabrication. Porter, too, perceived that Texas suffered from a gap between academic and industrial R&D compared to Silicon Valley, and he set out to insert TEES into that gap. By 1983 he increased its R&D budget from $3 million to $30 million, mostly with private funds, and TEES had some 800 people pursuing R&D.[19]

Oil money and the Silicon Valley model also met in the person of George P. Mitchell. A graduate of Texas A&M in 1940 with degrees in geology and petrochemical engineering, Mitchell had entered the oil business and eventually bought into a small firm that discovered a gas field in North Texas. Mitchell Energy and Development Corporation went public in 1972 and by the early 1980s was a *Fortune* 500 company, with several thousand employees. By that time Mitchell had made more money in real estate, including his Woodlands development outside Houston. He gave a lot of that money to Texas A&M, eventually becoming the school's largest benefactor with donations topping $95 million. He also developed a keen interest in environmental sustainability, sponsoring several conferences and prizes on the subject in the 1970s. His family called his combination of oil drilling and environmentalism "the Mitchell paradox"—a contradiction compounded by Mitchell's dogged pursuit of hydraulic fracturing for oil and gas, which finally paid off in the 1990s and made him known as "the father of fracking."[20]

Mitchell was also very interested in Texas's economic turn from oil and gas to high technology, and he, too, viewed Silicon Valley as a model, along with the Research Triangle, which was especially intriguing for its collaboration among several universities. In 1982 Mitchell created a Texas version, the Houston Area Research Center (HARC), at his Woodlands development. HARC was a nonprofit organization to manage cooperative research among Texas A&M, the University of Houston, and Rice. (UT-Austin joined the group in 1984.) Mitchell intended it to become a magnet for high-tech firms, and its research focused on fields with technological promise, such as superconductivity, supercomputing, lasers, digital image processing, and seismic imaging.[21]

The Microelectronics and Computer Consortium

The University of Texas meanwhile benefited from another development. In the early 1980s several major US electronics firms proposed a Microelectronics and Computer Consortium (MCC), under which the firms would pool resources and technologies in order to compete with Japanese and European companies. The MCC announced a national competition for the site, and Texas high-tech boosters saw an opportunity. They agreed that Texas needed to bid for it, and they settled on Austin as the site.

Out of fifty-seven cities in the original competition, Austin emerged as one of four finalists, alongside Raleigh-Durham, Atlanta, and San Diego. The Texas bid was viewed as a long shot, but it had the concerted backing of state and local governments, business executives and chambers of commerce, and the academic communities at UT and Texas A&M. Pike Powers, chief of staff to Texas governor Mark White, coordinated the effort, and tech executive Ross Perot—founder of Electronic Data Systems in Dallas—loaned his Lear jet to fly MCC representatives to Austin, where they were treated to a helicopter tour. On their tour the MCC team saw no suburban sprawl or belching factory smokestacks, but they did see the modern electronics plants of Tracor, Texas Instruments, IBM, and Motorola, along with the twenty empty acres of UT's Balcones Research Center awaiting the MCC.[22]

The campaign worked. The MCC's selection of the dark horse from Texas provoked grumbling from the other finalist cities. Mayor Andrew Young of Atlanta said, "I gather Texas bought it." He wasn't wrong. Texas offered a much larger package of financial support, including free use of the UT land; $23 million in low-cost mortgages and $3 million in bridge loans for MCC staff, provided by local businesses; and a Lear jet for recruiting. UT-Austin also committed to triple its microelectronics R&D through a combination of state and private funds, and Austin's congressman, Jake Pickle, and fellow Texan Jim Wright, Speaker of the House, promised legislative support against any antitrust barriers raised against the consortium.[23]

The MCC repaid the investment. Consortium members contributed between $30 and $70 million a year to the center, which supported 400 researchers and administrators in information technology, chip design, and software; many of them landed faculty or fellowship positions at UT. MCC's presence helped Austin win subsequent competitions for 3M's research lab in 1984 and, in 1988, the Semiconductor Manufacturing Technology consortium, or Sematech, a group of fourteen major US chip makers that aimed to revitalize American semiconductor production. Austin then attracted branch plants for Applied Materials, HP, Apple, Cisco, and Samsung in the 1990s.[24]

The MCC campaign also galvanized a major new source of support for UT-Austin. Not all of the money in Texas came from oil. Peter O'Donnell was a Dallas native who made a fortune in the securities business, and he donated much of it to Republican political candidates. He was chair of the Texas Republican Party in the 1960s and retained influence in Republican politics, albeit behind the scenes. In the early 1980s O'Donnell surveyed the Texas economy and concluded that agriculture and oil would not sustain it in the long run. The MCC campaign inspired him to embrace science and technology. In 1984 he pledged $8 million to UT, to be matched by $8 million from other donors and $16 million from UT itself, to endow thirty-two chairs in science and engineering. The chairs attracted many new prominent faculty, and students followed: by 1986 the UT-Austin graduate program in computer science was receiving three times as many applications as it had four years earlier. O'Donnell would go on to donate $135 million to UT, especially to its Institute for Computational

Engineering and Sciences, which eventually boasted eighty-seven faculty members.[25]

The MCC itself later fizzled, but it left a legacy in the thirty-two science and engineering chairs at UT, the associated thriving program in computer science, and the string of high-tech firms that followed the MCC to Austin. HARC, too, eventually declined in influence. Before they faded, however, both the MCC and HARC contributed to the major symbol of American science in the '80s.

The Superconducting Super Collider

The Lone Star State prided itself on bigness: Big Hats on men, Big Hair on women. When Texas embraced science in the '80s, it naturally backed Big Science. (Texans' scientific bent was evident in the Big Hair trend, which was known locally as the Big Bangs Theory.[26]) At the time, the biggest science involved the next accelerator for high-energy physics, a successor to the Tevatron proton collider at Fermilab. In spring 1983 the Department of Energy, the federal agency overseeing accelerators, decided to pursue a Superconducting Super Collider (SSC).

Texas had tried before to get into the accelerator business. In 1956 Texas A&M, Rice, the University of Texas, and the Welch Foundation lobbied the Atomic Energy Commission (AEC) to create a national laboratory in Houston, a counterpart to the existing national labs at Berkeley, Brookhaven, and elsewhere. The modest proposal included a particle accelerator alongside a nuclear reactor, high-speed computer, and medical center. The AEC, not wanting to multiply national labs, shot it down.[27]

The 1980s context encouraged Texans to try again, and they began angling for the SSC long before the Department of Energy (DOE) opened the site selection to a competition. The successful bid for the MCC provided inspiration, and HARC provided an institutional home.[28] In December 1983 the DOE announced that $20 million for initial SSC design studies would go to four existing accelerator labs (Fermilab, Brookhaven, Cornell, and SLAC) and the "Texas Accelerator Center" (TAC). The TAC was a new branch of the Houston Area Research Center, funded by George Mitchell and led by Peter McIntyre, a young accelerator physicist recently hired at Texas A&M after stints at Harvard and Fermilab. The TAC continued as one of the main SSC design sites through 1986.[29]

When the DOE in spring 1987 announced a national competition for the SSC site, Texas boosters again swung into action. The Texas bid centered on a site south of Dallas called Waxahachie, and it emerged as a leading candidate among the seven finalists, alongside an Illinois site near Fermilab. Texas again organized a high-powered campaign, coordinated by political consultant Karl Rove. Governor William Clements, a Republican, appointed a Texas National Research Laboratory Commission to spearhead the campaign, led by Peter Flawn, recently retired as UT president (and now working for Peter O'Donnell), and J. Fred Bucy, former CEO of Texas Instruments.

What would be called the "Texas collider cartel" ultimately enlisted dozens of state and federal lawmakers, faculty and administrators at several Texas universities, and business leaders. The preexisting technological base in Dallas provided one advantage. The state legislature provided another: seconded by Texas voters, it approved $1 billion in bonds, a huge local subsidy to cut federal costs. For the DOE's site visit, the Texas boosters organized a lavish parade followed by a dinner hosted, yes, by Ross Perot. This local enthusiasm may have provided the pivotal difference, since the Fermilab site suffered from some NIMBY-ist opposition from local residents. In November 1988 the DOE announced that Texas had won the SSC, prompting a *Dallas Morning News* headline celebrating the "Extraordinary Sales Job."[30]

The SSC announcement came two days after the election that saw Texan George H. W. Bush win the presidency. Together with the presence of Rep. Jim Wright (from Fort Worth) as Speaker of the House, Bush's election encouraged a perception that politics won it for Texas. That political leverage, however, perhaps came more into play subsequently, as the issue of funding the SSC shifted to Congress. In the ensuing years, spiraling cost estimates and schedule slips sparked doubts about the project's management, and then the demise of the Soviet Union removed the motivation of Cold War competition, putting the project funding in jeopardy. Texas's political clout helped keep the SSC alive for a time but also aroused resentment in other states, and in 1993 Congress voted to kill the project.[31]

In the meantime, though, the award of the SSC to Texas had provided the clearest signal yet of the Lone Star State's arrival on the map of American science and technology. Texas won the SSC not because a Texan won the White House and not because of Waxahachie's wide-open prairie or tunnel-friendly chalk geology. It won because Texans had already embraced science and technology: they had created institutions (such as HARC) that underpinned their claim to expertise, they had organized campaigns for new institutions (such as the MCC) that helped them mobilize for the bid, and they then rallied their political, economic, and sci/tech resources in pursuit of this new symbol of scientific leadership. The SSC decision was a fruit of the earlier activity.

The National Academy of Sciences

Texas's scientific and political footprint in the 1980s extended to the pinnacle of American science, the National Academy of Sciences. Cecil Green, one of TI's cofounders, had long been a prolific donor to the Academy. In 1983 the NAS formed an Academy-Industry Program to expand ties to industry, and NAS president Frank Press asked Green to join it. Green agreed, donated funds to it, and in November 1984 organized a lunch meeting between NAS representatives and thirty business leaders at the Dallas Petroleum Club to grease the effort.[32]

Around the same time, some friends of George Mitchell introduced him to Press so that the Academy could hear his views on sustainability and forecasting. Mitchell came away impressed and persuaded Press to create another avenue for fundraising and advice, called the President's Circle. Mitchell helped organize another meeting in Dallas with business executives, including a number of oil executives as well as Peter O'Donnell. Press insisted that they all put away their checkbooks; he just wanted their help getting the Academy's advisory reports to a broader audience in the business world. They listened to the second request but not the first: O'Donnell promptly wrote a million-dollar check to the Academy, and Mitchell later provided $1 million for an Academy study of sustainability called the Global Commons Project.[33]

What did each side hope to gain from the relationship? Why did Texans like Green and O'Donnell and Mitchell give loads of money to the NAS? One reason could be public spirit: as believers in the benefits of science and technology, they thought the Academy performed a valuable public service to American science. Another reason might be that they wanted the Academy's help on science policy issues relevant to their interests. One possibility for the oilmen was climate change, on which the

Academy had conducted several high-profile studies. The evidence here seems to be negative, and Frank Press himself later in the '80s came out strongly and publicly on the threat of human-driven global warming. One might also wonder whether the Academy's Texas connections influenced the SSC site selection, for which an Academy committee selected the seven finalists. The standard SSC history declares that the consensus among the interested parties was that political considerations did not influence the NAS panel in picking the Texas site. This view is supported by evidence from the NAS archives that Press and the NAS Council in fact wanted to keep the Academy *out* of the SSC site selection process, for fear both of appearing to endorse the SSC (which Press did not like) and of antagonizing scientists from rejected sites.[34]

What did the Academy, for its part, want from Texas? Money, certainly, but Press was also a savvy politico—he had served as President Jimmy Carter's science advisor—and he no doubt recognized that having friends in Texas might help the Academy politically in Washington. His political foresight paid off twice. The Academy conducted most of its advisory studies through the National Research Council (NRC), which was perpetuated by an executive order by President Woodrow Wilson in 1918. In the early 1990s the Academy requested a new executive order, one that would recognize the affiliated National Academy of Engineering and the Institute of Medicine (neither of which existed in 1918) and also, perhaps more important, confirm the NRC's sole-source status for federal contracts. In January 1993 the new executive order was sitting on President Bush's desk as his term wound down, and the Academy feared having to start over with a new administration. Press talked to Peter O'Donnell, who knew Bush well enough to call the White House and get the president to sign the order on his last day in office.[35]

O'Donnell also intervened in an even more critical episode. In the late 1990s a Supreme Court ruling threatened to subject the NAS/NRC to the Federal Advisory Committee Act (FACA). NAS/NRC advisory panels had historically worked in private, allowing committee members to debate technical and policy issues without fear of public criticism or political oversight. The Academy viewed FACA as an existential threat to its advisory function and fiercely resisted it, at one point threatening that the Academy would simply stop providing advice to the government. The Academy turned to Congress for help, but it again faced a deadline: it was crucial to pass the bill before Congress went on a two-month recess. The House majority leader was Dick Armey from Texas, who, like President Bush, was in O'Donnell's rolodex. O'Donnell told Armey how important the bill was for the Academy, and Armey brought the bill to a vote at one in the morning the night before the recess. The Senate then quickly passed it, and the Academy had won a major reprieve thanks to its Texas connections.[36]

Neoliberalism, the State, and Sci/Tech

The presidential election of 1992 featured two Texans: the incumbent George H. W. Bush, an old-school Republican, and Ross Perot, the high-tech executive and Texas tech booster whose populist platform attacked globalization and free trade. The Texans lost to Bill Clinton, whose support of a post-Fordist high-tech economy marked him as an "Atari Democrat." The Clinton administration, which encouraged and celebrated the fantastic growth of Silicon Valley and the internet economy, eventually marked the consolidation and triumph of neoliberalism.[37]

A central question for greedy science is the role of neoliberalism, an ill-defined concept that was nevertheless a defining feature of the 1980s.[38] If groovy science in the 1960s and 1970s was often about how the left-leaning romantic counterculture reconciled with modern science and technology, greedy science in the 1980s was often about how conservatives came to reconcile with modernity—and with science and technology—through the ascendant ideology of neoliberalism.

Neoliberalism was coterminous with conservative politics in its support of deregulation and free markets, manifested in policies undermining government social programs and the labor movement. But one might also see in neoliberalism an enthusiasm for new technologies, especially information technology, as a potent force for both political liberation and entrepreneurial economic activity. This embrace of technological innovation and progress differentiated neoliberalism from traditional conservatism.[39]

The efforts in Texas to encourage high-tech entrepreneurial activity seem another example of the incorporation of science and technology into neoliberalism. This vision, however, relied on a substantial state presence. Although greed was good in the '80s, so was public funding. Texas remained in the top two or three states for DOD prime contracts with around $10 billion per year, mostly from the Air Force.[40] Defense dollars kept flowing to Texas thanks in part to its recent institution building in science and technology. The very first HARC research program in 1983 was on lasers and materials for the Strategic Defense Initiative, better known as Star Wars, President Ronald Reagan's signature defense program for missile defense.[41] The UT system meanwhile became SDI's second biggest university contractor, behind the Utah state system. From 1983 to 1986 the UT system received $33 million in SDI R&D funds, for programs such as an electromagnetic railgun and a gamma-ray laser, and Texas Tech another $4 million.[42] Texas universities seem not to have shared the qualms of major counterparts, such as Cornell or Caltech, where faculty mounted anti-SDI campaigns and pressed their schools to refuse SDI contracts.[43]

Texas science in the '80s benefited further from an array of state and federal subsidies: $1 billion in state bonds to win the SSC plus tens of millions of dollars in federal R&D contracts for the SSC, as well as tens of millions in state funds to universities for endowed chairs and the MCC bid. By the time Congress killed the SSC, Texas taxpayers had contributed over $500 million to the project, some but not all of which the federal government later repaid in cash and land. One might also note that UT was in one sense a free rider, benefiting from the more than $100 million flowing into its endowment every year from oil leases—money derived from appropriated land.

Conclusion: Putting the Tech in Texas

In 1992, the year of the presidential contest, a new company entered the *Fortune* 500, a computer firm started in Austin by a young UT dropout named Michael Dell. The flourishing high-tech industry represented by Dell became the symbol of the New Texas in the 1990s, and it helped Austin win *Fortune*'s ranking as the top US city for entrepreneurship in 2000.[44] The more recent decisions by HP, Oracle, Meta, and Tesla to move major operations to Austin have signaled a shift in the center of gravity of high tech in the United States, reinforced by the relocation to Texas of the New Space firms SpaceX and Blue Origin, founded respectively by tech titans Elon Musk and Jeff Bezos. Thus Bezos sauntered to his first space flight in a Blue Origin vehicle, launched from West Texas in July 2021, in cowboy boots and a cowboy hat.[45]

These seemingly recent trends in fact derived from the vision of Texas's sci/tech boosters of the 1980s. The disparate developments described here—the entrepreneurial activity of universities to strengthen their science programs, the creation of HARC and MCC, the SSC decision, the engagement with the National Academy of Sciences—reveal a pattern: a concerted effort by academic, business, and political leaders to support science institutions and attract leading scientists and high-tech firms. The SSC of course came to naught, leaving Waxahachie with only a fine hole in the ground, and the MCC and HARC similarly ended up fizzling. Before they declined, however, they announced the arrival of Texas as a major player in academic science and high-tech industry.[46]

Austin is sometimes referred to these days as "Silicon Valley 2.0."[47] The comparison may have deeper historical roots than is usually recognized. Both Silicon Valley 1.0 and 2.0 trace back to natural resources, which led to extractive industries, which in turn provided financial capital and scientific and technological expertise for later sectors, in a process that economic geographer Richard Walker has called "resource industrialization." (This is not to argue for environmental determinism. Natural resources were necessary but not sufficient: extractive industries relied on a private property regime, capital and financial infrastructure, state subsidies and state enforcement of laws and regulations, labor scarcity, and so on.) Thus in California, mining, timber, agriculture, and oil encouraged the science and technology required to extract resources and process them; these capabilities then enabled new industries, from aviation and aerospace to electronics and computing—including Silicon Valley.[48] In Texas, oil provided primarily capital but also technological capabilities—including the electronics that led to Texas Instruments.

Texas itself was not a unitary object, as there were differences within it. Houston meant oil, space, and later biotech; Dallas was aerospace and electronics; Austin was electronics. In this period the center of gravity shifted from Houston to Austin, although Houston partly compensated with health care and biotech. And not all Texans jumped on the '80s sci/tech bandwagon. During the MCC competition one group was prominently missing from the Texas team of high-tech boosters: Austin's local government, especially the city council, which included slow-growth advocates. Growth was a contentious issue in Austin politics, and many local residents dreaded an influx of high-tech industry; winning the MCC probably cost the pro-growth mayor his job two years later.[49] Austin now shares the skyrocketing housing costs, gentrification, and gridlock of the original Silicon Valley.[50]

These costs suggest that we ask not just how and why Texas embraced science and technology but also what that meant for Texas. The sci/tech boosters seem to have considered only the economic effects of high tech. What did the influx of scientists and engineers mean for a culture committed to country music and football?

High-tech Austin is now viewed politically as a blue island in a sea of red, and the city's unofficial motto—"Keep Austin weird"—suggests that scientists and engineers conveyed a countercultural sensibility. For Texas at large, however, this period coincided with a rightward turn in politics. In a gradual process that started after World War II and accelerated with the Civil Rights Act, the previously Democratic Texas had by the 1990s turned solidly Republican. The oil industry encouraged the trend; it created great wealth but wasn't labor intensive and thus did not create a large working class. But high-tech industries may have contributed: many of Texas's new conservatives were not rural populists but rather middle and upper classes in cities and suburbs.[51]

Texas may thus provide a test case for the politics of high-tech suburbs. Lisa McGirr has described a political trajectory toward conservatism for engineers in the aerospace suburbs of Orange County, California, after 1960, while Lily Geismer observed the opposite trajectory for high-tech suburbs around Boston in the same period. Texas seems to lean more toward McGirr than Geismer, with its conservative scientists and engineers representing not blue-collar populism but rather middle-class modernity.[52] And the recent influx of Silicon Valley refugees to Austin seems to include a libertarian contingent that has made the region redder politically.[53]

Populism certainly persisted, evident in Texas's support of creationism in public schools. But business leaders in Texas generally opposed the creationists, for fear of hampering the recruitment of highly educated tech workers.[54] This suggests a final characteristic of the Texas sci/tech movement, a pragmatic bent that perhaps provides a final comparison to California. The so-called Californian ideology of Silicon Valley featured a mix of counterculture idealism, neoliberal economic ideology, and technological determinism (although that ideology often ignored historical realities, including the fundamental role of the federal government).[55] In Texas, figures such as Peter O'Donnell, apparently a Rockefeller Republican, or George Mitchell, with dual enthusiasms for environmental sustainability and gas fracking, were no doubt free-marketers but not techno-utopians;

their California counterpart would be David Packard, not Stewart Brand. They were also more oriented to longer-term scientific research—for instance, funding academic chairs at UT and Texas A&M—than to only short-term technological payoffs. But as businessmen they had a sharp eye on the bottom line, on the economic benefits—in short, the money—sure to flow to the Lone Star State from their support of science and technology.

NOTES

I thank Ruth Schwartz Cowan, Angela N. H. Creager, Michael D. Gordin, W. Patrick McCray, Cyrus C. M. Mody, Peter Neushul, and the conference participants for comments and suggestions.

1. See, for example, AnnaLee Saxenian, *Regional Advantage: Culture and Competition in Silicon Valley and Route 128* (Cambridge, MA: Harvard University Press, 1994); Chong-Moon Lee, William F. Miller, Marguerite Gong Hancock, and Henry S. Rowen, eds., *The Silicon Valley Edge: A Habitat for Innovation and Entrepreneurship* (Stanford, CA: Stanford University Press, 2000); Timothy Bresnahan and Alfonso Gambardella, *Building High-Tech Clusters: Silicon Valley and Beyond* (Cambridge, UK: Cambridge University Press, 2004); Martin Kenney and Donald Patton, "The Coevolution of Technologies and Institutions: Silicon Valley as the Iconic High-Technology Cluster," in *Cluster Genesis: Technology-Based Industrial Development*, ed. Maryann P. Feldman and Pontus Braunerhjelm (New York: Oxford University Press, 2005), 38–60; Christophe Lécuyer, *Making Silicon Valley: Innovation and the Growth of High Tech, 1930–1970* (Cambridge, MA: MIT Press, 2006); Michael Storper, *Keys to the City: How Economics, Institutions, Social Interaction, and Politics Shape Development* (Princeton, NJ: Princeton University Press, 2013).

2. Loren Graham, *What Have We Learned about Science and Technology from the Russian Experience?* (Stanford, CA: Stanford University Press, 1988).

3. Tyler Priest, "Shell to Houston," *The Houston Review* 3, no. 1 (2005): 10–11; Jon Kutner Jr., "Schlumberger," *Handbook of Texas*, Texas State Historical Association, www.tshaonline.org/handbook/entries/schlumberger; Mark Odintz, "Baker Hughes," *Handbook of Texas*, Texas State Historical Association, www.tshaonline.org/handbook/entries/baker-hughes.

4. Robert R. Shrock, *Cecil and Ida Green: Philanthropists Extraordinary* (Cambridge, MA: MIT Press, 1989).

5. John Merwin, "Sam Wyly's Biggest Gamble," *D Magazine*, December 1975.

6. David Gibson and Michael Oden, "The Launch and Evolution of a Technology-Based Economy: The Case of Austin, Texas," *Growth and Change* 50, no. 3 (2019): 952; Diana J. Kleiner, "Tracor," *Handbook of Texas*, Texas State Historical Association, www.tshaonline.org/handbook/entries/tracor.

7. Gibson and Oden, "Launch and Evolution."

8. William Glenn Cunningham, *The Aircraft Industry: A Study in Industrial Location* (Los Angeles: L. L. Morrison, 1951), 87, 224–226.

9. E. J. Malecki, "Military Spending and the US Defense Industry: Regional Patterns of Military Contracts and Subcontracts," *Environment and Planning C: Government and Policy* 2 (1984): 31–44; Ann R. Markusen and Scott Campbell, *The Rise of the Gunbelt: The Military Remapping of Industrial America* (New York: Oxford University Press, 1991), 13, table 2.1.

10. David Courtney, "Have Texans Played a Big Role in the Space Program?" *Texas Monthly*, July 2019.

11. Shrock, *Cecil and Ida Green*, 153–162.

12. US Congress, Office of Technology Assessment, *Technology, Innovation, and Regional Development: Census of State Government Initiatives for High-Technology Industrial Development* (Washington, DC: Office of Technology Assessment, 1983), 9.

13. Elizabeth Popp Berman, *Creating the Market University* (Princeton, NJ: Princeton University Press, 2012), 27–28, 34, 150–151; Margaret Pugh O'Mara, *Cities of Knowledge: Cold War Science and the Search for the Next Silicon Valley* (Princeton, NJ: Princeton University Press, 2005); US Congress, Office of Technology Assessment, *Technology, Innovation, and Regional Development*; US Congress, Office of Technology Assessment, *Technology, Innovation, and Regional Economic Development: Encouraging High-Technology Development* (Washington, DC: Office of Technology Assessment, 1984).

14. Jerry Olson, "Checking the Forecast: A Look Back at the Texas 2000 Commission," *Texas Business Review*, August 2000. On the oil boom: Bryan Burrough, *The Big Rich: The Rise and Fall of the Greatest Texas Oil Fortunes* (New York: Penguin, 2009), 335–386.

15. Yergin, *The Prize*, chaps. 35–36; Olson, "Checking the Forecast." On the bust in Texas: Burrough, *The Big Rich*, 406–432.

16. On appropriated land and land-grant universities: Robert Lee and Tristan Ahtone, "Land-Grab Universities," *High Country News*, March 30, 2020. Unlike the land-grant universities described in this article, the Texas "University Lands" grant came not from the federal government under the Morrill Act but rather from the state of Texas.

17. Gibson and Oden, "Launch and Evolution," 951–952; Joseph P. Kahn, "The Isosceles of Texas Is Upon Us," *Inc.*, October 1, 1983. On the oil well: Diana J. Kleiner, "J.J. Pickle Research Campus," and Julia Cauble Smith, "Santa Rita #1," both in Texas State Historical Association, *Handbook of Texas*, www.tshaonline.org/handbook/.

18. Marye Anne Fox, "Norman Hackerman," *Proceedings of the American Philosophical Society*, 153, no. 3 (September 2009): 351–357.

19. US Congress, Office of Technology Assessment, *Technology, Innovation, and Regional Development*, 69–70; W. Arthur Porter, *The Knowledge Seekers: Creating Centers for the Performing Sciences* (Austin: IC² Institute, The University of Texas, 1998), 10–12, 31.

20. Jurgen Schmandt, *George P. Mitchell and the Idea of Sustainability* (College Station: Texas A&M University Press, 2010), 1–5; "The Father of Fracking," *The Economist*, August 3, 2013. The fracking technique had been around since the 1940s, but Mitchell demonstrated its economic feasibility.

21. Schmandt, *George P. Mitchell*, 58–59; Porter, *The Knowledge Seekers*, 75–77. The name changed in 1990 to Houston Advanced Research Center.

22. Porter, *The Knowledge Seekers*, 15–19; Kirk Ladendorf, "30 Years Ago, MCC Consortium Helped 'Put Austin on the Technology Map,'" *Austin American-Statesman*, May 25, 2013; Kahn, "The Isosceles of Texas"; Gibson and Oden, "Launch"; "The Town That Won the Pennant: A Short History of Austin's Economic Development," *Austin Environmental Directory*, 2003, environmentaldirectory.info/a-short-history-of-austins-economic-development/.

23. Kahn, "The Isosceles of Texas"; "The Town That Won the Pennant."

24. Porter, *The Knowledge Seekers*, 19; Gibson and Oden, "Launch"; Ladendorf, "30 Years Ago." The first MCC director was Bobby Ray Inman, a native Texan and UT grad, former head of the National Security Agency, and just retired as deputy director of CIA. On Sematech, see Larry D. Browning and Judy C. Sheller, *Sematech: Saving the U.S. Semiconductor Industry* (College Station: Texas A&M University Press, 2000).

25. Ralph Haurwitz, "UT's 'Mr. Anonymous' a Force Behind Research," *Austin Statesman*, September 1, 2012; Gibson and Oden, "Launch," 957; Ladendorf, "30 Years Ago." Another non-coincidence: UT president Peter Flawn, upon retiring, turned up as a well-paid consultant to O'Donnell. Mary Huber, "Peter Flawn Helped Lead UT to National Prominence," *Austin Statesman*, May 8, 2017.

26. Skip Hollandsworth, "Hooray for Big Hair!" *Texas Monthly*, December 1992.

27. Peter Westwick, *The National Labs: Science in an American System, 1947–1962* (Cambridge, MA: Harvard University Press, 2003), 182.

28. On MMC as inspiration: Porter, *The Knowledge Seekers*, 19.

29. Michael Riordan, Lillian Hoddeson, and Adrienne W. Kolb, *Tunnel Visions: The Rise and Fall of the Superconducting Super Collider* (Chicago: University of Chicago Press, 2015), 36; Schmandt, *George P. Mitchell*, 59.

30. Riordan, Hoddeson, and Kolb, *Tunnel Visions*, 101–115, 150 ("sales job" on 115, "cartel" on 150); Mary Huber, "Peter Flawn." See also Peter Flawn, *The Story of the Texas National Research Laboratory Commission and the Superconducting Super Collider: How Texas Won . . . and Lost* (Austin: The University of Texas, Bureau of Economic Geology, 2003).

31. Riordan, Hoddeson, and Kolb, *Tunnel Visions*, 149–153, 168–169; Daniel J. Kevles, "Preface, 1995: The Death of the Superconducting Super Collider in the Life of American Physics," in *The Physicists: The History of a Scientific Community in Modern America*, 6th ed. (Cambridge, MA: Harvard University Press, 1995).

32. Shrock, *Cecil and Ida Green*, 180–181, 329.

33. Ken Fulton interview with author, May 22, 2018; NAS Council Minutes, NAS archives, June 18, 1993; Schmandt, *George P. Mitchell*, 84–85.

34. Riordan, Hoddeson, and Kolb, *Tunnel Visions*, 104; NAS Council Minutes, February 13, 1987. Steven Weinberg was on the NAS/NAE site committee, the only one among twenty-one members representing a Texas institution.

35. NAS Council Minutes, February 27, 1992, and February 25, 1993; Executive Order 12832, January 19, 1993, www.nationalacademies.org/nasem/na_053087.html; Frank Press interview by author, July 19, 2014; Ken Fulton, interview by the author, May 22, 2018.

36. Hearings on FACA amendments, November 9, 1997, and minutes of conference call, NAS Council Executive Committee, November 12, 1997 (NAS Council: Meetings: Agenda December 1997); William Colglazier interview by author, May 16, 2016; Jim Jensen, interview by Ruth Schwarz Cowan and the author, October 19, 2017.

37. Gary Gerstle, *The Rise and Fall of the Neoliberal Order: America and the World in the Free Market Era* (New York: Oxford University Press, 2022), chap. 5.

38. On the profuse and diffuse definitions: Daniel Rodgers, "The Uses and Abuses of 'Neoliberalism,'" *Dissent* 65, no. 1 (2018): 78–87.

39. Gerstle, *Rise and Fall*, especially 4–5, 8–9, 103–104; Fred Turner, *From Counterculture to Cyberculture: Stewart Brand, the Whole Earth Network, and the Rise of Digital Utopianism* (Chicago: University of Chicago Press, 2006). Cf. David Harvey, *A Brief History of Neoliberalism* (New York: Oxford University Press, 2005), 3–4, which has a more instrumental and less ideological view of the role of information technology in neoliberalism. On the origins of neoliberal ideology in Cold War game theory: S. M. Amadae, *Rationalizing Capitalist Democracy: The Cold War Origins of Rational Choice Liberalism* (Chicago: University of Chicago Press, 2003).

40. Department of Defense, *Prime Contract Awards by State, Fiscal Year 1985* (Washington, DC: Department of Defense, 1985); Department of Defense, *Prime Contract Awards by Region and State, Fiscal Years 1987, 1986, 1985* (Washington, DC: Department of Defense, 1987; Department of Defense, *Prime Contract Awards by State, Fiscal Year 1988* (Washington, DC: Department of Defense, 1988. These figures do not include money for military bases.

41. On SDI as Keynesian economic stimulus: Peter Westwick, "The International History of the Strategic Defense Initiative: American Influence and Economic Competition in the Late Cold War," *Centaurus* 52 (Fall 2010): 338–351.

42. Council on Economic Priorities, *Star Wars: The Economic Fallout* (Cambridge: Ballinger, 1988), 84–87; Porter, *The Knowledge Seekers*, 19; "Star Wars Research at UTD," *D Magazine*, February 1987; "Groans of Academe," *The Economist*, November 15, 1986. The ranking of SDI contractors does not include SDI work at the national labs at Los Alamos and Livermore, operated by the University of California, or at Lincoln Lab, operated by MIT.

43. For example: at Caltech, whose Jet Propulsion Laboratory engaged in SDI work, a group of faculty organized a "Committee against Space Weapons," and an anti-SDI petition garnered 500 signatures on campus, including those of six Nobel laureates. Peter Westwick, *Into the Black: JPL and the American Space Program* (New Haven, CT: Yale University Press, 2007), 128–130.

44. Burrough, *The Big Rich*, 435; Gibson and Oden, "Launch," 964n1.

45. Robin Givhan, "Astronaut Wally and the Cowboy Hat that Rocketed into Space," *Washington Post*, July 20, 2021. The astronaut/test pilot as cowboy on the western frontier is a familiar trope: Peter Westwick, "Photoessay: An Album of Early Southern California Aviation," in *Blue Sky Metropolis: The Aerospace Century in Southern California*, ed. Peter Westwick (Berkeley: University of California Press, 2012), 24.

46. On the decline of HARC and MCC: Schmandt, *George P. Mitchell*, 96–97, and Ladendorf, "30 Years Ago." One sector where Texas continued to lag, at least through the 1990s, was biotech: Meredith M. Walker, "Biotech Bonanza: Prospects for Texas," *Southwest Economy* 4 (July/August 1999): 1–7; Sarah Williams, "Could Austin—the 'Silicon Hills' of Texas—Become the Next Biotech Hub?" *STAT*, March 9, 2016.

47. Don Lee, "A Nascent Silicon Valley 2.0 in Texas," *Los Angeles Times*, February 10, 2022.

48. Richard A. Walker, "California's Golden Road to Riches: Natural Resources and Regional Capitalism, 1848–1940," *Annals of the Association of American Geographers* 91, no. 1 (2001): 167–199. For example, centrifugal pump technology for farm irrigation was subsequently developed for oil wells, harbor and water infrastructure, and rocket motors: Steven Usselman, "Revolutionary Machine: How Pumps Shaped Modern California," talk at Huntington Library, June 12, 2019.

49. Thomas C. Tuttle, *Growing Jobs: Transforming the Way We Approach Economic Development* (Santa Barbara, CA: Praeger, 2016), 67–70.

50. Elizabeth Findell and Konrad Putzier, " 'Startup City': Breakneck Growth Strains Austin," *Wall Street Journal*, December 27, 2020; "Silicon Valley Is Flooding into a Reluctant Austin," *Bloomberg News*, April 8, 2021.

51. T. R. Fehrenbach, *Lone Star: A History of Texas and the Texans*, 2nd ed. (Boston: Hachette, 2000), 660–661, 669, 704–707; Dominic Sandbrook, *Mad as Hell: The Crisis of the 1970s and the Rise of the Populist Right* (New York: Knopf Doubleday, 2011), 133.

52. Lisa McGirr, *Suburban Warriors: The Origins of the American Right* (Princeton, NJ: Princeton University Press, 2001); Lily Geismer, *Don't Blame Us: Suburban Liberals and the Transformation of the Democratic Party* (Princeton, NJ: Princeton University Press, 2014). On conservatism and modernity, see M. J. Heale, review of McGirr in *Journal of American Studies* 30, no. 2 (2003): 497.

53. Lawrence Wright, "No City Limits," *New Yorker*, February 13 and 20, 2023, 32–49, on 41.

54. James McKinley Jr., "In Texas, a Line in the Curriculum Revives Evolution Debate," *New York Times*, January 21, 2009; Tom Bartlett, "Evolution of the Specious," *Texas Monthly*, March 2014.

55. Richard Barbrook and Andy Cameron, "The Californian Ideology," *Mute* 1, no. 3 (1995); on the reality, as opposed to the mythology, see, for example, Margaret O'Mara, *The Code: Silicon Valley and the Remaking of America* (New York: Penguin, 2019); W. Patrick McCray, "Silicon Valley: A Region High on Historical Amnesia," *Los Angeles Review of Books*, September 19, 2019.

Part IV / Speculations and Spectacles

"The Required Allocations Grew Considerably"

Soviet Science, Military Imperatives, and
the Ambivalent Response to Reagan's Star Wars

Asif Siddiqi

Introduction

Beginning in 1981 and through the rest of the decade, the US Department
of Defense annually issued a lavishly illustrated publication with the im-
posing title *Soviet Military Power*. Distributed free of charge to all federal
officials but principally aimed at the general public, the high-quality soft-
cover book had a single overriding message: the evil empire was armed to
its teeth and it was only going to get worse if the free world didn't respond
in kind. As one of the many rhetorical instruments justifying Ronald Reagan's military buildup, each edition of *Soviet Military Power*, which grew
in size, ominousness, and gravity over the decade, seemed to couch Soviet
weapons in near-apocalyptic terms, implying that their existence could be
a threat to all existence. Matt Stanley notes in chapter 12 of this volume
that "the 1980s were an intensely apocalyptic time that had certain expectations and requirements for talking about the end of the world."[1] One
of these requirements, at least for *Soviet Military Power*, was to lay the problem at the feet of the Soviets. In his preface to the publication, Reagan's

secretary of defense Caspar Weinberger suggested that the book was "designed to assist informed citizens in free nations everywhere to make the choices required to provide for the defense and security necessary to safeguard freedom."[2] Many saw through the language and imagery of the books: most famously, Tom Gervasi published a broadside against *Soviet Military Power* helpfully subtitled *The Pentagon's Propaganda Document, annotated and corrected.*[3] But the hyperbole was the point. What *Soviet Military Power* did was to traffic in both spectacle and speculation, the former expressed in the many vivid artists' renderings of existing or planned Soviet weapons in frozen moments of combat, and the latter characterized by the obvious fact that none of the information in the booklet was sourced or cited; it was just declared as being true. Yet, given that the Soviets themselves wrapped their military operations in a veil of draconian secrecy, *Soviet Military Power*, with its charts, projections, budget numbers, and beautiful paintings, could easily slip from speculative to authoritative depending on the context.

The marriage of spectacle and speculation was manifest not only in the Reagan administration's declarations about Soviet military power but also, not uncoincidentally, in the crown jewel of the Reagan military buildup, the Strategic Defense Initiative (SDI). Advanced as a solution to the bipolar nuclear stalemate, the narrative of SDI was difficult for the public to apprehend without some form of spectacle ("Star Wars!") or speculation ("we *can* win a nuclear war!"). Even experts deeply involved with the implementation of SDI were not immune to these modes of discourse about the program since many of the technologies proposed for SDI—beam weapons, high-energy lasers, kinetic energy armaments—evoked more science fiction than sobering fact. And because SDI's very raison d'être rested on the premise of its effectiveness—basically an unknowable outcome—much of the discussion demanded a leap of faith from the public about what was largely a set of speculative claims about the efficacy of the program.

Spectacle and speculation were notably part of a large volume of commentary on SDI focused on a single question with a seemingly unknowable answer: How would the Soviet Union respond to Reagan's Star Wars program? Tapping into a clear demand to make sense of what the enemy might be doing, political scientists, security studies scholars, think-tank analysts, and political actors offered their versions of the inner workings of the Soviet military and space programs. Given the enormous levels of secrecy surrounding Soviet military activities, most of this writing was rather speculative, depending less on hard information than informed guesses married with bits of rumor or leaked intelligence information.[4] Not surprisingly, this canon disappeared rather abruptly in the early 1990s, coinciding with another disappearance, that of the Soviet Union itself. Subsequently, if anyone seriously considered the Soviet response to SDI, it was more as fodder for the broader argument: did Reagan's commitment to a military buildup bankrupt the Soviet Union and hasten the collapse?[5] Answering this question has largely been the domain of diplomatic historians interested in high politics and gamesmanship between Ronald Reagan and Mikhail Gorbachev (and, sometimes, Margaret Thatcher).[6]

In this chapter, I focus specifically on the Soviet response to Reagan's SDI with a view to foregrounding particularly the stance of powerful elites within the Soviet scientific community who were able to significantly shape that response. There are some good reasons to revisit the nature of the Soviet response to SDI at this juncture, centering less on the high politics of Geneva or Reykjavik than on the Soviet scientific community's ambivalence. First, we now have considerable evidence—much of it archival in nature—to understand the material actions undertaken by the Soviets in the aftermath of Reagan's speech announcing SDI in 1983. Second, while our understanding of Soviet science and technology has been enrichened by the considerable focus on harrowing episodes such as the atomic bomb project, Lysenkoism, and large-scale engineering projects such as the space program and computing, almost all of this focus has been limited in time to the early Cold War.[7] We lack any deep understanding of the forces at play shaping Soviet science and technology in the waning days of the Brezhnev era of stagnation let alone into the Gorbachev era of *perestroika* and *glasnost'*. Soviet considerations of SDI and the possible responses to it involved two overlapping expert communities—Soviet physicists and the designers in charge of the Soviet space program—who, in many ways, were at the peak of their authority in the 1980s. Extending our studies to the 1980s offers deep insights into the practices of these communities in the period just before the collapse of the Soviet Union and thus into their nature in the postsocialist era.

Finally, and most important, revisiting the Soviet scientific response to SDI adds to our understanding of the ways in which the greed for resources was a significant factor in how the Soviets responded to Reagan's challenge at a time when growing defense budgets across the West and East forged more intimate relationships between science and defense. A study of the Soviet response to SDI offers a striking case where the possible paths were articulated in terms such as "symmetrical" and "asymmetrical," which also mapped very clearly onto "very expensive" and "not expensive," respectively. We find here a clear context where powerful stakeholders in the Soviet military–industrial complex, used to big budgets and rapacious appetites, sought big contracts and more funding. Yet, as I show, their feelings of entitlement for increased resources, a kind of greed cultivated for decades during the Cold War, came into direct conflict with the position of elite scientists who ironically, until that point, had been their biggest cheerleaders.

The basic story that I present here covers a brief period between Reagan's initial speech on SDI, given in March 1983, and the final Soviet decision on a plan to respond to SDI, taken in January 1986. Throughout this period, the basic consensus among the Soviet scientific community was that SDI was unworkable and possible to blunt through countermeasures as part of an "asymmetric response." Leaders within the military and its research and development arm in industry, on the other hand, favored producing a Soviet copy of SDI—a "symmetrical" response—that could be combined with an asymmetric one. In December 1984, as part of the next five-year plan, Soviet industry approved increased funding for several existing projects related to missile defense. Unhappy with the modest reaction, military–industrial stakeholders pushed for much bigger action. They packaged their arguments in a language that was as much speculation as it was spectacle, sufficient to convince decision-makers. As a result, a massive party and government decree in July 1985 was signed by Mikhail Gorbachev foregrounding a large symmetrical response known under the code name *Alpha*. Yet soon, the wind appeared to shift once again through the remainder of the year. Six months later, in January 1986, after an extraordinary set of discussions led by Soviet scientists, *Alpha* was downgraded to a less expensive and largely asymmetrical set of initiatives. These programs essentially formed the backbone of Soviet efforts in the waning days of the Cold War to implement a coherent opposition to Reagan's SDI.

In recovering some of this history, we find first and foremost that the Soviet response, as it played out behind closed doors, was marked by an unusual level of ambivalence, delay, and instability. One aspect of this difficult process was common across both the secret and the open worlds: the position of elite Soviet physicists, who maintained from the very beginning that SDI was fundamentally unworkable and ultimately a fool's errand. As I show here, this consistency across the secret and open worlds destabilized the normative functioning of Soviet decision-making in the Soviet space program. In earlier times, upon assessment of new US military or space projects, the space program was usually allocated considerable funds to implement parallel, expensive, and symmetrical projects—that is, greed for military systems was typically rewarded, even when there was considerable doubt about the usefulness of certain systems or dysfunctional infighting among weapons designers.[8] But in the case of SDI, Soviet scientists' openly stated insistence that the American SDI was basically useless—a position that effectively negated both the spectacular and the speculative aspects of SDI—also implied that a domestic SDI would be equally useless. This position naturally posed a problem for those in the military who desired (and fully expected) a much bigger symmetrical response.

Typically, the military and the defense industry could push through their own demands, but in this case, they were hobbled by two additional contextual factors. First, there was a sudden disjuncture in the leadership of the Soviet military–industrial complex that interrupted a level of continuity established decades before. This changing-of-the-guard introduced uncertainty into the process of formulating an appropriate response to Reagan's SDI. By the time the Politburo approved an all-out program to counter SDI in January 1986, its scope had changed so many times that the new leadership was essentially not even sure of the scope of the response. Second, the Soviets had already invested significant resources prior to 1983 on SDI-type programs, which complicated the ability of the party and government to act quickly *after* 1983, since institutional inertia favored the activation of *new* programs to respond to SDI, not to resurrect old ones. There was greed for novelty, not extending existing projects. The result was a kind of organizational vacuum that allowed Soviet scientists to resist pressure to accede to the appetite of those who sought a more expensive symmetrical response.

The chapter is organized in three parts. First, I reconstruct the public response of Soviet physicists to Ronald Reagan's speech in 1983 announcing the SDI program. Here, I am interested especially in how their positions in both public and private were quite similar (i.e., that SDI was untenable and the only appropriate response was an asymmetrical one). In the second section, I explore those factors within the Soviet space and defense industry that impeded a clear and unambiguous response to SDI, including the considerable work carried out on space-based missile defense programs in the Soviet Union *before* Reagan's speech in 1983. In the third and final part, I show how all of these factors shaped the actual Soviet response at a granular level, ultimately deviating from precedent, which would have favored a much more exorbitant response reflecting the culture of greed prevalent in the Soviet defense industry by the 1980s.

The Phantom Menace

In a televised address to the nation made on March 23, 1983, President Ronald Reagan made a bold announcement that seemingly and suddenly shifted the terms and some would say equilibrium of the bipolar nuclear arms race by proposing a massive new project, the Strategic Defense Initiative (SDI). In simple terms, the idea was to build a shield against Soviet nuclear weapons, a kind of metaphorical umbrella that would protect the American landmass from incoming nuclear missiles. Reagan couched his call in utopian terms, challenging the American "scientific community who gave us nuclear weapons to turn their great talents to the cause of mankind and world peace; to give us the means of rendering these nuclear weapons impotent and obsolete." This would be accomplished with a defensive network "that . . . could intercept and destroy strategic ballistic missiles before they reached [US] soil."[9] Reagan's speech and the subsequent program evoked a spectrum of reactions both nationally and internationally. Many saw it as fundamentally destabilizing, since the creation of a credible missile shield would undermine one of the basic tenets of the nuclear arms race, the notion of mutually assured destruction (MAD) that held the two superpowers in a stalemate. SDI made the unthinkable possible: a survivable first strike against the Soviet Union. Others believed that the program would escalate the nuclear arms race into outer space since much of the activity of defending incoming rockets would take place beyond the Earth's atmosphere. Finally, some saw the program as foolhardy since its success would depend on a total blackout—in other words, even a single missile passing through the shield would render the whole project a failure. Many American physicists believed that absolute success was virtually impossible and that the whole project would be a monumental waste of money.[10]

The official and public Soviet response to Reagan's speech was striking. For decades, the official Soviet media—newspapers, popular journals, specialized publications—adopted a critical tone on alleged American militarism across the globe, which was framed as a kind of modern form of empire building now in the guise of global capitalism supported by American military bases. In the case of the space program, Soviet commentators, often cosmonauts but more frequently unidentified writers, published regular critiques of American spy satellites or supposed plans for weapons in space. These appeared in important mouthpieces of the military and space programs, such as *Aviatsiia i kosmonavtika* (Aviation and Cosmonautics), *Kryl'ia rodiny* (Wings of the Motherland), or *Krasnaia zvezda* (Red Star), and reproduced a number of usual talking points, contrasting American militarism with the peaceful internationalism of the Soviet space program. Their regularity, moderated tone, and relatively small numbers by the late 1970s ensured that they were largely ignored or filed away by Western analysts.[11]

The response to Reagan's speech, however, differed markedly in two respects. First there was the issue of scale. The ubiquity and relentless nature of Soviet critiques of SDI produced an unavoidable feeling that the contents of the speech had produced a substantive reaction among the elites of the Soviet military–industrial complex. For about two years, the Soviets published an avalanche of commentary on SDI from the highest party and government officials in the country to military officers and prominent journalists. In her exhaustive analysis of this discourse, the late Mary C. FitzGerald found some broad thematic obsessions, including, in particular, the belief among Soviet officials that SDI was really an offensive program hiding beneath the veneer of defense, that it would allow the United States the option to initiate a first strike against the Soviet Union, and that although based in space, its ultimate goal was to strike Soviet ground targets, such as cities.[12] Overall, one message was communicated unequivocally: SDI would fundamentally undercut strategic parity, which had been achieved at great cost by the Soviets.[13]

There was another aspect of the anti-SDI critique: the major role played by Soviet scientists in advancing commentary on Star Wars. From the Soviet perspective, SDI was substantively different from other American military or foreign policy initiatives that were perceived as militaristic. Unlike, say, US aid to dictators in Latin America or the development of the strategic B-2 bomber, there was one fundamental aspect of SDI—its grounding in various applied physics disciplines—that made it an appropriate subject for commentary from professional physicists from both sides of the ideological divide. As became evident over time, physicists in the American scientific community, with some key exceptions, were largely critical of SDI. The opinion of Wolfgang K. H. Panofsky, the director of the Stanford Linear Accelerator, who argued that in addition to raising "false hopes," SDI "poses grave dangers to national and world security," was probably representative of a large number of physicists.[14] American physicists critical of SDI also couched some of their opprobrium toward colleagues who they saw as money-grubbing scientists too happy to draw funding from SDI-related projects. In the eyes of many, this kind of greed, coupled with a Manichean view of good versus evil, was manifested most famously in the person of Edward Teller, the theoretical physicist who had contributed to the development of the hydrogen bomb and who repeatedly and aggressively sold the benefits of SDI.[15]

In the Soviet case, prominent physicists as well as specialists in other disciplines weighed in on SDI. In fact, their public engagement on the issue was unusually prolific and vibrant, putting them not only in the public eye within the Soviet Union but also especially beyond, in the Western media. There was no clear precedent for this kind of foregrounding of scientific expertise in public forums. Soviet academics and specialists previously had adopted two modes of commentary on US large-scale scientific and technical programs. They wrote esoteric accounts of the technical aspects of such projects—this was especially true in the 1950s when secrecy prevented a similar recounting of Soviet projects; speaking of American programs thus became the principal way to communicate to the public that similar work was going on at home without revealing the details.[16] A second mode was to speak in generalities of American achievements with an implicit acknowledgment that Soviet science was superior—this was especially true in the 1960s, for example, when Soviet scientists downplayed the achievements of Apollo while suggesting that their robotic lunar sample collector was qualitatively superior.[17]

While not organized with any kind of coherent deliberation, SDI forced a reformulation of these modes. A large complement of elite Soviet scientists, including physicists (Evgenii Velikhov, Roal'd Sagdeev, Aleksandr Aleksandrov), mathematicians (Gurii Marchuk), aerodynamicists (Vsevolod Avduevskii), and control systems experts (Boris Raushenbakh), offered comment on SDI in prominent public venues. The main message was impossible to miss: SDI would be fundamentally unworkable. Two points were germane here. First, in order for SDI to be successful, the missile shield had to be absolutely perfect, impeding any incoming missiles from piercing through to their targets. But, they argued, the kind of weapons systems proposed for SDI—directed energy weapons (including particle beams, microwaves, and various types of lasers) and kinetic energy systems (involving mass accelerators or electromagnetic guns, for example)—could not be perfected to such a degree as to be workable. Second, they argued that even if they were fully impermeable, the Soviet Union could, with relatively less investment than the United States, mount an "asymmetric" response to completely neutralize the SDI shield.[18]

In the wake of Reagan's SDI announcement, academician Evgenii Pavlovich Velikhov (1935–), one of the preeminent active physicists in the USSR and an expert in just the kind of weapons proposed by SDI, assumed a leading position on behalf of Soviet scientists on SDI. An up-and-coming star in Soviet science, Velikhov had a wide range of appointments and obligations, including vice president of the Soviet Academy of Sciences and a senior position of the famous Kurchatov Institute of Atomic Energy. He reached the peak of his career in the late 1980s as a science advisor to Mikhail Gorbachev while still firmly fitting into the party-state structure, exemplified by his election as a deputy in the Supreme Soviet (in 1980) and his service as a candidate member of the Central Committee (from 1986). Less well known at the time, Velikhov had been involved, since the late 1960s, in a number of highly secret weapons programs to develop ground-based lasers under the code names *Omega* and *Terra-3*.[19] His experience with these programs gave him unusual authority within the Soviet military–industrial complex.

As one would expect, Velikhov was critical of the US position but always couched his critique in technical terms, avoiding both speculative claims and spectacular imagery and, in fact, embracing the exact opposite rhetoric of emphatic and sobering claims. In 1985, to the newspaper *Sovetskaia rossiia*, he argued that "the plans for creating an absolutely impregnable anti-missile defense system with space-based components are an illusion not confirmed by any modern scientific or technical notions. The scientists' conclusions, based on profound knowledge of the fundamental laws of nature and a comprehensive assessment of the state of and prospects for developing the technology, rule out all variant readings and different interpretations. They are categorical and the arguments are conclusive."[20]

A second point about countermeasures was made equally forcefully. Here, in a major article in *Pravda*, he notes that "even if we assume that it will be possible to solve certain tricky (from the scientific-technical viewpoint) 'defense' problems, the anti-missile defense system will turn out to be highly vulnerable to various countermeasures. The point is that there are always simpler and cheaper methods of overcoming the most sophisticated 'defense systems.' The creation of a space anti-missile defense system [such as SDI] would quickly lead to offensive means to overcome it."[21] All of this ended on a grim note: "The hopes of U.S. strategists who consider that their anti-missile defense system could save the United States from an annihilating retaliatory strike, are . . . illusory."[22]

Velikhov and other Soviet physicists came to this position not without serious inquiry. Two months after Reagan's speech, Soviet scientists had formed the so-called Committee for the Defense of Peace against the Nuclear Threat, which involved at its inception twenty-five elite Soviet scientists from many different disciplines, including the social sciences.[23] The members, who were skewed toward disciplines that had connections to space science, very quickly established contacts with American physicists who were similarly skeptical of SDI and sought to forge an international alliance that, at least, in theory if not practice sought to present a universalist scientific position that was above "politics." Most crucially, they produced a detailed and exhaustive analysis of SDI using openly published data from American sources and issued a series of publications, originally only in Russian but then translated into English with titles designed to alarm.[24] These prolific publications, edited by Velikhov and coauthored often with Sagdeev and a well-known political scientist, Andrei Kokoshin, were designed to convince a Western audience, but they were not simply cynical exercises in propaganda. Their conclusions, in fact, represented concurrent studies conducted in secret by many of the same Soviet scientists on the Committee for the Defense of Peace against the Nuclear Threat on the efficacy of the American SDI program. And this—a transgression of the open and secret worlds—proved to be one of the key factors that undermined and complicated the Soviet military response to SDI.

The Force Awakens

If the official and public Soviet response to SDI was one of near-hysteria, the internal and secret one was not so different. A slew of recently available information has filled in in some key details on the causes and nature of this alarm. Peter Westwick has persuasively argued that the Soviets' reaction to the SDI announcement was particularly intense—frenzied and obsessed even—due principally to the perception among key Soviet officials of the offensive nature of SDI systems. Specifically, he identifies a particular fixation on the notion that SDI would open the way for "space-strike weapons," which to the Soviets represented "a new generation of space-based beam weapons that could instantly strike targets on Soviet territory at any time."[25] In his telling, this was one of the reasons that, early on at least, Soviet arms control negotiators linked SDI to any discussion over reductions in strategic weapons. Pavel Podvig too has offered trenchant analysis on the Soviet response to SDI, making use of the papers of V. L. Kataev, a party apparatchik who worked in the Central Committee unit dedicated to strategic weapons.[26] In illuminating the dizzying array of projects that were part of the response to SDI, Podvig shows that the Soviet defense industry "took advantage of the situation created by the SDI initiative to increase the levels of funding and get access to additional resources for its programs." Greed, in other words, won the day, as the Soviet Union implemented, in secret, a combination of asymmetrical and symmetrical projects in the 1985–1986 period.[27] But this is only part of the story.

A hitherto unused source—an official three-volume history of the Soviet and Russian military space forces—adds a further dimension to understanding the vicissitudes of the Soviet response.[28] We find here that the Soviet response was actually marred by chaotic decision-making that, despite the best efforts of stakeholders from the military, the defense industry, the space program, and the scientific community, was never a fully committed one.

Why the chaos? First, the position of the scientific community, led by Velikhov and Sagdeev, was one maintained both in public and in secret consistently since Reagan's speech in 1983. The Velikhov commission's public positions were well known, having been published widely. But we now have confirmed evidence that he and his colleagues were saying the exact same thing in private. A secret commission under Velikhov had, already by late 1983, come to these conclusions.[29] A later iteration, known as an "Interbranch Working Group," set up by decree of the Military–Industrial Commission on July 6, 1985, and composed of representatives of the industry, the Academy of Sciences, and the Ministry of Defense, was unequivocal in maintaining the position that SDI would ultimately be ineffective. In a classified and surprisingly candid report prepared for the personal use of Mikhail Gorbachev before his meeting with Reagan in Reykjavik in October 1986, the assessment of Velikhov's commission was thus put: "An evaluation of the possibilities of the USA's ABM [anti-ballistic missile] system, as created under the SDI program, showed, that a 100-percent interception of all missiles and warheads is practically impossible to achieve. An analysis, carried out by the Ministry of Defense together with industry (under the scientific-research themes *Duel'-2*, *Vekha-2*, [and] *Protivodeistvie*), showed that even with the creation of some [SDI] echelons, the effectiveness of the USA's ABM system in the 2010–2020 timeframe could be around 0.99 (i.e., 1 percent of warheads could penetrate to the strike objects), and that's only theoretically."[30] With this position maintained persistently from 1983 to 1986 by the scientific community, the powerful Soviet defense industrial constituency, who favored a symmetrical response, found a very powerful adversary. The more the Velikhov commission pushed the line that SDI was basically useless, as they did in public repeatedly, the more it undercut the preference of Soviet defense industrialists in private.

The defense industry's inability to fully overcome the implications of the scientific community's hardline critique of the effectiveness of SDI was related to a second factor: the profound changing-of-the-guard at the apex of the Soviet military–industrial complex from 1983 to 1985. In those three years, three of the most important architects of Soviet military power in the postwar era, Dmitrii Ustinov, Leonid Smirnov, and Sergei Afanas'ev, left their posts. Ustinov, who had been appointed by Stalin to be the founding manager of the postwar ballistic missile program in 1946, presided over several generations of weapons development before moving to a policy position in the party hierarchy in 1965, thus overseeing the entire defense and space sectors for about two decades.[31] Under him, Smirnov had been serving, since 1963, as the chairman of the Military–Industrial Commission, a mysterious body that had a name worthy of a supervillain organization but in fact coordinated all contracts for Soviet military research, development, and production. A woefully understudied and underappreciated personage, Smirnov's signature is in the archives of every single weapon developed in the Soviet Union in the Brezhnev era.[32] Finally, under Smirnov was Afanas'ev, the minister in charge of all missile and space developments in the Soviet Union. As the head of the obviously named Ministry of General Machine Building (what if anything was a spaceship but a "general machine"?), Afanas'ev acquired the name "Big Hammer," for both his physical stature and his terrifying personality.[33] Together, Ustinov, Smirnov, and Afanas'ev produced a kind of equilibrium in Soviet Big Science that prized overfunding, overproduction, and a semi-chaotic competitive environment that kept weapons designers constantly under pressure.[34] Here, the elite scientific community performed an important but largely advisory role in producing the applied research necessary to maintain Soviet power. But barring a few exceptions, scientists rarely acted against the interests of the military–industrial complex. In a few cases where scientific input was crucial—such as the decision to develop a copy of NASA's Space Shuttle or the decision to build the next generation of intercontinental ballistic missiles (ICBMs) with multiple independently targetable reentry vehicles—dissent was mild or suppressed outright.

Afanas'ev was fired from his job in 1983. Ustinov died in 1984. And in a culling of the old guard, Smirnov was dismissed by Gorbachev in 1985.[35] The loss of these three architects of the Soviet military–industrial complex in such a brief time span was unprecedented and disruptive, producing ruptures in the so-called "administrative-command system" of normative weapons development. An official history of the Russian military space forces lamented that the decision-making for the response to SDI (and, in general, all military space activities) "had to proceed without Dmitrii Fedorovich" who "had promised [further] space systems and the formation of future military space assets."[36] Ustinov's replacement as the Communist Party's secretary in charge of space (i.e., the de facto head of the Soviet space program) was an old Leningrad party hack, Grigorii Romanov, who had zero experience in the sector and, according to Sagdeev, was "the most anti-intellectual kind of party apparatchik" whose "main idea was to undermine the role of [the late] Ustinov."[37] Afanas'ev, who had been heavily invested in maintaining his grip on the generous disbursements to his pet programs, including both the Buran space shuttle and the ongoing Salyut space station missions, was replaced by Oleg Baklanov, who evidently supported a massive symmetrical response to Reagan's SDI with whole new projects. The overall effect was a dissipation of the united front usually presented by the elite of the Soviet military–industrial complex, who could be trusted to present and select options for countering US military power relatively quickly and with large investments spread among multiple design bureaus and research institutes. The scientific community played an advisory role in this equation, serving on various ad hoc advisory councils that were conjured up and dissolved as soon they put their imprimatur on some already preagreed decision. In this context, the Velikhov commission's claim, made both in secret and in public, at home and abroad, that a direct symmetrical response made little sense, was difficult to counter, despite considerable motivation among actors in the military and industry to do so.

One more factor—a counterintuitive one—complicated and ultimately delayed a direct symmetrical Soviet response to SDI: the Soviet defense industry had already initiated a whole host of SDI-type programs long before Reagan gave his famous speech in 1983. For example, alarmed at the prospect of the US building a space shuttle supposedly capable of sudden (and possibly invulnerable) strikes against ground targets in the Soviet Union, engineers at one of the largest spacecraft design firms in the Soviet Union, NPO Energiia, were tasked, in 1976, with coming up with a range of options to respond in kind. One of these was a Soviet space shuttle (later called "Buran"), looking much like the American one that would be launched on a super-rocket (later named "Energiia" after the organization).[38] Attached to this massive initiative were a couple of smaller research and development profiles that resembled elements of Reagan's (later) SDI project: "integrated research was carried out on defining the possible ways of creating space assets with the ability to the solve the goals of destroying military spacecraft, ballistic missiles in flight, and also special important, air, naval, and ground targets."[39] The research also involved reequipping the "civilian" Salyut space station, which in the late 1970s was playing host to visiting cosmonauts from brotherly socialist nations, into two "battle" versions, one to fire kinetic energy projectiles (known as *Kaskad*, after the Russian word for "cascade") and another to fire lasers (known as *Skif* after "Scythian"). For the Soviets at the time, war in space was an imminent reality, not an imagined one.

There were more investments in SDI-type systems in the early 1980s. It's now clear that the Soviets had already began to discuss a significant investment in space weapons a year *before* Reagan's speech, in response to a statement from the Reagan administration on the militarization of space, the so-called National Space Policy issued on July 4, 1982, on the same day that Reagan welcomed home the crew of the fourth mission of the Space Shuttle *Columbia*. Also known as the National Security Decision Directive 42 (NSDD 42), the document reiterated a commitment to a robust space program with the Space Shuttle playing a central role in the future of American national security. The classified version also emphasized the commitment to various space weapons systems, including anti-satellite weapons. It gave overall control of Reagan's space policy over to the National Security Council rather than any civilian advisory body.[40]

The text of the directive—and it appears that the Soviets had access to the classified version rather than the one released to the press—clearly alarmed Soviet Minister of Defense Dmitrii Ustinov, the seventy-three-year-old eminence of the Soviet military–industrial complex. He immediately instructed the powerful industrial manager in charge of all Soviet defense research and production, Leonid Smirnov, to prepare, in two months, a report analyzing US intentions and capabilities in space while also suggesting proposals for implementation at home that focused on "accelerating the transition of a number of existing research projects from [the research stage to experimental stage], [as well as] paying special attention to the creation of new promising space assets to solve the problems of the Ministry of Defense."[41] Representatives from the nine primary ministries involved in the defense industry as well service officers were involved in articulating the problem, which was framed in terms of the already-in-play eleventh Five-Year Plan, covering 1981 to 1985. The report, produced by September 1, 1982, on behalf of the chief of the general staff, Marshal Nikolai Ogarkov, acknowledged that the Reagan statement signaled a fundamental shift in American posture in military space activities. The "new goal" of the Soviet space program, the authors of the report proposed, should be "to actively oppose the aspirations of the United States to gain dominance in space." The report recommended "the achievement of parity and prevention of U.S. superiority in the use of outer space for military purposes" with expenditures on military space to be increased by 35 percent in the twelfth Five-Year Plan (in 1986–1990) and 50 percent in the subsequent plan.[42] Most ominously, there was acknowledgment that these efforts would have to be undertaken "at the expense of other, less important [programs]," meaning the routine civilian Salyut space missions.[43]

These were only recommendations, but Reagan's stance on the weaponization of space clearly alarmed a large cross section of leading figures in the Soviet military and space programs. The result was that already approved space-based missile defense programs assumed a higher profile in discussions of militarization. The Ministry of the Radio Industry—the center for all radar and ABM efforts in the Soviet Union—had already, in the late 1970s, initiated a very large research program under the name of *Fon-1* (*fon* is the Russian word for "background") to explore theoretical research on "different kinds of beam weapons, electromagnetic rail guns, [and] anti-ballistic missiles."[44] Meanwhile, General Designer Valentin Glushko, in charge of the civilian Salyut long-duration station program, also foregrounded his pet space projects to be more welcoming to space defense experiments. Specifically, he reoriented the experimental work on a large science platform that was supposed to be linked to the Salyut-7 "civilian" space station to host a larger number of military experiments to support space-based defense. In August 1982, the minister in charge of the Soviet space program, Sergei Afanas'ev, signed an order to add a special suite of instruments (called *Pion-K*) to a future Salyut mission that cosmonauts would use as part of a program known as *Oktant* to advance space-based antiballistic missile defense and enemy satellite detection.[45] All this happened before Reagan said a single word about the Strategic Defense Initiative.

But when it came time to respond to Reagan's SDI, these "old" programs proved inadequate. Instead, stakeholders expressed a desire, a greed, for *novelty*. Andrei Kokoshin, one of the leading members on both the public and private Velikhov commissions, recalled later that "instinctively, many state, party, and military leaders, as in previous years, were more inclined to respond 'point to point' (*ostrie protiv ostriia*), to adopt a strategy of symmetric measures."[46] There was, in other words, considerable pressure to initiate a program-to-program response, which was typically the precedent in the Soviet defense and space worlds. We see this pattern played out, for example, with many high-profile space projects since the mid-1950s, including the first Soviet satellite project, approved ten days after the Eisenhower administration announced their intention to launch one; the Soviet military space station program, Almaz, an analog to the US Air Force's Manned Orbiting Laboratory (MOL); and the crewed lunar landing program, N1–L3, a direct response to Apollo.[47] But in the case of responding to SDI, in a setting marred by a deep discontinuity in decision-making and the existence of older anti-SDI projects, the scientific community's full-on critique of the American program continuously hobbled an appropriate response.

The Empire Strikes Back

So what *did* the Soviets end up doing after Reagan's speech? Their actions from 1984 to 1986 show a great degree of ambivalence and indecision, with the scientific community seemingly at odds with powerful actors in the military–industrial complex. Initially at least, there was pressure for an overwhelming response to SDI. One of the first evaluations of the efficacy of Reagan's SDI was carried out at a secret defense industry thinktank known as "TsNII-50," a research institute in the Moscow suburb of Bol'shevo whose mandate was to develop strategy for future military actions in space. Their study showed that SDI was "a real threat on the part of the United States to undermine existing parity in the field of strategic arms and made it imperative to take timely response measures."[48] This initial evaluation, that Star Wars was a potentially catastrophic intervention into the arms race, prompted a set of initiatives that were folded into a general plan of rearmament covering the period from 1986 to 1995 and approved by the Soviet party and government in December 1984.[49] The plan included a number of projects "reactivated" by Dmitrii Ustinov, including a large space-based gas dynamic laser installation known as *Skif* to be launched by Buran, the Soviet version of the Space Shuttle, as well as a new anti-satellite project known as *Kontakt*, which would use a MiG-31 jet fighter to shoot satellites out of orbit.[50] The older *Fon-1* research theme was transitioned into *Fon-2*, now focused on asymmetrical responses to SDI (i.e., to neutralize the power of SDI through "non-lethal weapons" such as electromagnetic pulses, lasers, and microwave devices).[51]

Yet, already by the end of 1984 it was clear to some highly placed military and industrial figures that these initial measures would be insufficient to deter SDI and that a further consolidated plan was required. Through early 1985, driven largely by the momentum of the TsNII-50 report, which essentially called for a massive symmetrical response, a large number of players from the Soviet military–industrial complex were able to mobilize and insert their pet projects into a new integrated plan. The ideology at this point was a full-on symmetric response—a Soviet version of SDI—driven by Ustinov's preferred practice of "point-to-point" responses. The first step in moving all of this into a higher footing was a joint party–government decree issued and signed by Mikhail Gorbachev (representing the party) and old-timer Nikolai Tikhonov (representing the government) on July 15, 1985.[52]

Although the decree was focused on both symmetric and asymmetric measures, the focus was clearly on the former. The envisioned objective, which was still in the future and lacking a firm commitment, was "to create a single defensive system in space and from space [under the code name] *Al'fa* [or *Alpha*]. As part of this system," the idea was to "create 'space-to-space' [and] 'space-to-earth' space strike assets [as well as] a space-based ABM echelon."[53] There is the suggestion that this stage of research could be shifted into a more aggressive posture as needed within ten years: "The implementation of comprehensive long-term missile defense programs will make it possible by 1995 to create a technical and technological reserve in case it is necessary to deploy a multi-layer missile defense system." Finally, the decree included some modest concessions to an asymmetrical response to blunt the power of SDI: "The nation will also deploy research projects to improve domestic strategic offensive weapons, aimed at reducing the effectiveness of the SDI system (the *Protivodeistvie* [or *Counteraction*] program)."[54]

Four large broad-ranged research and development programs involving a considerable amount of basic research, testing facilities, and experimental models were approved. Two focused on symmetric responses (D-20 and SK-1000) and two on asymmetric ones (*Kontseptsie-R* and *Protivodeistvie*). The symmetric responses were split according to ministry assignments, a standard Soviet industrial policy to distribute big juicy projects according to ministry rather than a more logical cross-industrial consortium. D-20, assigned to the Ministry of the Radio Industry, was largely aimed at producing ground-based systems to fight incoming missiles, while the more ambitious SK-1000, assigned to the "space" ministry (the Ministry of General Machine Building), also involved space-based weapons. Many of the weapons systems brought under these two symmetric umbrella profiles were already ongoing, such as various ground-based antiballistic missile projects under D-20 or the mainstream Soviet space program involving the Salyut and Mir space stations, now consolidated under SK-1000.[55] Glushko's Energiia firm, already responsible for the bulk of the Soviet space program—cosmonauts, space stations, space shuttles, and so on—was contracted to be the prime contractor for *Alpha*, the Soviet version of SDI.[56]

Taken in isolation, the July 1985 decision approving *Alpha* displayed, in Podvig's words, "a major commitment to [the] development of a broad range of missile defense and space weapons technologies."[57] Yet it's clear now that, even at the moment of its issuance, this "decision" was fully contingent on further elaboration and not binding beyond the broad contours of the effort. At this point, normally, the Military–Industrial Commission, the coordinating authority over the entire Soviet defense industry, would issue specific contracts to organizations to carry out their part of the work. But Ustinov was no longer there and Afanas'ev was gone too. Smirnov, the last remaining member of the military–industrial troika, was close to retirement and already sidelined. There was another problem related to timing: the July 1985 decision was issued out of sync with standard five-year planning and, in fact, required a kind of delinking from the normative rhythms of Soviet research and development due to the urgency of the matter.

Still, the work had to be done: how to put all this into practice in the form of contracts, deliverables, and so on? In the fall of 1985, a large number of organizations representing the scientific community, the defense industry, think tanks, and several leading space organizations, including Energiia, broke down the July 1985 decree and systematically examined each option presented. These considerations included, foremost, an accounting of the costs of each, the need to develop extensive ground-testing infrastructure, links to other projects, possible locations of projects in an organizational matrix, the need for more fundamental research, and, as was typical of Soviet management practices, the need to link fielded systems into "a single defense system in space and from space in terms of creation time, control, communications, transmissions, observation of space, and access to space."[58]

This period set off alarm bells to many. First, the analysis showed that, because of the many prior existing programs, now supplemented with new ones, there was considerable "duplication of topics." Second, because of the need for substantive basic research on directed energy weapons, advanced computing, and command and control, "the required allocations grew considerably." When the status of the programs was examined first by Chief of the General Staff Marshal Sergei Akhromeev and then sent up to the new minister of defense, Marshal Sergei Sokolov, both "expressed concerns about the too high-level of funding required and the slow pace of alignment with other programs," with the latter a reference to the problem of duplication. When Sokolov ordered a complete report of the road ahead to be prepared by December 7, 1985, the Military–Industrial Commission moved into action. By this time, the final member of the old troika, Leonid Smirnov, had been forced into retirement, and continuity with the "old regime" was irrevocably severed.

In this context, Smirnov's replacement, Iurii Masliukov, who had no connection or interest in the space industry, ordered a crash review of *Alpha*, but with one key element reordered: he put physicist Evgenii Velikhov, then vice president of the USSR Academy of Sciences, in charge of the preparing the final plan to respond to SDI. Velikhov had, of course, chaired other commissions and councils dealing with SDI, but this one included a bevy of gray-faced bureaucrats from the Soviet defense industry, Communist Party, and military, including prominent deputy ministers, various generals, economic planning officials, and leading space scientists and space program chief designers. But the insertion of the scientific community's opinions at such a high level virtually ensured that the balance between symmetry and asymmetry in the response to SDI would tip from the former to the latter. For a whole week, from December 4 to 11, 1985, the commission convened in a remote country location supposedly isolated from the rest of the world ("without telephones") and considered the July 1985 decree and all the subsequent analysis of its potential. The attendees flew through three iterations of a plan before a fourth one was agreed upon, with key input and compromise between leaders of the space industry, particularly Valentin Glushko, and leaders of the Soviet military space forces.[59]

The outcome of all this was another government decree, issued in January 1986, that essentially superseded the one from six months before and downgraded *Alpha*, reverting back to some of the more modest research initiatives approved much earlier in December 1984—ones that had been considered "insufficient" at the time. With the agreement of forty-eight different branches of the Soviet government, two major aspects of the July 1985 decision were downgraded. First, it was decided to "significantly reduce the amount of funding." Many of the major projects under the umbrella of a symmetric response were downgraded or eliminated, with the focus being more on asymmetric responses and fundamental research. Second, any decision was no longer tied to the five-year plan horizon but to be "establish[ed] by *annual* decisions of the government," meaning that everything was contingent and potentially under threat of cancellation. The program would now include two broad *research* directions: one focused on "systems" and another on "fundamental research," carried out in two phases, 1986–1990 and 1991–1995. In terms of experimental work, all of it would be directed by Glushko's Energiia organization and performed on their already ongoing programs such as the Salyut and Mir space stations and the in-development Buran space shuttle. Finally, there would be investments in "the creation of [other] space complexes" and their integration into the larger space program.[60] The projected cost for all this? The disbursements for the first period, 1986–1990, were estimated to be 30.7 billion rubles (roughly $34.2 billion in 1985 amounts), downgraded from an estimated 50 billion rubles.[61]

This final decision in January 1986 was both a realignment and a tipping point. There were a number of further twists and turns in the saga of the Soviet response to SDI, but from this point on, the scientific community's agency, largely led by Velikhov, Sagdeev, and the new president of the Academy of Sciences, mathematician Gurii Marchik, was to insist, behind closed doors, on a measured response to the perceived power of SDI. This was not always a straight line, but the access that Velikhov had to Mikhail Gorbachev played a significant role in ensuring that there were no major reversals of this policy. As Sagdeev later noted, Velikhov, having "firmly launched himself in the front line of opposition to SDI," was a central obstacle to increased military budgets at home. Military–industrial leaders, Sagdeev wrote, "did not know yet how much Velikhov's efforts were to eventually undermine their interests in getting budgets and providing a stable and secure program for themselves."[62]

Conclusions

On both sides of the Iron Curtain, discussions surrounding SDI were frequently framed in language that drew on both the speculative and the spectacular. This was partly by design since SDI involved technical systems that were still in the realm of conjecture whose complete effectiveness could protect each nation from the catastrophe of nuclear holocaust. Those advocating for SDI (or, in the Soviet case, a full-scale symmetrical response) found such assessments unhelpful to their cause since they inevitably offered qualifications, cast doubt, and suggested caution. In exploring the Soviet response to SDI and putting the Soviet scientific community at its center, we find that their position, one of supposed "objective" science, put them at loggerheads with the prevailing sentiment of the Soviet military–industrial complex. In his memoirs, Evgenii Velikhov pointed to this very problem but more as a point of pride. Remembering plans to publish an evaluation of SDI jointly with his American counterparts in both the Soviet and American media, Velikhov noted that "we decided to write and publish it independently with the main goal of educating the public, since there was practically no information in the press in the USSR, while in the USA there was the usual mixture of military-industrial *advertising*, the schizophrenic *fantasy* of General [George J.] Keegan and all sorts of hack writers, painfully familiar to us today. We proceeded from the idea that the laws of physics in the U.S.A. and the USSR are the same, and [our] analyses were based only on physics, without using any secret information."[63] The implication here was that because this was an assessment free of "advertising" and "fantasy"—which we might interpret as forms of speculation and spectacle—it would not only offer a corrective to prevailing wisdom on SDI but also possibly shape the Soviet response. And as we see here, this is exactly what happened: the intervention of Soviet scientists to advance an "objective" view of the situation caused a crisis within the Soviet program and eventually enervated the power of an effective response.

In practical terms, Soviet scientists, by assessing SDI as untenable, essentially endorsed a domestic response that was strictly asymmetrical in nature, since they believed that through relatively inexpensive and less sophisticated means, any missile shield could be penetrated. Such a view went against the principal actors of the Soviet military–industrial complex, who, accustomed to a culture free of resource constraints where for decades they could be as greedy as necessary, were determined to launch a symmetrical response, a Soviet version of SDI that could generate new contracts and new work for years if not decades. Typically, as a constituency, Soviet scientists, while playing an important advisory role in the Soviet space and military programs, had never fundamentally dislodged an existing commitment to a certain path in space or military development. Until SDI.

Why were Soviet scientists able to essentially derail a commitment by the Soviet military–industrial complex to mount a symmetric response? Two factors stand out. The first was the discontinuity at the top leadership of the defense industry and the loss of three key actors in the space of three years from 1983 to 1985. This instability not only opened up a period of zigzagging decision-making but also allowed Velikhov and his allies to seize the high ground in terms of their advisory role on decisions in support of the SDI response. Undoubtedly, Gorbachev's ascendance to power produced a welcoming climate for Velikhov, but the relative deemphasis of the symmetric option was fostered more directly by the disruption caused by the loss of Ustinov. Smirnov, and Afanas'ev. Second, within this moment of instability, the military were unable to mount a sufficient defense of the much more expensive symmetric option because of the plethora of missile defense projects already ongoing in the Soviet Union. Here, a deep awareness of the multifaceted problems of missile defense, dating back to the late 1960s, paradoxically hampered rather than helped decide on the specific symmetric and asymmetric options against SDI. In other words, *there was not only a greed for resources but also a greed for novelty.* The result was a poorly coordinated set of projects that seemed more chaotic than thought through. For a brief period, the big beneficiary of all of this turned out to be Valentin Glushko's Energiia corporation, given a bevy of new contracts related to militarizing the Mir space station. Ultimately, the speed and pace with which all this fell apart was remarkable: Energiia (the rocket) and Buran (the shuttle) flew once in 1987 and once in 1988, respectively, and that was it. The following year, Glushko died. In two years, the Soviet space program became the Russian space program, and Glushko's Mir, which might have been the heart of an anti-SDI program, suddenly became the host not to visiting Buran shuttles with weapons on them but, as Margaret Weitekamp points out in chapter 5 in this

volume, to cosmonauts selling advertisements for Pepsi and MTV, spectacle and speculation of an entirely different sort than that of SDI.

NOTES

1. Matthew Stanley, chapter 12 in this volume.

2. Caspar W. Weinberger, "Preface" in *Soviet Military Power 1987* (Washington, DC: Department of Defense, 1988), 5.

3. Tom Gervasi, *Soviet Military Power: The Pentagon's Propaganda Document, annotated and
corrected* (New York: Random House, 1987).

4. For a small sampling of this work, see David Holloway, "The Strategic Defense Initia-
tive and the Soviet Union," *Daedalus* 114, no. 3 (1985): 257–278; Bruce Parrott, "The Soviet
Debate on Missile Defense," *Bulletin of the Atomic Scientists* 43, no. 3 (1987): 9–12, 257–278;
Stephen M. Meyer, "Soviet Strategic Programmes and the US," *Global Politics and Strategy* 27,
no. 6 (1985): 274–292; Benjamin Lambeth and Kevin Lewis, "The Kremlin and SDI," *Foreign
Affairs* 66, no. 4 (1988): 755–770; Mary C. FitzGerald, "The Soviet Military on SDI," *Studies in
Comparative Communism* 19, nos. 3–4 (1986): 177–191; Daniel Gouré, "The Impact of the SDI
on Soviet National Security Policy," in *The Technology, Strategy, and Politics of SDI*, ed. Ste-
phen J. Cimbala (New York: Avalon, 1987); Eric Stubbs and Rosy Nimroody, "The Soviet
Response to Star Wars," *Challenge* 30, no. 1 (1987): 21–27.

5. There is, likewise, a massive body of work in this vein. For a good selection of the recent works in this genre, see the notes in Melvyn P. Leffler, "Ronald Reagan and the Cold War:
What Mattered Most," *Texas National Security Review* 1, no. 3 (2018): 76–89.

6. For a more nuanced take on this triangular relationship, see Aaron Bateman, "Intelligence and Alliance Politics: America, Britain, and the Strategic Defense Initiative," *Intelligence
and National Security* 36, no. 7 (2021): 941–960. For the international context more broadly, see
Peter Westwick, "The International History of the Strategic Defense Initiative: American Influence and Economic Competition in the Late Cold War," *Centaurus* 52 (2010): 338–351.

7. For works on the atomic bomb, see David Holloway, *Stalin and the Bomb: The Soviet
Union and Atomic Energy, 1939–1956* (New Haven, CT: Yale University Press, 1994); Mi-
chael D. Gordin, *Red Cloud at Dawn: Truman, Stalin, and the End of the Atomic Monopoly* (New
York: Macmillan, 2010). The canon on Lysenko is vast. For a good start and an excellent
bibliography, see Michael D. Gordin, "Lysenkoism," *Encyclopedia of the History of Science* (on-
line), ed. Christopher J. Phillips (Pittsburgh: Carnegie Mellon University Libraries, 2022).
For the space program, see Asif Siddiqi, *Challenge to Apollo: The Soviet Union and the Space
Race, 1945–1974* (Washington, DC: NASA, 2000). For computing, see Ben Peters, *How Not to
Network a Nation: The Uneasy History of the Soviet Internet* (Cambridge, MA: MIT Press, 2016).

8. For chaotic infighting in the Soviet space program, see Asif Siddiqi, "Fighting Each
Other: The N-1, Soviet Big Science, and the Cold War at Home," in *Science and Technology in
the Global Cold War*, ed. Naomi Oreskes and John Krige (Cambridge, MA: MIT Press, 2014),
189–225.

9. Ronald Reagan, "Address to the Nation on Defense and National Security," March 23,
1983, https://www.reaganlibrary.gov/archives/speech/address-nation-defense-and-national
-security.

10. There is a considerable body of literature on the contours, responses to, and debates
over SDI. Much of it was produced by policymakers, scientists, or journalists *during* the high
period of SDI, in the 1980s. Probably the best summation of the contemporaneous literature

is Steven W. Guerrier and Wayne C. Thompson, eds., *Perspectives on Strategic Defense* (Boulder, CO: Westview Press, 1987), which includes statements from proponents and opponents of SDI as well as many other primary sources, such as statements by Reagan and Gorbachev. Scholarly historical treatments are surprisingly rare, although SDI features quite prominently in a huge market of books on Reagan's role in the end of the Cold War. For trenchant post facto works focused exclusively on SDI, see Aaron Bateman, *Weapons in Space: Technology, Politics, and the Rise and Fall of the Strategic Defense Initiative* (Cambridge, MA: MIT Press, 2024); Frances FitzGerald, *Way Out There in the Blue: Reagan, Star Wars and the End of the Cold War* (New York: Simon & Schuster, 2000); Columba Peoples, *Justifying Ballistic Missile Defense: Technology, Security and Culture* (Cambridge, UK: Cambridge University Press, 2010); Mira Duric, *The Strategic Defence Initiative: US Policy and the Soviet Union* (London: Routledge, 2003); Donald R. Baucom, *The Origins of SDI, 1944–1983* (Lawrence: University Press of Kansas, 1992); Luc-André Brunet, ed., *NATO and the Strategic Defence Initiative: A Transatlantic History of the Star Wars Programme* (London: Routledge, 2023); Peter J. Westwick, "From the Club of Rome to Star Wars: The Era of Limits, Space Colonization and the Origins of SDI," in *Limiting Outer Space: Astroculture after Apollo*, ed. Alexander C. T. Geppert (London: Palgrave Macmillan, 2018), 283–302.

11. For an excellent summary of Soviet official statements about the US space militarization in the late 1970s, see Congressional Research Service, *Soviet Space Programs: 1976–80, Part 1: Supporting Vehicles and Launch Vehicles, Political Goals and Purposes, International Cooperation in Space, Administration, Resource Burden, Future Outlook* (Washington, DC: GPO, 1982), 184–190.

12. The concern with a first strike was made evident in an oft-shared article under the title "SDI—A Strategy of First Strike" with the subheading "Imperialism—The Enemy of the People." See M. Aleksandrov, "SOI—strategiia pervogo udara," *Aviatsiia i kosmonavtika* no. 10 (1985): 42–43.

13. Mary C. FitzGerald, *Soviet Views on SDI*, The Carl Beck Papers in Russian and East European Studies, no. 601 (May 1987), University of Pittsburgh Center for Russian and East European Studies. For another contemporary summary, see Rip Bulkeley, "Soviet Military Responses to the Strategic Defense Initiative," *Current Research on Peace and Violence* 10, no. 4 (1987): 129–142. For a summary of translated Soviet pronouncements on SDI, see *Selected Soviet Statements on Countermeasures Against SDI*, Special Memorandum, Foreign Broadcast Information Service, FB 85-10050, October 15, 1985.

14. Wolfgang K. H. Panofsky, "The Strategic Defense Initiative: Perception vs Reality," *Physics Today* 36, no. 6 (1985): 34. See also Hans A. Bethe, Richard L. Garwin, Kurt Gottfried, and Henry W. Kendall, "Space-Based Ballistic-Missile Defence," *Scientific American* 251, no. 4 (1984): 39–49. For a sociological study of American physicists and their stance toward SDI, see Michael R. Nusbaumer, "Scientists, Activism and Arms: The Strategic Case of Star Wars," *Sociological Focus* 29, no. 2 (1996): 89–105.

15. William J. Broad, *Teller's War: The Top-Secret Story Behind the Star Wars Deception* (New York: Simon & Schuster, 1992). For bios, see Peter Goodchild, *Edward Teller: The Real Dr. Strangelove* (Cambridge, MA: Harvard University Press, 2004); Istvan Hargittai, *Judging Edward Teller: A Closer Look at One of the Most Influential Scientists of the Twentieth Century* (Amherst, NY: Prometheus Books, 2010).

16. For secrecy and Soviet science in the postwar period, see Asif Siddiqi, "Soviet Secrecy: Toward a Social Map of Knowledge." *American Historical Review* 126, no. 3 (2021): 1046–1071.

17. Siddiqi, *Challenge to Apollo*, 740.

18. A. Kokoshin, "Asymmetric Response vs. Strategic Defense Initiative," *International Affairs: A Russian Journal of World Politics, Diplomacy, and International Relations* 53, no. 5

(2007). Original Russian published as " 'Assimetricheskii otvet' vs. 'Strategicheskoi oboronnoi initsiativy,' " *Mezhdunarodnaia zhizn'* 8 (2007): 29–42.

19. See F. V. Bunkin, E. P. Velikhov, P. P. Pashinin, and E. M. Sukharev, "Istoriia razrabotki i sozdaniia moshchnykh lazerov oboronnogo primeneniia," in *Sozdateli rossiiskikh lazerov: nauchnoe izdanie*, ed. Iu. V. Rubanenko and E. V. Mozhelev (Moscow: Stolichnaia entsiklopedia, 2016), 9–22.

20. E. Velikhov, "Illiuzii 'Zvezdnykh voin,' " *Sovetskaia rossiia*, April 21, 1985, 5. This quote is also translated and excerpted in FitzGerald, *Soviet Views on SDI*, 22.

21. E. Velikhov, "Kosmicheskie ambitsii—zemnye ugrozy," *Pravda*, April 20, 1984, 6. Quote also in FitzGerald, *Soviet Views on SDI*, 43.

22. Velikhov, "Kosmicheskie ambitsii—zemnye ugrozy."

23. For an official history and full membership list of the committee, see B. A. Kontarev, "Komitet Sovetskikh uchenykh v zashchitu mira, protiv iadernoi ugrozy (kratkaia sprakva)," in *Klimaticheskie i biologicheskie posledstviia iadernoi voiny*, ed. E. P. Velikhov (Moscow: Nauka, 1986), 168–170.

24. These included Yevgeni [Evgenii] Velikhov, ed., *The Night After . . . : Climactic and Biological Consequences of a Nuclear War, Scientists' Warning* (Moscow: Mir Publishers, 1985); *"Star Wars": Delusions and Dangers* (Moscow: Military Publishing House, 1985); *The Large-Scale Anti-Missile System and International Security: Report of the Committee of Soviet Scientists for Peace against the Nuclear Threat* (Moscow: Novosti Press Agency, 1986); Yevgeni Velikhov, Roald Sagdeev, and Andrei Kokoshin, eds., *Weaponry in Space: The Dilemma of Security* (Moscow: Mir Publishers, 1986). A translated extract from "Space-Strike Arms and International Security" (authored by Roal'd Sagdeev and Andrei Kokoshin) was reproduced in *Perspectives on Strategic Defense*, 297–321.

25. Westwick, " 'Space-Strike Weapons,' " 958.

26. These papers are stored at the Hoover Institution. See "Vitalii Leonidovich Kataev papers," Hoover Institution Library & Archives, https://digitalcollections.hoover.org/objects/39/vitalii-leonidovich-kataev-papers, accessed June 23, 2024.

27. Podvig, "Did Star Wars Help End the Cold War?"

28. V. V. Favorskii and I. V. Meshcheriakov, *Voenno-kosmicheskie sily (voenno-istoricheskii trud): kniga II: stanovlenie voenno-kosmicheskikh sil* (Moscow: Voenno-kosmicheskie sily MO RF, 1998).

29. Podvig implies the Velikhov commission was set up "shortly after the US announcement" in March 1983. See Podvig, "Did Star Wars Help End the Cold War?" 7.

30. Vitalii Leonidovich Kataev papers (henceforth Kataev papers), Box 7, Folder: Vitalii Kataev Arms Control 7-4, "Materials on the Question of Nuclear-Space Armaments for Preparations to Meet with R. Reagan" (undated, but probably October 1986), leaf 8. The Russian words *duel'*, *vekha*, and *protivodeistvie* in English can be rendered as "duel," "milestone," and "counteraction."

31. For biographical treatments of Ustinov, see Iu. S. Ustinov, *Narkom ministr marshal* (Moscow: Patriot, 2002); Iu. S. Ustinov, *Trizhdy Geroi i Kavaler 11 ordenov Lenina*, 2 vols. (Moscow: Geroi otechestva, 2005).

32. A. V. Degtiarev, ed., *Smirnov: U istokov raketostroeniia* (Kiev: Speis-Inform, 2016).

33. Ia. V. Nechesa, *Pervyi raketno-kosmicheskii ministr* (Moscow: 'Restart,' 2008); Ia. V. Nechesa, *S. A. Afanas'ev: Pervyi raketno-kosmicheskii ministr* (Moscow: Art-poligrafiia, 2010). Sagdeev reveals his nickname in *The Making of a Soviet Scientist*, 199.

34. These points are elaborated in Siddiqi, "Fighting Each Other." For the work of these three men as a single "pressure group" during the Cold War, see Siddiqi, *Challenge to Apollo*, 428–436.

35. For Afanas'ev, see Sagdeev, *The Making of a Soviet Scientist*, 257–258. For Ustinov, see Serge Schmemann, "Defense Minister of Soviet Union Is Dead at Age 76," *New York Times*, December 22, 1984, 1. For Smirnov's dismissal, see Degtiarev, *Smirnov*, 73.

36. Favorskii and Meshcheriakov, *Voenno-kosmicheskie sily (voenno-istoricheskii trud)*, 109.

37. Sagdeev, *The Making of a Soviet Scientist*, 258–259.

38. The project, with the overall name "1K11K25," was given the green light in February 1976 with a joint decree of the Party Central Committee and Council of Ministers, siphoning off a huge amount of NPO Energiia's resources in the subsequent decade. There are plenty of official or semi-official accounts of the history of the Energiia rocket and the Buran space shuttle. For the best English-language book, see Bart Hendrickx and Bert Vis, *Energiya-Buran: The Soviet Space Shuttle* (Chichester, UK: Springer-Praxis, 2007). For somewhat technical but reliable Russian-language accounts, see Iu. P. Semenov, G. E. Lozino-Lozinskii, V. L. Lapygin, and V. A. Timchenko, *Mnogorazovyi orbital-nyi korabl' "Buran"* (Moscow: Mashinostroenie, 1995); Iu. P. Semenov, ed., *Raketno-kosmicheskaia korporatsiia "Energiia" imeni S. P. Koroleva, 1946–1996* (Moscow: RKK 'Energiia,' 1996), 362–399; V. E. Nesterov, V. A. Omel'ko, S. G. Samusenko, and V. F. Cherniavskii, eds., *Mnogorazovaia kosmicheskaia Sistema "Energiia-Buran"* (Moscow: NPP "OmV-Luch," 2004).

39. Semenov, *Raketno-kosmicheskaia korporatsiia "Energiia" imeni S. P. Koroleva*, 419.

40. "National Security Decision Directive Number 42, 'National Space Policy,' July 4, 1982," https://nsarchive.gwu.edu/document/21095-document-1-19820704-nsd-directive-42, accessed June 23, 2024. See also John M. Logsdon, *Ronald Reagan and the Space Frontier* (Cham, Switzerland: Palgrave Macmillan, 2019), 50–51; Howell Raines, "Reagan Affirms Support for U.S. Space Program," *New York Times*, July 5, 1982, 8.

41. Favorskii and Meshcheriakov, *Voenno-kosmicheskie sily (voenno-istoricheskii trud)*, 86.

42. Favorskii and Meshcheriakov, *Voenno-kosmicheskie sily (voenno-istoricheskii trud)*, 89.

43. Favorskii and Meshcheriakov, *Voenno-kosmicheskie sily (voenno-istoricheskii trud)*, 89.

44. Dmitrii Tikhanov, "Esli zavtra budet voina" [Interview with Col.-Gen. Iurii Votintsev, Part 2 of 3], *Vechernyi almaty*, June 2, 1993, 3.

45. S. Shamsutdinov, "'Almaznye' kosmonavty," *Novosti kosmonavtiki* no. 12 (2000): 78–81; Asif A. Siddiqi, "The Almaz Space Station Complex: A History, 1964–1992, Part 2: 1976–1992," *Journal of the British Interplanetary Society* 55 (2002): 35–67.

46. Andrei Kokoshin, "'Assimetricheskii otvet' vs. 'Strategicheskoi oboronnoi initsiativy,'" *Mezhdunarodnaia zhizn'* 8 (2007), https://interaffairs.ru/jauthor/material/1650, accessed June 23, 2024.

47. On Sputnik, see Asif Siddiqi, "Sputnik 50 Years Later: New Evidence on Its Origins," *Acta Astronautica* 63 (2008): 529–539. For Almaz, see Asif Siddiqi, "The Almaz Space Station Complex: A History, 1964–1992, Part 1: 1964–1976," *Journal of the British Interplanetary Society* 54 (2001): 389–416. For the N1 lunar decision, see Asif Siddiqi, "A Secret Uncovered: The Soviet Decision to Land Cosmonauts on the Moon," *Spaceflight* 46 (2004): 205–213.

48. V. V. Favorskii and I. V. Meshcheriakov, eds., *Kosmonavtika i raketno-kosmicheskaia promyshlennost': razvitie otrasli (1976–1992): kn. 2* (Moscow: Mashinostroenie, 2003), 95. For an official history of TsNII-50, see E. V. Alekseev, V. A. Men'shikov, and I. V. Meshcheriakov, *Na peredovykh rubezhakh: ocherki istorii 50 TsNII MO imeni. M. K. Tikhonravova* (Iubileinyi: NII KS im. A. A. Maksimova, 2008). The name of the study evaluating SDI was *Klokot* (or *Bubbling*). See 10–11.

49. Favorskii and Meshcheriakov, *Voenno-kosmicheskie sily (voenno-istoricheskii trud)*, 106.

50. The decree on *Skif* was issued by the Central Committee and Council of Ministers on December 25, 1984, for a space laser with a range of 100 kilometers, developed by the firm NPO Astrofizika to be installed on the *Skif* spacecraft. The first flight of *Skif* was set for 1991.

See Kataev papers, Box 7, Folder: Vitalii Kataev Arms Control 7–15, "Nuclear and Space Missile Talks Ballistic Missile Defense Information 'The real situation re: US claims . . .'" (undated, but probably November 1985), leaf 6. The *Kontakt* system, to target satellites to altitudes of 600 kilometers, was approved on November 27, 1984. The first test was set for 1989. See same source, leaf 14.

51. Tikhanov, "Esli zavtra budet voina."

52. Kataev papers, Box 7, Folder: Vitalii Kataev Arms Control 7–15, leaf 24.

53. Kataev papers, Box 7, Folder: Vitalii Kataev Arms Control 7–15, leaf 24.

54. Kataev Papers, Box 7, Folder: Vitalii Kataev Arms Control 7–15, leaf 18.

55. Kataev provides a useful one-page summary of the four programs in Kataev Papers, Box 1, Folder: Vitalii Kataev Biographical File 1–5 ("Diaries 1974–1989"), leaf 68. Details of each of these are provided in leaves 70–75.

56. Favorskii and Meshcheriakov, *Voenno-kosmicheskie sily (voenno-istoricheskii trud)*, 107.

57. Podvig, "Did Star Wars Help End the Cold War?" 11.

58. Favorskii and Meshcheriakov, *Voenno-kosmicheskie sily (voenno-istoricheskii trud)*, 106–107.

59. Favorskii and Meshcheriakov, *Voenno-kosmicheskie sily (voenno-istoricheskii trud)*, 107–108.

60. Favorskii and Meshcheriakov, *Voenno-kosmicheskie sily (voenno-istoricheskii trud)*, 108.

61. The 50 billion figure is from Savranskaya, "Soviet Response to the Strategic Defense Initiative," 44, citing Andrey Grachev, *Gorbachev's Gamble* (Cambridge, UK: Polity Press, 1998), 94. Conversion rates of 0.8975 rubles to $1 from "Treasury Reporting Rates of Exchange as of March 31, 1985," https://www.govinfo.gov/content/pkg/GOVPUB-T63_100-7eddaf427 3a66e80b5be6d4909795a8e/pdf/GOVPUB-T63_100-7eddaf4273a66e80b5be6d4909795a8e. pdf, accessed June 23, 2024.

62. Sagdeev, *The Making of a Soviet Scientist*, 261.

63. Evgenii Velikhov, *Moi put': ia na valenkakh poedu v 35-i god* (Moscow: ACT, 2017), 156. An earlier version of the biography also includes the same quote. See E. P. Velikhov, *Ia na va-lenkakh poedu v 35-i god: vospominaniia* (Moscow: Astrel' AST, 2010), 215–216, emphasis added.

CHAPTER TWELVE

Extinction, Insurance, or a New Weapons Industry

Asteroid Impacts and the Triumph of the Apocalyptic Lobbyist

Matthew Stanley

At 7:14 P.M. Eastern Time on September 26, 2022, NASA intentionally crashed a spacecraft into an asteroid. This was the culmination of forty years of efforts to convince humanity that we, as a species, were in existential danger from a rogue cosmic body on a collision course with our planet. The spacecraft mission, known as DART (the Double Asteroid Redirection Test), was a trial run of technology to see how we might nudge a threatening asteroid onto a safer path. The $300 million mission was a statement of scientific, political, and bureaucratic confidence in scientific predictions made in the early 1980s.

At the time, though, those predictions seemed self-evidently ridiculous. A small group of scientists worked diligently to persuade their colleagues, the public, and the US government that the sky was literally falling and that action was required. For the most part, they failed. Precise calculations, empirical evidence, and interdisciplinary investigations were not enough. Even before the forty years to DART, it took a decade between the discovery of what astronomers saw as their smoking-gun evidence and the first meager government response to it.[1]

I argue here that this time lag was largely due to those scientists having difficulty grappling with the context of the late, resurgent Cold War. Specifically, the 1980s were both an intensely apocalyptic time that had certain expectations for talking about the end of the world and an intensely greedy time that had certain expectations for what would motivate government action. Public attention is a scarce resource, and it is not always obvious how to turn a scientific prediction into a political cause.[2]

In the Reagan era, simply claiming technical expertise was not sufficient to spur the government. Rather, the hoped-for action needed to be framed in that era's revamped understanding of the role of government as supporting private initiative and profit. If these scientists wanted Congress to spend money on their project, they needed to speak the language of the military–industrial complex, not merely the public good. They tried to stress the obligation of the government to protect its citizens, and failed. They tried to make the case for cost-effective regulation, and failed. In the end, their handmaiden was the industrial lobbyist, that much-despised but apparently irreplaceable agent for explaining how government dollars could fit the 1980s vision of limited government. The asteroid scientists' success came through adaptation to a specific institutional pathway that accepted the strange realities of American politics at the end of the Cold War: the road to the apocalypse needed to be lined with dollars.

Warnings of Total Destruction

This story really begins with some father–son bonding time. Luis Alvarez, winner of the Nobel Prize for his inventions in particle physics and scientific observer of the Hiroshima bombing, had begun working with his geologist son Walter on the problem of dating certain rock layers. He and his son's team were trying to use meteoric iridium as a dating technique—perhaps the deposition of iridium dust from disintegrating meteorites could provide a steady clock. But instead of a gradual deposition, they discovered an enormous amount of iridium in deep-sea limestones formed about sixty-five million years ago. They suspected this thirtyfold increase over background amounts was an anomaly of some sort, so they checked the same layer hundreds of miles away. There, an increase of 160 times. Elsewhere, twenty times. There seemed to be a layer of iridium, a substance rare on Earth but common in meteorites, spread across the entire globe.

This limestone layer happened to mark the end of the Cretaceous period (at the time called the K-T boundary, today often described as the K-Pg boundary), best known for the mysterious disappearance of the dinosaurs. Alvarez's group decided that it could not be a coincidence that a mass extinction was associated exactly in the geological record with a huge amount of meteoric material. Their conclusion: an asteroid around 10 kilometers wide had struck the Earth, hurling ejecta into the atmosphere, causing global climactic effects that wiped out *Tyrannosaurus rex* and friends.[3] This so-called "Alvarez hypothesis," published in 1980, was enormously controversial but was relatively quickly supported by other analyses and, eventually, the discovery of the Chicxulub crater.[4]

Part of the difficulty of the ensuing scientific discussion was that there were not many people who knew much about asteroids. Astronomers who studied those lonely rocks were part of the subdiscipline of planetary science, at the time marginal to mainstream astronomy—perhaps because close studies of space rocks and solar system surfaces still reminded people of Lowell and his imaginary canals.[5] For many years, the American Astronomical Society's Division for Planetary Sciences had to hold its own meetings separate from the main society gathering.

To bring this marginal group into what was now an extremely important conversation, a conference was organized to discuss the impact hypothesis and get geologists, astronomers, and paleontologists up to speed with each other's fields. Their focus was squarely on the dinosaurs and whether their sudden absence was in fact best explained by a space rock.

There was another conference spurred by the Alvarez hypothesis, however, with a slightly different emphasis. This was for a group who were less concerned with the K-T extinction itself and more worried that if it happened to the dinosaurs, it might also happen to *us*. This meeting was led by Gene Shoemaker, the geologist who taught NASA how to think about rocks and trained the Apollo astronauts inside Meteor Crater. Shoemaker was the first to identify convincingly that depression as an impact site and was one of the very few scientists to have been considering the possibility of asteroid impacts on Earth's surface at all. He persuaded skeptics that Meteor Crater had indeed been formed by a meteor by comparing it closely to the craters left behind by years of nuclear testing. Those sites proved him with a great deal of information on compression mechanics, impact hydrodynamics, and crater formation.[6] Decades later, he was praised for carrying out "kind-of-heroic calculations based on almost no data."[7]

His interest in cosmic impacts started an entire new career for him as virtually the only scientist focused on finding threatening asteroids before they hit. He first worked with Eleanor Helin, with whom he had a falling out, and for the rest of his life with his wife Carolyn. Helin went on to start her own program, and the Shoemakers continued with their tedious and largely unrewarded task of searching for Earth-crossing asteroids (ECAs).[8] Parallel to them was Tom Gehrels, "Mr. Asteroid," who had been studying the rocks for purely scientific interest since the 1960s. Gehrels, a Dutch-born astronomer who parachuted into Nazi-occupied Europe, said he was woken up to the impact hazard by the Alvarez hypothesis. Asteroids would no longer be charming lights in his telescope; they would be a constant reminder that what happened to the dinosaurs might happen again.

Shoemaker, along with the NASA administrator Bill Brunk, put together a weeklong informal workshop titled "Collision of Asteroids and Comets with the Earth" in Snowmass, Colorado. One of the participants, Clark Chapman of the Planetary Science Institute, remembered that it "included a few people from the military, asteroid scientists like myself, people who knew about orbits . . . and some NASA mission-planning-type people. It was just a wide-ranging discussion—it included the nature of the threat, the possible damage to the environment."[9] Under the intellectual leadership of Shoemaker, the group came to consensus on the likelihood of future impacts and what effects it might have on the planet. They calculated population numbers (how many asteroids of what sizes were present in the solar system), approach velocities, and how often we could expect a K-T–style disaster.

Unlike the other Alvarez conference, though, this group pondered what such an impact would be like *now*, on a planet inhabited by humans and their technological civilization. Mike Gaffey, an astronomer at the meeting who grew up on a farm in Iowa, pointed out that the dust clouds that shattered the dinosaur's food web would also destroy the sprawling agricultural systems on which the billions of modern humans depended.[10] Both of these were essentially direct effects of the impact.

Someone (no one seems to remember who), however, also pointed out that the world of the 1980s brought with it other sorts of indirect dangers from an impact. The stalemate of the Cold War meant that nations heavily armed with nuclear weapons were watching very carefully for surprise explosions and were prepared to release those weapons with the briefest of consideration. What if a relatively small asteroid were to impact and be mistaken for a first strike, releasing the hair trigger? One scenario: what if the asteroid that caused the Tunguska explosion in Siberia had arrived in 1980 instead of 1908? The Soviets would surely have seen it as an American attack, leading to a massive nuclear war. So the asteroid danger not only was a natural one but also brought with it an "indirect risk of misinterpretation of otherwise relatively innocuous events by nuclear nations that could possibly lead to man-made catastrophes," as Shoemaker later wrote.[11] The '80s, as it turned out, would be a particularly bad time to be hit by an asteroid (although the dinosaurs might disagree).

The Cold War background to the Snowmass conference brought other issues to the fore, such as the possibility that the Soviets might intentionally use an asteroid as a first-strike weapon (the so-called Ivan's Hammer scenario). Shoemaker's group took this possibility seriously enough to consider the effects on specific targets such as missile silos or submarines. Both this threat and the natural one raised the question of what to do if an asteroid was on a collision course with Earth. It takes a lot of energy to move a celestial body, and the merciless laws of physics meant that nuclear weapons were the only plausible way to get that energy to where it needed to be. The workshop developed detailed calculations for what kinds of weapons to use, how many, and where they should be placed on or within the rock.

Shoemaker compiled all the work from the conference into a hundred-page report . . . which he then declined to show anyone. Eventually, someone stole it from his desk and informal copies began circulating.[12] Shoemaker never explained why he kept the report—in some sense, the culmination of his life's work—to himself. Other participants in the Snowmass meeting have a loose consensus that the problem was the nukes. Specifically that NASA, as the sponsor of the workshop, did not want to be associated with any recommendations for the use of nuclear weapons in space. Or perhaps they felt that the warnings of a cosmic impact triggering a nuclear war had an implicit critique of mutually assured destruction that would be politically unacceptable in the early Reagan years. Another possibility was that Shoemaker, by all accounts an antiwar leftist, was anxious about recommending the use of nuclear weapons in any context.[13]

No one asked him why he did not publish the report, and it seems that his stature in the asteroid community was such that everyone trusted his judgment.

If the problem was the nukes, everyone's concerns were probably justified. While Shoemaker's report never came under public scrutiny, some related calculations did, and it did not go well. In a published paper estimating the climactic effects of an asteroid impact, Shoemaker referenced Brian Toon's 1982 paper that described how an injection of dust into the stratosphere could remain aloft for months or years, blocking sunlight and dropping temperatures by tens of degrees. Shoemaker did not mention that Toon's work on this was part of research on the so-called nuclear winter phenomenon. Working with Carl Sagan and a handful of other scientists, Toon made the case that a full-scale nuclear war between the United States and the Soviet Union would have global consequences, possibly even destroying all life on Earth.[14] Whether the dust was lofted by asteroid collisions or nuclear fire, the atmospheric physics was the same and the equations just as apocalyptic. The death of the dinosaurs became a model for what humanity might do to itself.

Much of the work on nuclear winter was done by Jim Pollock, along with Brian Toon and Tom Ackerman, at the NASA Ames Research Center. The political implications of that work were noted quickly by anxious NASA higher-ups, and that trio's boss, David Morrison, had to intervene to protect them. Morrison and Pollock had both been graduate students under Carl Sagan at Harvard in the 1960s and were quite committed to promoting science of public importance even if it was politically dangerous.[15]

Sagan tried to leverage the research to influence nuclear policy within the US government (and, to a smaller degree, in the USSR). He argued that the nuclear winter theory showed the need for an immediate nuclear freeze and the impossibility of a winnable nuclear war even with SDI (the Soviet reaction to SDI was quite complicated; see Asif Saddiqi's discussion in chapter 11 of this volume). Sagan made a conscious decision to move beyond conventional modes of science communication to speak to the public directly through his celebrity status and, it was hoped, government leaders through them. He published essays in *Parade* and *Foreign Affairs*, appeared on *Nightline* and numerous other television shows, and sent copies of the papers directly to, among other leaders, McGeorge Bundy, Robert McNamara, and Brent Scowcroft.[16] He was relentless in describing the spectacular terror and civilization-wide destruction that nuclear winter would entail.

As Myrna Perez's story of Stephen Jay Gould in chapter 2 of this volume shows, Sagan was not alone in using his status as a high-profile scientist to push back against the resurgent right-wing politics of the 1980s. While Gould battled creationists, Sagan contested the new Reaganite militarism. Both gained a great deal of public attention, although reactions from their colleagues were mixed. Gould was largely celebrated for defending the integrity of science while Sagan was attacked for apparently abandoning long-standing scientific norms about modesty of expression, humility of claims, and value-free science. By advocating for specific policy proposals, he had jeopardized the authority of scientists as objective intellectuals. His tendency to lean into the worst possible implications of the theory did not help either; it felt to many like exaggeration. He was trying to intervene in the national conversation about nuclear war and set himself up as an expert. In the end, he failed. He did not manage to influence government policy. The nuclear winter computer models came under scrutiny for their precision and consistency. Sagan came under criticism for overstepping the bounds of proper scientific conversation.

Although it was the same equations that predicted the apocalypse from asteroids and from nuclear winter, the latter received more attention. Indeed, the first mention of any danger from cosmic impacts on the floor of Congress was a side note during William Proxmire's 1983 speech warning about nuclear winter.[17] He read into the record the World Health Organization (WHO) report on the effects of nuclear war on human health, along with two articles by Sam Iker, an environmental reporter for *Time*, titled "No Place to Hide" and "Death from the Sky." In explaining how nuclear war would lead to massive crop failures, he quoted Iker's article: "Night will descend on much of the northern hemisphere—a funereal pall of the kind which may have blanked out the sun some 65 million years ago and sent the dinosaurs into oblivion. At that time, some scientists theorize, a 10-kilometer-wide asteroid collided with the Earth and hurled vast quantities of fine dust into the atmosphere . . . it's believed that the frigid darkness lasted long enough to cause mass extinctions of mammals and plants."[18] With so much at stake (from nuclear war, not asteroids), Proxmire urged his colleagues to vote for a nuclear freeze.

Asteroids appeared in the political conversation of the mid-1980s only as minor points in the dynamics of the Cold War. For Proxmire, the K-T impactor was a stand-in for nuclear winter. Hawai'i senator Spark Matsunaga suggested that addressing the threat of asteroid impacts would be an excellent opportunity for the United States and the USSR to undertake joint missions in space and reduce tensions.[19] Representative George Brown Jr. of California, while speaking against the militarization of space with mass drivers, made an apparent reference to the Snowmass conference: "These problems seem comfortably distant, yet a recent NASA conference on diverting asteroids whose orbits cross with Earth's recalled the military's long-standing ambition to learn how to bombard Earth with such objects."[20] As befits these marginal mentions, none of these points were followed up on. And indicating the realities of the greedy decade, the government's only stated interest in asteroids was whether we would one day mine them: small references to possibly harvesting asteroids appeared regularly in NASA appropriation bills through the decade, without any actions taken. This profit-oriented entry point into getting government attention remained unexploited by the scientists.

Instead, in these early years, asteroid scientists emphasized the close connections between their apocalypse and nuclear war. They played up the totality of destruction accompanying each and how one could trigger the other. The same equations and modeling were shared by both. The reasoning seemed to be that government leaders did heed the threat of nuclear war, so if similarity could be established, they would also pay attention to asteroids. In both cases, it should be the government's responsibility to protect its citizens from destruction. But the strategy backfired. Touching on nuclear policy was dangerous and seemed unscientific, as the nuclear winter story makes clear. And even the suggestion of using nuclear weapons made the Snowmass report unpublishable. Invoking the specter of nuclear war did not make asteroids worthy of more attention; it made them unmentionable.

Warnings of Precise Danger

Not much changed in asteroid science during the middle part of the decade. There were few technical publications about them, and even fewer about impacts. David Morrison and Clark Chapman helped revitalize the conversation in October 1988 at a conference in Snowbird, Utah (coincidentally, where the Alvarez meeting had been held in 1981). NASA had put together the meeting on "Global Catastrophes in Earth History" to look at volcanoes, earthquakes, and other mass mortality events (perhaps spurred by the 1987 earthquake in Southern California).

It was an explicitly interdisciplinary conference, a framework that Morrison and Chapman took seriously. They wanted to bring asteroid impacts into the conversation and followed Shoemaker's guidance from the Snowmass meeting: a proper consideration of the asteroid threat would require an analysis of "the physical, economic, sociological, and psychological implications." So while most of the papers at the conference dealt with specific technical issues (e.g., what variance of carbon isotope anomalies should be expected in geological boundaries), this pair presented a forward-looking paper on "Risk to Civilization."[21] They had been thinking about writing a popular science book together for a while, and catastrophes seemed like a good topic for attracting an audience.

The conference paper was presented as an updating of the 1981 Snowmass conference to include recent research. Chief among that research was the progress that nuclear winter studies made in understanding atmospheric dust effects. Chapman and Morrison's contention was simple: "The risk that civilization might be destroyed by impact with an as-yet-undiscovered asteroid or comet exceeds risk levels that are sometimes deemed unacceptable by modern societies in other contexts. Yet these impact risks have gone almost undiscussed and undebated."[22]

"Risk" here meant danger in the classical sense of car insurance: cost per event (dollars or lives) multiplied by number of events. That number was then how much a society should spend to ameliorate that danger. But for low-frequency, high-cost events like asteroid impacts, the calculation became a little strange. The cost was all of humanity and the global economy, and the frequency was once every ten million years or so. The resulting number meant that, statistically, more people died every year from asteroids than from plane crashes. This was, of course, untrue (there is no confirmed record of any human being killed by a cosmic body).[23] Nonetheless, it was mathematically sound for Chapman and Morrison to say that the chance of dying from an impact was ten times greater than the danger allowed by the regulation of nuclear power plants and five thousand times greater than the chance of dying from exposure to TCE (the industrial pollutant of the day). These were all concerns that American society was willing to spend money to guard against, so asteroid defense should get a budget too.[24] This was a long-standing technique for bringing technical expertise to bear on social problems, and it seemed a reasonable way to approach the matter.

Unfortunately for this strategy, the 1980s was precisely the time when classical risk assessment was becoming less and less persuasive. Kahneman and Tversky's prospect theory was undermining conventional wisdom about profit and loss in the markets.[25] Their critiques of how we thought about investments grew into broader concerns about confidence in risk assessment. Ulrich Beck's influential 1986 book *Risk Society* argued that the emergence of global, incalculable risks like environmental destruction had fundamentally altered public willingness to accept such precise quantitative predictions.[26] The Chernobyl disaster occurring between his writing the book and its appearance in print made him seem prescient. In a similar attack on quantification, Charles Perrow wrote in 1984 that considering deaths only in terms of numbers led to absurdities (was the life of a child really equivalent to a check for $300,000?). Instead, one must consider how views about risk relied on "social ties, family continuity, a distinctive culture and valued human traditions."[27] The emergence of the possibility of catastrophic disasters (humanmade or otherwise) destabilized simple frequency-cost estimates.[28]

Chapman and Morrison ran up against these difficulties immediately and looked for resources outside astronomy. Chapman happened to hear a presentation by the psychologist Paul Slovic at an American Association for the Advancement of Science meeting that explored the factors involved in what made people perceive a given situation as dangerous.[29] Slovic and his colleagues had noticed years before that people were wildly inconsistent with how they thought about risk (they were willing to tolerate vastly more toxins in the air than in food, for instance). He developed a new research project called the "psychometric paradigm" that sought to empirically measure what people *actually* found frightening, rather than calculating dangers in advance and telling them what they *should* find frightening.[30] He concluded that these apparent inconsistencies actually came from heuristics of danger that were quite consistent; they simply did not follow the lives-per-year-per-dollar calculations of classical risk. Rather, Slovic found that events were seen as more risky if they involved unknown causes or what he called "dread": a fear of things that were newly recognized (industrial pollutants) or due to unobservable agents (radiation). This internal logic was, he argued, steady and predictable, even if it disrupted the bureaucratic approach to calculating danger. In some of his earliest work on this problem, he contended that this could help us understand why society addressed some dangers but not others: "Because society is limited in its capacity to worry about risk and to provide resources to reduce it, hazards must compete with each other for public attention."[31]

This seemed to be exactly what Chapman and Morrison were looking for. An explanation for why their warnings were not being well accepted, as well as tools for making their predictions more persuasive. The asteroid threat matched the characteristics for dread rather well, and they began framing it in those terms. They also took Slovic's (and Beck's) warning that simply delivering instructions from a position of scientific authority was no longer tenable in the 1980s. They claimed that they were merely calling for more research, not dictating what should be done.[32]

Many of the papers from the 1988 conference were published in a proceedings volume by the Geological Society of America, although not Chapman and Morrison's. Chapman had written some popular science books and he took the lead on expanding their paper into an entire trade book, *Cosmic Catastrophes*, in 1989.[33] This was their first attempt to spread widely the warnings about possible impacts, in a somewhat technical but rather readable book. Its core argument was that accepting the truth of the impact hazard required a complete shift in the way scientists thought about the universe.

The problem, they wrote, was that scientists were under the spell of Charles Lyell. That Victorian geologist's principle of uniformitarianism—that the world should be explained through gradual, regular effects—had made it impossible to realize that "sudden, immense, often unpredictable forces—in short, catastrophes—have helped to shape the stars, the planets, and the Earth."[34] They acknowledged that there were good historical reasons for this prejudice. Catastrophism had traditionally been associated with biblical literalists and fringe characters like Immanuel Velikovsky. Velikovsky's bizarre (but very popular) theories of planets careening and smashing their way around the solar system in accordance with ancient scriptures had been causing headaches for astronomers like Sagan for years.[35] Chapman and Morrison conceded that any talk of celestial collisions put their colleagues on edge: "The very term 'catastrophe' carries a negative connotation to many scientists. It suggests the irrational or un-predictable and conjures up images of the 'acts of God' much beloved by newspaper writers."[36] Science's emphasis on natural laws, causality, and re-producibility made it very difficult to acknowledge the importance of random or unique events such as asteroid impacts. Even deeper, the human psyche has "difficulty with randomness and irregularity. We fear unex-pected catastrophe. As human beings, scientists also have difficulty com-ing to grips with rare or unusual events, with chaotic processes."[37] We needed to be wary of ascribing to nature the kind of processes that we found pleasant to think about.

Astronomers had changed their mind in the previous decades, partly due to patient accumulation of data. Photos of Mercury's pitted surface and cracks on Phobos had taught them to be humble before a violent uni-verse.[38] Chapman and Morrison spent a large portion of *Cosmic Catastro-phes* defending their new catastrophism as scientific. One chapter discussed Velikovsky head-on and why his story of cosmic pinball was different from their own. Another looked at creation science's flood geology and why it was different as well. They knew that impact science could be dismissed as pseudoscience, and they wanted to draw the boundaries clearly. Unlike Velikovsky, they said, their warnings were based on accumulation of facts, observation of the solar system, laboratory experiments, and detailed cal-culations on massively powerful computers. It was only as a result of this kind of evidence that they concluded that "in some ways the universe is inherently chaotic, and that catastrophes are inevitable. . . . The scientific process itself, not mere arm-waving or reliance on ancient texts, has led inescapably to this conclusion."[39] This new catastrophism was not a reli-gion, and catastrophes were not the proper explanation for every mystery in science. One should only endorse catastrophic explanations that pass "rigorous scientific tests."[40] Unpredictability did not mean total ignorance, but it did mean that probability and statistics were the best that could be done.

This turned out to be a challengingly narrow path to negotiate. They found themselves in what Alvin Weinberg called "the regulator's di-lemma."[41] Weinberg, the nuclear physicist who ran Oak Ridge National Laboratory from the Manhattan Project until his death, was grappling with the limitations of what science seemed to be able to tell us about future dangers such as nuclear disasters. He worried that the strength of science came from its focus on the regularities of experience, which was barely helpful for infrequent events and their attendant uncertainty. Scientists might be able to understand a unique happening in retrospect, but they were expected to characterize those in advance. Science was being "asked a question that lies beyond its power; the question is trans-scientific." The example Weinberg chose in 1985 to illustrate this was telling: "Although science can often analyze a rare event after the fact—for example, the extinction of dinosaurs during the Cretaceous-Tertiary period following the presumed collision of the earth and an asteroid—it has great difficulty predicting when such an uncommon event will occur."[42] Weinberg worried that the reliance on catastrophism proposed by Chapman and Morrison made it essentially impossible to rely on science for this kind of decision-making.

Nonetheless, the asteroid scientists were frustrated that their authority on celestial collisions was not carrying the day. They wondered why political leaders were ignoring the threat even though "experts *have* considered the matter, and, in fact, they concluded that the threat is serious enough to warrant awareness at the highest levels of government." The astronomers' "thought-provoking conclusions apparently remain unappreciated and unheeded." Leaders had been warned that there was a real threat and had chosen to do nothing.[43] Despite their earlier disavowal of technocracy, Chapman and Morrison offered some specific considerations about what should be done.

The first suggestion was just to think more about the problem. This was cheap: 1 percent of NASA's budget would find all the likely impactors within a decade. There was no rush, per se, but "to let this slide for decades, without action, would be to treat our whole society the way a smoker treats his or her life: irrationally. If it is worthwhile to buy state lottery tickets or home fire insurance, it is sensible to invest in further knowledge about the threat from the skies." If we find nothing, then we have lost nothing. "If, instead, a doomsday impact is discovered and averted, then we will have made one of the most important contributions to human civilization in recorded history."[44] This call was stirring and dramatic, although it was clear that the authors were ambivalent about becoming prophets of apocalypse: "As the millennial year of 2000 approaches, we can expect an increase in the predictions of imminent catastrophe. There is something is such concepts that appeals to people. . . . The universe is indeed a big and powerful place, and there are real dangers to consider, as described throughout this book."[45]

Cosmic Catastrophes was a clear product of the Cold War. Its language, structure, and arguments were heavily embedded in Reagan-era America. The book opened with the same thought experiment that shaped the Snowmass conference years before: a Tunguska-style explosion near a Soviet military base leading to a massive nuclear exchange. The hair-trigger of the Cold War became a metaphor for our unstable planetary environment. One small accident could lead to the deaths of billions. The authors were also clear about the influence of the nuclear winter debate on their thinking. They saw themselves as following in the footsteps of Sagan: scientists bringing a politically unwelcome warning necessary to save the world.[46] They argued that nuclear winter and impact extinctions were closely linked not just by computer models and stratospheric physics but also conceptually: "these recent advances in understanding the horrible consequences for our planet have come from research by planetary scientists familiar with studying the other planets. Perhaps it takes a cosmic perspective to appreciate, and deal with, the global consequences of such a man-made Armageddon."[47]

Simply framing the impact threat in terms of nuclear war had not worked, however, nor had their attempts to provide both numerical and psychological risk assessments for the threat. Surely part of the issue was that the Reagan era was not a good time to call for something that was, essentially, a new kind of regulation. It may also be that their neo-catastrophism undermined the fundamentals of both classical and psychometric risk perception underlying such regulations. The global, unprecedented nature of the threat made it difficult for anyone to know exactly how to react to it. Indeed, this is precisely what Beck argued happened to the way risk was conceived of during the decade—the novel dangers on the world scene were so disruptive to conventional understandings that not only were regulators unable to cope, but public trust in scientific authority was damaged on a deep level. Chapman and Morrison thought cosmic disaster should seem even scarier in the wake of Chernobyl; Beck showed that Chernobyl destroyed the very conceptual tools that one would have used to decide whether or not to be scared.

Warnings of Long-Term Government Projects

Chapman and Morrison were competing for apocalyptic attention. The 1980s were an intensely apocalyptic decade, and there was only so much social energy that could be spent on such problems. The hazard theorist Robert Kates in 1983 tried to understand why some apparently clear dangers were not being directly addressed. He contended that one explanation was that individuals and societies have a "small, relatively fixed stock of worry beads to dispense on the myriad threats of the world."[48] The same year, Mary Douglas and Aaron Wildavsky argued that societies had to choose which dangers were deserving of attention.[49] This choice, they said, was inherently cultural and political, and thus filtered through our institutions.

And in the United States, the most important institution for naming danger was Congress. There were many groups trying to get the House and Senate to give them a "worry bead" or two. So when asteroid scientists predicted the end of the world, they were part of a crowded field. When *Cosmic Catastrophes* was being published, John Dingell was already holding hearings before the House Subcommittee on Energy and Commerce on global warming.[50] In testimony that seems distressingly familiar today, officials from the Environmental Protection Agency, the National Oceanic and Atmospheric Administration, and others tried to establish the facts of global warming, likely consequences, and policy options. Stephen Schneider from the National Center for Atmospheric Research explicitly grappled with the questions raised by these sorts of apocalyptic predictions: what kind of policy response is appropriate for the huge scientific uncertainties? He concluded that this was "a value judgment, not an issue resolvable by scientific methods." It was worthwhile to ensure against possible catastrophes, however, depending on how likely they seemed. "In my estimation, space invaders are no more likely than the collision of the Earth with a large asteroid—which would be a catastrophe!" The odds of that collision he reported as 1 in 10 million each year, whereas the probability of several degrees of warming was a better than even bet.[51]

It must have been dispiriting for asteroid scientists hoping to persuade government leaders to have their predictions invoked as the very example of something that would never happen, along with literal science fiction scenarios. Fortunately, Chapman and Morrison received some help from the heavens and from the military–industrial complex. In March 1989, an asteroid was spotted within 400,000 miles of Earth, a near record. Close, but not outrageously so—less than twice the distance to the Moon. Interest grew when its trajectory was calculated and it was realized that it had shortly before passed directly through the Earth's orbit. The rock, known technically as 1989 FC and later named Asclepius, had hurtled through the space where the Earth had been just six hours before. It was 300 meters across and would have yielded an explosion of at least hundreds of megatons. This was exactly the sort of danger Chapman and Morrison were warning about—and we only spotted it *after* it had passed. It was still an event largely of academic interest until "an exaggerated—actually erroneous—press release by NASA placed the 'what-if-it-had-hit' issue on the front pages of newspapers around the world."[52] The *Wall Street Journal* complained, "Now they tell us! But where would you run, anyway?" The *Toronto Star* declared "NASA Caught Napping as Asteroid Buzzes By."[53]

One might expect that such an evocative demonstration of the asteroid danger, just as two of the world experts on that danger published a new book, would have galvanized policymakers. In fact, it did not. Even the media coverage was surprisingly short-lived, with most articles simply finger-wagging at NASA without any sense that there was a deeper problem involved. The media seemed somewhat unsure how to follow up on the story as well (there was not one interview with Chapman or Morrison). So the intuitively obvious scenario—sudden danger makes leaders pay attention to experts—did not happen. One more factor was needed for a proper intervention: greed. Specifically, greedy lobbyists.

The Asclepius close call drew the attention of Johan Benson, administrator of public policy at the American Institute of Aeronautics and Astronautics (AIAA), an industrial lobbying group. He asked Edward Tagliaferri, an AIAA committee chair and physicist by training, to look into it. Tagliaferri was startled to discover that a half-dozen searchers were responsible for almost all our knowledge of near-Earth asteroids and contacted them all directly to get up to speed on impact science. He received many suggestions from the Shoemakers and was deeply impressed by Chapman and Morrison's *Cosmic Catastrophes*, which led to "many conversations" with Chapman.[54]

Tagliaferri took away three conclusions from these conversations and his literature search: (1) the threat of asteroid impacts was real, (2) we were almost totally ignorant of where the threatening asteroids were, and (3) no one had any clue how we would cope with an impending impact.[55] He wrangled all his political skills to get his conclusions and recommendation approved by the AIAA board, which would normally take about two years. The board was not entirely sure how to approach the issue, especially because of the "giggle factor." The AIAA was a place for serious business, aircraft, and missiles, not Chicken Little warnings about the sky falling. There was a further issue too, still lingering from Snowmass. If they were going to talk about actual defense against an asteroid, that meant nuclear weapons and related technology: "here, one not only needed an international team of knowledgeable scientists and engineers, but participation from the military community as well, since most of the technology needed to cope with such a threat would have been developed during the US Strategic Defense Initiative and its Soviet counterpart."[56] In the end, the board approved the position paper by an 11-to-10 vote.

The paper was short, clear, and designed to be read by laypeople. It warned that if an asteroid collided with our planet, it would be "a disaster of unprecedented proportions." The probability was small, but the consequences were enormous. The risk was therefore finite.[57] It decried the nearly zero resources spent on detecting threatening objects, which meant that any of these objects could strike "without warning."[58] Using Chapman and Morrison's methods for risk analysis, it concluded that the likelihood of being killed by an asteroid was about the same as dying in an airplane crash. Therefore, government intervention would be worthwhile. Even more, "since the risk of danger from the impact of an asteroid is global, it is clearly in the best interests of all the world's nations to meet those risks collectively." And since relations between the United States and the USSR had never been better, "this program provides an excellent opportunity for the two superpowers to demonstrate leadership in an area of global concern."[59] In 1990, the image of the superpowers cooperating in asteroid defense looked somewhat different than it did when Senator Matsunaga suggested it in 1984.

The paper concluded that the United States would be derelict if it did nothing and humanity went the way of the dinosaurs. It recommended that a "systematic and open program be established" to detect and define the orbits of asteroids. Additionally, there should be studies on deflection and destruction of a threatening body.[60]

Immediately the AIAA went to work publicizing their document. As a lobbying group, they had a great deal of experience getting noticed by government leaders and had an extensive network of contacts to exploit. Benson sent copies of it to every congressional representative and senator. He made appointments with "several high-level government decision makers," including Lt. Col. Peter Worden of the National Space Council. Worden was an astronomer by training and reportedly "immediately understood the implications." He had many contacts in the executive branch (he was later the director of technology for SDI) and got a copy of the paper into the hands of the vice president, Dan Quayle.[61] This turned out to be a mixed blessing, as Quayle's speech on the matter only cemented his reputation for malapropisms and general obliviousness. He was the last person you wanted to have warning that the sky was falling. The press ridiculed him as usual. Mike Royko of the *Chicago Tribune* suggested than when Quayle was done looking for asteroids, he could get started on UFOs and the abominable snowman.[62]

Despite that setback, Benson organized informal briefings for several members of Congress and their staffs, focusing on members of the Space Committee. The archival record on these briefings is spotty, but it seems that the key moment was when Benson arranged for David Morrison to present to about twenty or so staffers for the House Committee on Science, Space, and Technology.[63] One of the people there was Harry S. "Terry" Dawson, senior staffer to George Brown, who would become the next chairman of that committee. Dawson, along with Wernher von Braun, had been one of the founding members of the advocacy group the National Space Institute in 1974. He was extremely interested in ambitious space programs, particularly as the Cold War incentive for space exploration waned. It seems that he, rather than George Brown himself, became determined that the government would act on the AIAA paper.

This was all taking place against the backdrops of the hearings for the 1991 NASA authorization bill.[64] These hearings were held, unusually, in front of the full committee instead of merely the relevant subcommittee because this was seen as the first budget for NASA after the end of the Cold War, and it was radically unclear how the "peace dividend" would reorder priorities, and the legacy of the Reagan-era pressures to privatize space (as discussed by Margaret Weitekamp in chapter 5 of this volume) was still very strong. The hearings were overwhelmingly concerned with the space shuttle and the space station, with some mention of the Earth Observing System. Asteroids were not mentioned at all. The transcripts ran to some three hundred pages, plus four hundred pages of appendices. Those appendices included a six-page manifesto for "Scientific Literacy for the 21st Century," an eighteen-page pitch for the commercial development of space (including Reaganite provisions to support small businesses); the details of the budget for the renovation of the fireproofing of the third floor of the construct processing control center at the Kennedy Space Center; eleven pages from a software lobbying group hoping for more NASA contracts; and twenty pages from the AIAA generally supporting the proposed budget (but with no reference to asteroids). Separate from the general AIAA statement, and second to last in the four hundred pages of submitted documents, was Tagliaferri's eight-page position paper on the danger of asteroids. The distillation of a decade of work by asteroid scientists was available only to the most diligent and devoted of readers.

This was, however, enough for Dawson to get language into the final authorization bill that called for the government to fund two small conferences on the impact threat: one dedicated to the detection of potentially hazardous bodies and one dedicated to methods to deflect or destroy them. The bill became law, and very small sums of government money were intentionally spent, for the first time, with the purpose of keeping humanity from going the way of the dinosaurs.

David Morrison was the logical person to be put in charge of the detection workshop. He was one of the world experts on asteroids, an experienced administrator, and (so far) the only scientist to have talked to even the staff of a member of Congress about the impact threat. The second workshop, on possible mitigation measures, was given to John Rather, a space weapons expert at Lawrence Livermore National Laboratory. Even with the Cold War over, discussions of asteroid impacts simply could not seem to escape the shadow of nuclear weapons.

The sudden enthusiasm of the national labs for asteroid defense did not go unremarked. Chapman pointed out that it was surely not a coincidence that an aerospace lobbying group first brought the issue to Congress. As the Cold War petered out, he wrote, the defense industry "looked for new applications of their 'Star Wars' detection and interception technologies, and began pressing the government to take action against a new enemy: asteroids and comets."[65] As with oil companies in Cyrus Mody's chapter 1 of this volume, corporations whose income streams are endangered by geopolitical events feel strong pressures to diversify. The specter of a world at peace made impact mitigation a fruitful route to explore. The *Wall Street Journal* reported that NASA's apparent need for a whole new generation of weapons "moved a weapons scientist at Los Alamos to shout: 'Nukes forever.'"[66] The planetary scientists, many of whom had been conscientious objectors during the Vietnam War and were active in causes such as nuclear disarmament, were frustrated to have to suddenly be working with figures such as Edward Teller.

Carl Sagan spoke evocatively for this group, warning that the Department of Defense was just looking for a new task in the wake of the Soviet Union's collapse: "All bureaucracies attempt to maintain themselves when their primary mission fades. They invent new tasks, preferably urgent ones, and the resulting inertia becomes especially high when jobs and profits are at risk. . . . A credible, sufficiently dangerous enemy is a great convenience for politicians unable to real with proliferating domestic problems and potential discord. And if such an enemy doesn't exist, it's usually easy enough to arrange for one."[67] The Pentagon's sudden interest in asteroids and comets seemed to be simply self-preservation. Sagan contended that the weapons labs saw this as "a way to unite continuing nuclear weapons development with a permanent seat on the save-the-Earth bandwagon."

Conclusion

Sagan was clearly right that the weapons designers and their associated lobbyists were thrilled to have an opportunity to build and test new nuclear devices (Teller was reportedly particularly enthused about the prospect of nuking Ceres to fragments). But it is probably not an accident that the prophets of the asteroid apocalypse were almost completely unheard until the weapons industry became involved. The planetary scientists did not really have much more convincing data and predictions in 1991 than they did in 1981. Even the Asclepius near-miss did not seem to do much to move their cause forward. Instead, they had to wait for a lobbyist to get interested. They needed someone to translate their work and concerns into a form understandable by the people on Capitol Hill—to show how government dollars could make money for the private sector.

Stephen Hilgartner's analysis of how scientists function as advisors to the federal government can be helpful in understanding what happened here. He approaches the problem with a theoretical framework grounded in performance: at any time, there are many competing performers trying to establish themselves as credible and important. The key is that some have successful "stage management" and some do not. That is, some are able to structure an effective relationship between experts, public, and the government.[68] This is what lobbyists brought to the table. They already knew how to stage manage their experts in a way that drew the correct kind of government attention. And particularly important for the era of greed, they needed to show how this government attention could support a nation of industrialists and entrepreneurs.

For the decades running up to 1990, lobbyists had a very effective (and profitable) apocalyptic narrative along these lines: preparing for nuclear war. And there were clear experts about nuclear war, along with administrative infrastructure and rhetorical expectations. It was extremely difficult to wrest control of that narrative from the existing experts, as Sagan discovered with nuclear winter. Similarly, if asteroid scientists during the 1980s wanted the government to pay attention to their apocalypse, they had an even more difficult task. Beyond modifying an existing narrative as Sagan wanted to, they had to *replace* that narrative with an entirely new one. This turned out to be very challenging, and we have seen how at every turn they were engulfed, blocked, and entrapped by existing concerns and concepts of nuclear war. It was essentially impossible for them to propose a new kind of apocalypse when there was already one dominating every conversation. It was only when actors became involved who already knew how to stage manage the existing apocalypse that some forward movement appeared. The new apocalypse had to be put in terms recognizable by the old one—an existing performer could be given an innovative role, rather than bringing someone entirely fresh onto the stage. The successful prophets, then, are the ones who are able to reframe their new predictions via existing modes already understood by the power structure. And the most powerful tool for getting the attention of the power structures in this era was greed, perfectly distilled and deployed by Washington lobbyists.

This did not result in immediate large-scale action. The impact scientists still had to wait three decades for DART and had to wander a winding path shepherded by a shattered comet on Jupiter, Bruce Willis, and yet another meteor strike in Russia. But they were eventually successful in engendering a new government bureaucracy implementing their warnings (an achievement perhaps unthinkable in the Reagan years). Their achievement still bore the markings of its Cold War origins, though—DART used no nuclear technology, instead focusing on the much less reliable kinetic-impact method. The worry that interventions against an asteroid apocalypse might trigger a more terrestrial one is clearly still with us.

NOTES

1. On the general history, see Erik Conway, Donald Yeomans, and Meg Rosenburg, *A History of Near-Earth Objects Research* (Washington, DC: NASA Office of Communications, 2022), NASA SP-2022-4235; Felicity Mellor, "Negotiating Uncertainty: Asteroids, Risk, and the Media," *Public Understanding of Science* 19, no. 1 (2010): 16–33; Felicity Mellor, "Colliding Worlds: Asteroid Research and the Legitimization of War in Space," *Social Studies of Science* 37, no. 4 (August 2007): 499–531.

2. Stephen Hilgartner and Charles Bosk, "The Rise and Fall of Public Problems: A Public Arenas Model," *The American Journal of Sociology* 94, no. 1 (1988): 53–78.

3. Luis W. Alvarez, Walter Alvarez, Frank Asaro, and Helen V. Michel, "Extraterrestrial Cause for the Cretaceous–Tertiary Extinction," *Science* 208 (1980): 1095; Walter Alvarez, *T. Rex and the Crater of Doom* (Princeton, NJ: Princeton University Press, 1997).

4. Jan Smit and Jan Hertogen, "An Extraterrestrial Event at the Cretaceous–Tertiary Boundary," *Nature* 225 (1980): 198–200.

5. David Morrison, interview by author, September 17, 2021; Clark Chapman, interview by author, September 10, 2021.

6. Eugene Shoemaker, "Impact Mechanics at Meteor Crater" (PhD dissertation, Princeton University, Department of Geology, 1960). The core argument is presented in E. C. T. Chao, E. M. Shoemaker, and B. M. Madsen, "First Natural Occurrence of Coesite from Meteor Crater, Arizona." *Science* 132 (1960): 220–222; see also William Hoyt, *Coon Mountain Controversies* (Tucson: University of Arizona Press, 1987).

7. David Morrison, as quoted by David Chandler, "Planetary Science: The Burger Bar That Saved The World," *Nature* 453 (2008): 1164–1168.

8. Chandler, "Planetary Science." Dangerous celestial objects are also sometimes referred to as near-Earth asteroids (NEAs), near-Earth objects (NEOs), or potentially hazardous asteroids (PHAs).

9. Chandler, "Planetary Science."

10. Chapman interview.

11. Shoemaker, "Collision of Asteroids and Comets with the Earth: Physical and Human Consequences," NASA conference report, unpublished draft (1982), 8.

12. Shoemaker, "Collision of Asteroids and Comets with the Earth." Some of the conclusions of the workshop appeared in Eugene Shoemaker, "Asteroid and Comet Bombardment of the Earth," *Annual Review of Earth and Planetary Sciences* 11 (1983): 461–494.

13. Chapman and Morrison interviews.

14. On the history of nuclear winter, see Lawrence Badash, *A Nuclear Winter's Tale: Science and Politics in the 1980s* (Cambridge, MA: MIT Press, 2009); Lawrence Badash, "Nuclear Winter: Scientists in the Political Arena," *Physics in Perspective* 3 (2001): 76–105; Paul Rubinson, "Containing Science" (PhD dissertation, University of Texas at Austin, 2008); Matthias

Dörries, "The Politics of Atmospheric Sciences: 'Nuclear Winter' and Global Climate Change," *Osiris* 26, no. 1 (2011): 198–223; Naomi Oreskes and Erik M. Conway, *Merchants of Doubt: How a Handful of Scientists Obscured the Truth on Issues from Tobacco Smoke to Global Warming* (New York: Bloomsbury, 2010).

15. Morrison interview.

16. List of TTAPS recipients, Carl Sagan Papers Box 813, Folder 2, October 28, 1983.

17. Congressional Record, Senate, Thursday, September 15, 1983, 24353.

18. Congressional Record, Senate, 24354.

19. Congressional Record, Senate, Wednesday, October 10, 1984, 31159–31162.

20. Congressional Record, House of Representatives, February 29, 1984, 3925.

21. Abstracts for the conference can be found in "Global Catastrophes in Earth History: An Interdisciplinary Conference on Impacts, Volcanism, and Mass Mortality," Lunar and Planetary Institute Contribution No. 673 (1988), NASA-CR-183329. Some of the papers were published in Virgil Sharpton and Peter Ward, eds., "Global Catastrophes in Earth History: An Interdisciplinary Conference on Impacts, Volcanism, and Mass Mortality," *Geological Society of America Special Paper* 247 (1990).

22. Chapman and Morrison, "Risk to Civilization," NASA conference paper abstract, at Global Catastrophes in Earth History: An Interdisciplinary Conference on Impacts, Volcanism, and Mass Mortality (1988), 26.

23. There is room for disagreement on this point. One or two people may have died as a result of the Tunguska explosion, and it is possible to interpret some older records as describing deaths caused by meteorites.

24. Clark Chapman and David Morrison, *Cosmic Catastrophes* (New York: Plenum US, 1989), 283.

25. D. Kahneman and A. Tversky, "Prospect Theory: An Analysis of Decision under Risk," *Econometrica* 47 (1979): 263–291.

26. Ulrich Beck, *Risk Society: Towards a New Modernity*, trans. Mark Ritter (London: SAGE, 1992). Originally published in German as *Risikogesellschaft* (Frankfurt: Suhrkamp, 1986). For an overview of Beck and his legacy, see Gabe Mythen, *Ulrich Beck: A Critical Introduction to the Risk Society* (London: Pluto Press, 2004).

27. Charles Perrow, *Normal Accidents: Living with High-Risk Technologies* (New York: Basic Books, 1984), 308.

28. See also Robert Kates, "Managing Technological Hazards: Success, Strain, and Surprise," in *Hazards: Technology and Fairness*, ed. National Academy of Engineering (Washington, DC: National Academy Press, 1985).

29. Slovic's general theory is outlined in Paul Slovic, "Perception of Risk," *Science* 236, no. 4799 (1987): 280–285. He later carried out a survey specifically on fears of impacts as well. Paul Slovic and K. Peterson, "Perceived Risk of Asteroid Impact," unpublished manuscript (1993) as summarized in Paul Slovic, "Perception of Risk from Asteroid Impact," in *Comet/Asteroid Impacts and Human Society: An Interdisciplinary Approach*, ed. Peter T. Bobrowsky and Hans Rickman (Berlin: Springer, 2007), 369–382, 379.

30. Slovic, "Perception of Risk." Slovic and his colleagues laid out their research agenda in many places: Paul Slovic, "Informing and Educating the Public about Risk," *Risk Analysis* 4 (1986): 403–415; Baruch Fischhoff, Paul Slovic, and Sarah Lichtenstein, "Characterizing Perceived Risk," in *Perilous Progress: Managing the Hazards of Technology*, ed. R.W. Kates, C. Hohenemser, and J. X. Kasperson (Boulder, CO: Westview Press, 1985), 91–125.

31. Barbara Combs and Paul Slovic, "Newspaper Coverage of Cause of Death," *Journalism Quarterly* 56 (1978): 837.

32. Clark Chapman and David Morrison, "Risk to Civilization," NASA conference paper at Global Catastrophes in Earth History: An Interdisciplinary Conference on Impacts, Volcanism, and Mass Mortality (1988), 27.

33. Clark Chapman and David Morrison, *Cosmic Catastrophes* (New York: Plenum US, 1989). Hereafter *CC*.

34. *CC*, v.

35. Michael D. Gordin, *The Pseudoscience Wars: Immanuel Velikovsky and the Birth of the Modern Fringe* (Chicago: University of Chicago Press, 2012).

36. *CC*, 8.

37. *CC*, 34.

38. *CC*, 36.

39. *CC*, 38.

40. *CC*, 73.

41. Alvin Weinberg, "Science and Its Limits: The Regulator's Dilemma," *Issues in Science and Technology* 2 (1985): 59–72.

42. Weinberg, "Science and Its Limits," 60–61.

43. *CC*, 276.

44. *CC*, 286.

45. *CC*, 211. As it happened, Morrison became NASA's chief debunker for apocalyptic rumors in 2012.

46. *CC*, 113–114.

47. *CC*, 122.

48. Robert Kates, "Hazard Assessment: Art, Science, and Ideology," in Kates, Hohenemser, and Kasperson, *Perilous Progress*, 251–264.

49. Mary Douglas and Aaron Wildavsky, *Risk and Culture: An Essay on the Selection of Technological and Environmental Dangers* (Berkeley: University of California Press, 1983).

50. "Global Warming," Hearings before the Subcommittee on Energy and Commerce, House of Representatives, February 21 and May 4, 1989, Serial No. 101-31.

51. "Global Warming," 62.

52. Quote from Chapman, "What if . . ." in *The Great Comet Crash*, 103–104; NASA Press Release 89-52, "NASA Astronomer Discovers 'Near-Miss' Asteroid That Passed Earth," April 19, 1989. It is unclear what Chapman is referring to as being "incorrect."

53. *Wall Street Journal*, April 20, 1989; *Toronto Star*, April 21, 1989. Also see *St. Louis Post-Dispatch*, April 29, 1989; *Orlando Sentinel*, April 21, 1989; *Chicago Sun-Times*, April 21, 1989.

54. Edward Tagliaferri, "The History of AIAA's Interest in Planetary Defense," ET Space Systems, AIAA 96-4381, September 24–26, 1996, 1. Presented at the 1996 AIAA Space Programs and Technologies Conference.

55. Tagliaferri, "The History of AIAA's Interest in Planetary Defense," 2.

56. Tagliaferri, "The History of AIAA's Interest in Planetary Defense," 3.

57. "Dealing with the Threat of an Asteroid Striking the Earth," AIAA Position Paper, originated in the American Institute of Aeronautics and Astronautics Space Systems Technical Committee, chaired by Ed Tagliaferri (April 1990), 1.

58. "Dealing with the Threat," 3.

59. "Dealing with the Threat," 5.

60. "Dealing with the Threat," 6–7.

61. Tagliaferri, "The History of AIAA's Interest in Planetary Defense," 3.

62. Mike Royko, "Astronomical Task Perfect for Quayle," *Chicago Tribune*, June 6, 1990, C3.

63. The AIAA position paper has a reference to D. Morrison, "Cosmic Catastrophes: The New Catastrophism," presentation to US House of Representatives' Subcommittee on Space

Science and Applications, June 26, 1989. However, there is no official record of the subcommittee meeting on that day or any hearings that year that matched this description. Morrison himself does not remember the exact circumstances of his presentation.

64. "1991 NASA Authorization," Hearings before the Subcommittee on Space Science and Applications of the Committee on Science, Space, and Technology, House of Representatives, no. 129 vol. II, H 701-77, February 6, 8, 21, and 27, 1990.

65. Chapman, "What If," 104.

66. Bob Davis, "Never Mind the Peace Dividend, the Killer Asteroids Are Coming!" *Wall Street Journal*, March 25, 1992.

67. Carl Sagan, "Between Enemies," *Bulletin of the Atomic Scientists*, May 1992, 24.

68. Stephen Hilgartner, *Science on Stage: Expert Advice as Public Drama* (Stanford, CA: Stanford University Press, 2000), 7, 147.

CHAPTER THIRTEEN

Service with a Smile, Or, How Profit Made Japanese Robots Personal and Personable

Yulia Frumer

Visitors to the opening ceremony of the 1985 International Science and Technology Expo in Tsukuba were treated to a show. They were promised futuristic technologies from the giants of the Japanese electronics industry—Mitsubishi, Sony, Hitachi, Fujitsu—names that evoked at the same time both a covetous desire for the latest gadgets and pervasive fear of Japanese domination of the tech world. But the ceremony began with a more traditional-sounding classical music concert by the NHK (Japanese Broadcasting Corporation) Symphony Orchestra performing J. S. Bach's "Air on the G-string."[1] That is, until the soloist was revealed to the audience—a humanoid robot playing from the score in front of it on an electronic organ. The prototype for the robot was WABOT-2, developed by Katō Ichirō and his lab at Waseda University between 1980 and 1984. The name WABOT is a portmanteau of WAseda roBOT, although the sound "wa" in Japanese can also be interpreted to mean "Japanese." The robot at the 1985 expo differed from the original WABOT-2 by just a couple of tweaks to facilitate maintenance, but it was considered a whole new model and had a different name—WA-SU-BOT. The WA and the BOT were the same, but SU indicated something new. It was the presence of a sponsor: Sumitomo Electric Industry Ltd., an offshoot of the powerful conglomerate Sumitomo.

Why would this economic powerhouse associate its brand with what looked like a gimmick? The drivers behind the booming Japanese economy in the 1980s were the automotive and the electronics industry. Factories that manufactured the products that dominated markets worldwide were indeed occupied by robots, but these were industrial robots responsible for arc-welding and the assembly of cars, computers, cameras, and sound systems. WASUBOT, however, was incapable of manufacturing anything. It could read musical notation, render that music on an electronic organ, and respond to simple voice commands, but it did not produce *stuff* that could be sold. And yet, Sumitomo was clearly interested in it.

Today, Sumitomo's investment in humanoid robotics would hardly raise eyebrows. But humanoid robotics research was not always enthusiastically embraced in Japan. In the 1960s, Japanese engineers worked on "automation" research, not robots. The term "robot" was associated with the clunky, double antenna-crowned, square-headed tin cans one found in toy stores. Robots were for children, not grown-up engineers. They belonged to the realms of sci-fi, animation, and manga, not the worlds of research labs, corporations, and government industrial policy. And, even when around 1970 Japanese engineers did allow the word "robot" to enter their lexicon, they repeatedly had to explain that they were not talking about *those* kinds of robots—the humanoid ones—lest they be denigrated or dismissed for frivolously tinkering with childish playthings.[2] Humanoid robots may have generated profits in the entertainment industry, but what attracted government and corporate sponsorship in engineering at the time were industrial machines employed in manufacturing that were only sometimes vaguely anthropomorphic and not even remotely cute.

So, what happened? How did humanoid robotics research in Japan go from being an embarrassment devoid of funding to a potentially revenue-generating line item that corporate giants could not wait to lavish sizable investment upon? And how did Japanese robots go from box-like machines with industrial arms to cute, humanoid companions engaged in activities that, up until recently, had been considered the exclusive domain of *Homo sapiens*, such as creative expression?

The story below follows the evolution of Japanese robots from industrial, to personal, to personable. This evolution was not driven by random mutation but by Japanese engineers who adapted robots to the rapidly changing landscape of government policies, corporate greed, labor and gender relationships, and the historically situated, yet individual, needs of Japanese consumers. The central feature in this landscape was a particular Japanese take on the global phenomenon of emerging neoliberal economies: the adoration of techno-fixes and legislation that was friendly to corporations, combined with rhetoric that would provide the veneer of serving the "public good" and targeted campaigns to guilt-trip individuals into assuming government responsibilities for social welfare. As we shall see below, the seed for government and corporate investment in humanoid robotics had already been sown by the 1970s, with the emergence of the techno-utopian *Information Society* vision. By the 1980s, humanoid robots came to be envisioned as the perfect contrivance to address the real-life consequence of market-driven government policies. Robots were now supposed to take over service labor, especially within the household, and to satisfy citizens' emotional needs. For the government, funding research into "personable" robots served to absolve itself of responsibility for Japanese citizens' well-being and to mask the failures of its policies. For corporations, making robots personable answered a perceived market demand for a product that would alleviate emotional distress associated with the market-driven world order.

Tokyo, We Have a Problem

To an untrained eye, Japan of the late 1970s appeared to be thriving. The poverty and starvation of the immediate postwar period had been relegated to a distant past. Following a brilliant political move, Japan was allowed to pay reparations in kind, instead of currency, and to regain foreign markets in the guise of "technical aid."[3] By converting wartime manufacturing capacity to produce peacetime products and by relying on human capital developed during the war, Japanese corporations generated rapid industrial and economic growth, paving the way for the so-called "Japanese miracle."[4] The Japanese economy was so strong in the early seventies that it emerged from the global oil crisis relatively unscathed. Japanese products were dominating world markets, wages were growing, and unemployment—at least according to official statistics—was low. And Japanese citizens enjoyed their new cars, single-family homes, TV sets, and automatic rice cookers.

But there were troubling signs. While Tokyo was becoming a jungle of elevated highways and smog, rural areas were stagnating and shriveling. The public was becoming increasingly aware of the environmental costs of industrial growth: the Minamata mercury poisoning, the Yokkaichi asthma, and the Itai-itai disease, caused by cadmium contamination.[5] Then there was the rising inflation. By the late '60s, people had become skeptical of governmental policies and were taking to the streets to demand better working conditions and more investment in infrastructure and social welfare.[6] Students barricaded themselves on campuses, demanding changes to not only the curriculum but also the economic policies of the whole country. And finally, the so-called Nixon shock of 1971 created a cascade of new global economic trends that were increasingly hurting Japanese exports by later in the decade.[7]

What drove the Japanese policymakers to action, however, was neither the empathy for human suffering nor the fear of an angry mob, but greed. Victims of industrial pollution could be ignored or paid off; students, union leaders, and political protesters were handled with force. But what can one do against labor shortages that threaten to arrest the growth frenzy, cap the money-making capacity, and sink the GDP? *That* was something that kept business leaders and economists awake at night.[8]

Enter the Information Society

Realization of the looming crises spurred think-tanks into action as early as the late '60s, when the Japanese economy seemed to be going full tilt. Recognizing both the impending doom of an economy based on heavy industry, corporations paid "experts" to produce reports and predictions that would convince the government to invest in their industry. In 1969, one of them, Hayashi Yūjirō, came up with a tempting solution that excited both the policy and the corporate worlds—a so-called *Information Society*, which would not only open new economic markets but also create a new world order in which humans would no longer suffer from problems that were plaguing the Japanese society at the time—harsh working conditions, environmental damage, and political discord.[9]

As the phrase "information society" suggests, the path toward a new economic world order was predicated on "knowledge-intensive" industries such as computers, telecommunication, and biotech.[10] Human-less factories, run solely by industrial robots, would take care of all manufacturing, thus solving the problem of the dwindling workforce and emancipating working humans from harsh and hazardous working conditions.[11] Social services would be run by computerized systems. So perfect was this world that Yoneji Masuda uncynically referred to the coming of the information age as "Computopia." Further, he painted the coming Computopia in religious terms as a "theological synergism of man and a supreme being."[12]

The most obvious feature of this utopian vision was the insistence that technology would fix social problems. In 1981, a report issued by the Industrial Structure Council, appropriately titled *Guidelines for Increasingly Prosperous Informaticized Society*, claimed that "information is . . . effective in solving social problems" and that "technological innovation is the fountainhead of social progress."[13] And while the vision itself was utopian, it had very real consequences in the form of corporate and government investment. Similar to what is described by Cyrus C. M. Mody in chapter 1 of this volume, energy companies rebranded their image by investing in research on artificial intelligence (AI), robotics, and biotech.[14] The government began aggressive legislation that offered tax breaks and even funding to businesses that manufactured electronics. The 1978 Machine Information Law made it explicit that the funding of private corporations was a way to "improve citizen's welfare and create a society in which life is worth living."[15] The famous Fifth Generation Computer project that began in 1982 was catalyzed by the *Information Society* vision and explicitly framed not only as a technical challenge but as social solution.[16]

The focus on technological fixes absolved the government from investing in social welfare. Calls for the improvement of the welfare state were drowned out by the optimism of the information society vision, according to which welfare would just happen automatically with the introduction of automated information technology. The assumption was that the new information world would bring prosperity to *all*, creating a harmonious, class-less society and obviating the need for a massive welfare system. And because automation of office work promised to create an abundance of free time, individuals would be able to dedicate some of this time to "community, or voluntary activities [of the social] service type," and practice "self-control," so that "the dominant form of maintenance of public order . . . will be the *autonomous restraints exercised by the citizens themselves*."[17]

The insistence on techno-fixes, the belief in the power of the corporate world to regulate the economy and society, the desire to "shrink the government," and the government divestment of welfare—all sound like a neoliberal utopia. The rhetoric of Japanese policymakers inspired by the *Information Society* vision sometimes sounded very much like Margaret Thatcher, described by Jon Agar in chapter 4 earlier in this book. Welfare, according to one of them, was "obstructing the spirit of independence, self-help, and responsibility for oneself."[18]

Japanese proponents of the *Information Society* even cultivated its resonance with the American-style neoliberalism. The term coined by Hayashi Yūjirō's in his 1969 essay, *jōhōka shakai*, literally translates to "society that goes through informatization." With the proliferation of "information society" language in American social analysis in the 1970s, such as Daniel Bell's and Alvin Toffler's, however, Japanese authors began translating Hayashi's term as "information society."[19] It is not a coincidence that Masuda published an English-language version of his book—like many other Japanese theorists, he considered Japan to be at the forefront of the global technological movement. And there were good reasons to think this way: Toshiba's computerized health evaluation system, for example, was installed as early as 1970 and served as a model for American proponents of computerization of health care.[20]

But Japanese *Computopia* also had a very distinctive local flavor. The concept of a computer-based utopia tapped into the long Japanese tradition of techno-nationalism, promising to reinforce "the cultural values of the nation."[21] Offloading responsibility for the welfare industry onto individuals and their families was couched in the rhetoric of "Japaneseness." In this framing, Japanese "traditional" family structures and "Japanese traits" of social cohesiveness made it "natural" for families to take care of their relatives and for the elderly to do their best to avoid becoming "a burden on society."

Unlike their American counterparts, the Japanese government did not simply retreat from welfare, leaving to individuals the "freedom" to make their own choices. Instead, the government of Japan was actively involved in what Sheldon Garon called "the molding of Japanese minds," packaging computerization in terms of a "public good" and running public campaigns to convince individuals to take responsibility for social welfare.[22] The government perhaps "shrank" its investment in institutionalized welfare, but it did not disengage its active engineering of social dynamics by means of psychological manipulation.

The "Man" in the Man–Machine System

For the new economy to work, there had to be new markets for information technology products. But it was not enough for corporations to *create* new products; they also had to find ways to appeal to the consumer base, either by tapping into existing needs or by creating new ones. Surveying the market, corporate experts recognized a generational shift in consumer preferences.

The key descriptor of those new preferences was "personal." For many decades, "personal" was somewhat of a dirty word, when individuals were expected to suppress their personal needs and wants, first on the battlefield and then on the factory floor. With rising standards of living in the late 1950, however, people began searching for meaning beyond work. Recognizing the new need, corporations sought to turn "personal" into a commodity as a means of divesting consumers of their accumulated capital.[23] As cars became more affordable, car makers began targeting individual consumers. Toyota ads placed their sleek products in environments that were decidedly not work related: a ride through an enchanting forest, a view of moving clouds as seen through a car window, a young, fashionable female passenger with a dreamy look on her face. . . . Cars were not just functional but also reflected their owners' aspirations and personalities. In 1961, Hoshino Yoshirō, best described as a "visioneer," published an essay that described individual car ownership as a revolution.[24] The Japanese title of his essay, *Mai Kaa*, was a transliteration of the English phrase "my car"—a nod to the American automobile culture portrayed in Hollywood movies. In the essay, Hoshino linked "kaa" ownership to feeling free from constraints and living in the moment.[25] This message resonated with so many that the phrase *mai kaa* became an affectionate term for one's own personal automobile.

The emphasis on "my" was not limited to cars. Akio Morita, the founder and the former chair of the Sony corporation, tells the story of trying to promote miniaturized stereos in American markets. According to Morita, he was told that Americans like things that are big. Having big homes and big cars, why would they choose to buy a tiny stereo when they could get a large one? Morita's reply was that a small stereo allows everybody in the family to have their *individual* music in their own rooms.[26] This story—as many others Morita tells in his memoir—may be apocryphal. What we know for sure is that it was important to Morita, the salesperson, to portray his products as *personal*.

Perhaps nothing symbolizes personal electronics more than another Sony product—the Walkman, released in 1979. Pragmatically speaking, the Walkman was meant to enable an individual to carry their music wherever they went,[27] but it was more than listening to music. The Walkman allowed people to create a private bubble, protected from interruptions from the outside world.[28] At a certain point, Sony designed a Walkman with two audio jacks and marketed it as a dating device—sharing the music, sharing one's own private world was a sign of intimacy. Consumers may have wanted to be in their individual bubbles, but they did not want to be lonely—they also wanted personal companionship.

With the release of IBM's personal computer in 1981, "personal" became a brand. As Laine Nooney discusses elsewhere, including in chapter 3 of this volume, the 1970s saw a growing investment in microcomputing for personal (as opposed to corporate) use.[29] The Commodore Personal Electronic Transactor even contained "personal" in its name. With IBM's PC, however, "personal" became an eponym. In Japan, it was first referred to as *maikon*—a portmanteau combining the sound of the English words "my" and the first syllable of "MIcro," together with the beginning of the word "computer" (*konpyu-ta* in Japanese). But the brand name won over, and a shortened version of "personal" (*pa-sonaru*) replaced "my" in a word that is still used today as a stand-in for a "computer"—*pasokon*.

The commodifying of the "personal" allowed the visioneers of Japan's *Information Society* to appeal to people on the emotional level. The architect of the vision, Hayashi Yūjirō, promised that automation would create free time to pursue leisure, hobbies, education, and creativity.[30] Another spokesperson of the *Information Society* craze, Masuda Yoneji, argued that "knowledge-based" technologies satisfied the need for "self-actualization" and for "fulfillment."[31] "Free time," "leisure," "creativity," "self-actualization"— all evoked individuation and personal fulfillment. The fascinating, novel, leisure-related, and futuristic electronics products were reframed in "personal" terms and presented as a path toward self-realization.

The Limits of Corporate Techno-Utopia

Despite the Japanese government's efforts, however, the reality did not play out the way the prophets of the *Information Society* promised. By all parameters, the early '80s saw "the first stage of the shift towards the information society."[32] With the increasing automation of factories, soaring sales of electronics, and ever-growing funding for new technologies, problems plaguing the Japanese society were supposed to be resolved. Investment in electronics industries, according to the Ministry of International Trade and Industry (MITI), would result in a completely transformed society already in the '90s.[33] But the reality was different.

First of all, a computerized workplace was not easier but *harder* on the worker. Only relatively few workers got the chance to engage with computers creatively while the rest toiled in mindless tasks of information drudgery, such as data input. Working on computers caused a variety of professional illnesses, such as wrist injuries, neck muscle strain, eyestrain, and headaches, all of which contributed to workers' stress and depression. An exclusive focus on the screen also disrupted workspace social matrices, resulting in a growing sense of alienation in the workplace. This was particularly hard on women, most of whom were part-time employees who never had the chance to form strong social bonds with their coworkers.[34]

Workers at automated factories did not fare better. For example, Toyota's famous management system promised to increase worker satisfaction—and hence productivity—by allowing every employee a voice in the production process. In reality, the possibility of making an impact was illusory—dissenting employees were deliberately excluded from workplace satisfaction surveys.[35] Worse, even those workers who were overall satisfied with the workplace environment reported more psychological stress when working in roboticized factories.[36]

One of the most glaring departures from the utopian vision was related to time. The whole social system described by the *Information Society* proponents was predicated on the assumption that an automated workplace would free up human employees' time.[37] It was in their newly found free time that people were supposed to find self-actualization by engaging in creative hobbies and learning. It was free time that should have encouraged people to participate in "voluntary synergetic cooperation" and take upon themselves welfare responsibilities. And it was free time that would allow people to maintain meaningful social relationships. But in the early 1980s, there were no signs of increased free time. Quite the opposite—people were working *more*, not less.

Another oversight was workforce balance. With their emphasis on cultivating customers' "creativity" and "personal growth," corporations needed a constant supply of novelty—a phenomenon that Tessa Morris-Suzuki has described as "information capitalism" predicated on "perpetual innovation."[38] Industries driven by "perpetual innovation" lured workers with the promise of creative work that would satisfy consumers' intellectual needs rather than material necessities. This meant that when a growing economy created an increasing demand for service, there were fewer people willing to go into the service sector. More high-speed trains required more railroad employees, more travel meant higher demand for hotels and hence for hotel staff, and a larger volume of consumer electronics meant increased demands for store employees. This resulted in significant shortages in the service workforce.[39]

The vision of the *Information Society* failed to account for service work. The Japanese word for "service," *sa-bisu* (サービス), is sometimes used to indicate that something is "on the house"—such as free dish at a restaurant, or a free neck rub at a hairdresser. Of course, nothing is ever actually free, but lacking an explicit transaction, the customers are tempted to ignore the hidden costs. Similarly, proponents of the *Information Society* refused to consider the hidden costs of service. Enamored with the idea of automaticity, they assumed that alongside welfare and policing, service would just happen automatically. Treating service like magic, they even ignored the warnings of engineers—the ones that were supposed to make automation happen—that nothing was *really* automatic and that every device appearing to act "on its own" was enabled and maintained by the efforts of humans.[40]

Even more significant was the fact that women began rebelling against their government-assigned roles as welfare providers. They, too, demanded personhood and self-fulfillment. They pursued education for the purpose of career advancement and not just to improve their marriage prospects, sought work as a means of personal development and not just to supplement family income, and desired hobbies and societal investment in their personal interests. Some of these demands were finally answered with the passage of the Equal Employment Opportunity Law in 1985, which seemed to promise middle-class women an existence beyond the household.[41] But seeing that the influx of household appliances did not deliver on the promise of reduced housework, working women also began to demand a more equitable distribution of household labor, which experts warned might lead to an increased divorce rate.[42]

What prompted the Japanese government to realize that something had to change was not women's demands, however, but demographics. In the early 1980s, there was no way to deny that median life spans were rising, and birth rates were falling. If families maintained the ideal of the "traditional Japanese" three-generation household, soon, demographic trends showed, a couple would need to take care of *both* sets of parents. This meant that a middle-aged woman would now be taking care of four elderly people *on top* of childrearing, extra education, cooking, cleaning, comforting, entertaining, and so on. This would put the country into an economic double-bind. To do all this housework and provide caregiving, women would have to abandon their hopes of participating in the workforce, which would cause a further dwindling of the labor force in the caretaking professions such as nursing, as well as eliminate a significant portion of contribution to the pension funds. Pension funds, however, were increasingly needed to sustain the growing population of retirees, who were now expected to live around thirty years after retirement and often even more.[43]

The demographic crisis exposed the hypocrisy of the *Information Society* vision that presumed women would just pick up the slack at home. Care and service needs were not satisfied "automatically"—they were addressed through hidden and unacknowledged labor. And with women no longer willing or even capable of playing along, somebody—or something—had to do the job.

Robots to the Rescue

When, in 1994, Katō Ichirō was interviewed about his work in robotics, he pointed to his secretary (OL, in Japanese: Office Lady), who just brought cups of tea, and replied, "One day, I was chatting in the lab with a few colleagues of mine. Since there was no one who could make tea for us, we started to wonder how nice it would be if we had a robot that could bring us tea."[44]

One of the main figures in the history of Japanese robotics, Katō was a researcher at Waseda University. Early in his career, he made himself a name for developing myoelectric hand prosthetics. After hands and arms, he turned to investigating robotic legs and locomotion and, in 1975, unveiled a bipedal walking robot that became known to the world as the first WABOT. In the early 1980s, he moved on from the legs to the brain. The new robotic project may have been less of a mechanical engineering challenge than its bipedal predecessor, but it was inspired by the breakthroughs in artificial intelligence promised by the Fifth Generation Computer project and was supposed to result in a robot that had the capacity to learn and to think abstractly.[45] As a way to realize this goal, Katō decided to build a musical robot, since music was within the realm of humanity, of self-cultivation, of refinement. The new robot would be capable of reading musical notation, playing an electronic organ, and responding to voice commands.

Critics, however, accused Katō of being too greedy. Experimental research like Katō's was painstakingly slow, had vague goals, and could not promise concrete application. All the while, it relied on almost unlimited government funding and cost billions of taxpayers' yen, with no clear indication of how and when the investment would bring a return. There were mounting voices that called for redirecting funds to corporate labs, which would be able to turn the investment into a profit in just a couple of years.[46]

Katō claimed to have a broader economic vision. Similar to Soviet and American scientists and engineers, described by Asif Siddiqi (chapter 11) and Mathew Stanley (chapter 12) earlier in this book, Katō framed his research as critical and strategic. He pointed out that in the several decades after the end of World War II, the economic driver of the whole Japanese economy was "My Car." But, he warned, "the automobile industry is becoming saturated. With the number of passenger cars reaching 26–27 million, the growth period is over." The new area of economic growth, Katō noted, was "My Computer." But within ten years, "the number of computers was expected to reach 10 million, maybe more if society reached a place when there is more than one computer per household." What would be next?[47] The next big thing after "My Car" and "My Computer," Katō predicted, would be "My Robot."[48]

This would not be the robot you imagine, Katō warned in 1983. Remember, he implored his audience, how in the first years of the '70s engineers wanted to talk about industrial robots and everybody imagined a toy or something out of manga? Well, the situation in the '80s was reverse, and when one said "robot," everybody was imagining an industrial robot.[49] But this is not what "My Robot" was going to be. Industrial robots were used in the secondary industry for physical labor. My Robot, in contrast, was destined to the tertiary industry, namely, the service sector.[50]

And, per Katō, the service sector was the area robot makers should be thinking of. "Today is the golden age of industrial robots," he declared in 1983, stating that Japan owned 70 percent of the world's industrial robots.[51] Only a year later, Katō's tone changed. "Recently, Western countries began following Japan's lead, and it is noticeable that the number of industrial robots in the US, the UK, and France is rapidly increasing. Now, Japan's share is only as high as 50%."[52] Appealing to corporate greed, Katō implored companies to invest in "My Robot," which, he claimed, was "the only strategic product for the post-computer era."[53]

Katō's pitch appealed for economic but also emotional reasons. Unlike computing, robotics was a very young discipline that was developing during the Japanese economy's meteoric rise. Consequently, whereas software engineering was perceived in Japan and other countries as something imported from abroad, Japanese engineers could rightfully claim significant technological breakthroughs in robotics, making it a perfect candidate for a "national" technology.[54] The rapid advancement in Japanese robotics engineering from the late '60s to late '70s, combined with the futuristic image of robots in the media, made robots the epitome of "next-generation technology," so central to the *Information Society* vision. Robotics engineering promised the perpetual innovation required by an economic market predicated on satisfying the psychological need for creativity. If robots could be made to provide the dreaded household labor, the cost would be borne by the consumers, freeing the government from the need to invest in social welfare.

The idea that robots would take over women's household work was rather explicit. "Take dishwashing after dinner, for example," Katō pointed out. "I don't think it's a fun thing to do. Everyone else in the family is having fun after dinner, but *she* has to do it. . . . I am sure that every woman wants to do something about it, and for those who don't, it is probably because they have come to terms with the fact that it is unavoidable. *This is where women's burning need for My Robot is hidden.* My Robot, with its supple intradermal sensor, will easily do the cleaning without breaking a single plate," while the woman *can spend that time on something meaningful for herself.*"[55]

By now, it should not be difficult to see why corporations such as Sumitomo would take an interest in WABOT 2. The ability to play music, the capacity to recognize objects or people, and the use of voice communication were framed as a "symbol of a future with robots" and a proof that "tomorrow's science [will bring about] the enrichment of mankind."[56] In market terms, it was a technological breakthrough that charted the path toward a functioning service robot. The official guidebook of the 1985 Tsukuba Expo, where WASUBOT performed, opened with a story of "Alice in the Land of Wonders." In it, Alice is pictured visiting the future. In one imagined room, she spots a robot with a vacuum and exclaims, "Oh, wow. With this thing, I wouldn't have Mom on my back about every little bit of mess. I could lie back on the carpet and eat away at cookies and have the crumbs cleaned up just like that." Enamored with the idea but frustrated with reality, Alice throws a fit, wanting to bring the service robot back to the present.[57]

Other developments in robotics research similarly pointed in the direction of service. In the mid-1980s, multiple robotics labs were working on products with commercial potential. The Mechanical Engineering Lab of the Japanese Ministry of International Trade and Industry was working on MELDOG—a robotic guide dog that was supposed to enhance the sense of independence for the visually impaired.[58] An array of other labs were working on locomotion and gentle handling of delicate objects. In 1986 Honda began its own research on bipedal robots, with the intention of "develop[ing] a more viable mobility that would allow robots to help and live in harmony with people" in a "real world environment."[59]

Seeing yen signs, the Japan Industrial Robot Association (JIRA) began conducting surveys in an effort to assess the market potential of household robots, or "My Robots." In 1988, JIRA sent a questionnaire to about 200 households, particularly targeting housewives. The survey asked about the potential to entrust household tasks to robots, to engage in recreational activities with the robot, and to pay the steep price it would cost. We can get a glimpse of JIRA's technological optimism from the fact that the company's researchers were surprised to learn that women were not so keen on delegating their children's education to machines. Also, while survey participants enjoyed toying with the idea of having something like C-3PO in *Star Wars*, they were not thrilled about shedding large sums of money on it.[60] Still, Katō was encouraged by JIRA's questionnaire results. "When personal robots start spreading to households," he predicted, "'My Robot' will become a new core industry, just as automobiles (My Car) and computers (My Computer) once did."[61]

From Personal to Personable

In order to make the dreams of "My Robot" come true, however, there needed to be an overhaul in the whole manufacturing process of robots. One of the major concerns was safety. Industrial robots were considered so dangerous that safety protocols prohibited humans from approaching areas with actively working robots and forbade the activation of robots when humans were nearby.[62] Obviously, such dangerous robots could not be marketed for service functions in the domestic setting. Moreover, if robots were to be used in people's homes, they would need to deal with stairs, tatami mats, fragile objects, and an ever-changing, unpredictable environment with children, pets, moving chairs, and whatnot.

And there was something else, something that took engineers by surprise but would have been obvious to anybody who has experience providing caregiving labor. While robots were designed to perform specific tasks, care involved much more than cooking, cleaning, and helping others dress. To take care of the aged, do housework, participate in hobbies, and respond to finicky humans with their nuanced and ever-changing needs, service robots would have to be designed for what Katō called "information work."[63]

What does "information work" mean? The reader can probably guess that it has something to do with the *Information Society*. But there is no need to guess because the prophet of the computerized corporate utopia, Hayashi Yūjirō, explained "information work" as early as 1969. Hayashi noted that labor shortages in the service sector that accompanied the "miracle" economy resulted in rude and inattentive service. Overextended workers, he wrote, only provided what he called the "practical" side of the service one paid for, such as giving a customer a massage or getting a customer safely to their destination. But for Hayashi, service included the practical task as well as an "information function," which was "not only to achieve a goal but also to make customers feel good and generate positive mood." When receiving the functional part of the service, the customer should be greeted with an *invigorating* (*sawayaka*) "Good Morning!" and a *cheerful* (*nikoyaka*) and warm attitude.[64] For Hayashi, facilitating customers' good mood was an essential part of the service and hence part of the commodity. And so, long before the term was coined, Hayashi identified what we now call "emotional labor" and attached a yen sign to it.[65]

If robots were to be used to solve the woes of the neoliberal economy, they would need to be able to provide the whole package—both the practical and the emotional components of service. Robotics engineers often say that by designing a robot that is supposed to replace human functions, they learn a lot about what makes one human. What Japanese roboticists of the late '80s realized was that to replace housewives or friends in acts of service, robots would need to perform the emotional labor these people performed or, in Katō's terms, to "understand" humans, to "detect the state of [the] human mind," to attend to their human's unique "personality."[66] In other words, for robots to become truly personal, they needed to become *personable*.

What does personable mean in engineering terms? At first, Japanese engineers had thought that making personable robots was only a matter of a few external tweaks. Industrial-looking robots "did not fit the home décor," and a fully humanoid robot "could be mistaken for an intruder," so perhaps companion robots should be designed to resemble famous media characters, and everything would be fine.[67] In reality, appearance wasn't enough. Emotional labor involves socially and cognitively complex human functions that could not be programmed into a robot. But Katō realized that there was actually no need to engineer complex robotic brains; the easier way out was to trick the humans into *believing* that service robots are sentient. "If a My Robot *behaves* like a human and can *seemingly* guess human feelings, human psychology would kick in, and people will become emotionally involved, thinking 'ah, this robot has a mind.' "[68] A cursory read in psychology suggested that the solution was to identify behavioral proxies that humans associate with emotional connection, which could be quantified and programmed. For example, Katō suggested that a robot could learn to follow the human gaze, recognize their human's gait, or memorize their preferences.[69] Instead of engineering robotic minds, one could engineer human ones.

To learn how to engineer human minds, robotics engineers turned to a scientific discipline that capitalized on human cognitive tendencies to turn a profit—consumer psychology. The evolution of the discipline from ergonomics reflects dramatic shifts in attitudes that characterized different decades. The '60s, according to engineers' reflections, were an era of rapid economic growth, resource extraction, and the necessity to satisfy basic material needs. Toward the end of the decade, however, it became apparent that rapid growth imposed a heavy toll on humans and the environment. At the same time, rising standards of living allowed moving beyond strict satisfaction of material needs. The '70s, engineers envisioned, would be "an era of restoration of humanity,"[70] when humans would seek *"spiritual satisfaction and fulfillment."*[71] The leaders of the ergonomics association felt that the old name, Science of Labor (*rōdō no kagaku*), was too focused on the labor itself and did not reflect the emerging demand that the discipline be more "man-centered and man-oriented."[72] The new name was Human Engineering (*ningen kōgaku*, which could be understood as either "engineering *for* humans" or "engineering *of* humans"). The '80s were a game-changer altogether. "Today, we live in an age of food and clothing sufficiency, an age of maturity, abundance of goods, and individuality; today, we only buy the things we *want*. It is the age of affect," declared Matsuo Nagamachi as he promoted his disciplinary agenda—emotional engineering (*jōcho kōgaku*). Back in the era of rapid growth, he continued, "if you manufactured *something*, you'd sell it," but in the '80s, just manufacturing something was no longer enough; the only way for companies to survive was to identify people's emotional needs and design products that address those. "Emotional engineering," he wrote "translates psychological factors, such as feelings and images that people hold in their minds, into concrete design elements that express, or realize those feelings and images in matter."[73]

What does that mean, and how does this work? Nagamachi's favorite method in the '80s was C. E. Osgood's method of semantic differential.[74] The theoretical premise central to this method is that people assign meanings to objects through associations with various signifiers. So, when somebody went to buy a mug, for example, they weren't just looking for a vessel from which to drink a hot beverage but "a cup for two young people" to share a drink together or a "cup for a quiet drink by oneself." Or, in another example, when buying a car, people didn't just evaluate its performance by some kind of objective criteria but had an image of what a car would make them look and feel like, such as "a fast car that makes you feel carefree and makes your girlfriend faint with excitement." To engineer emotions, researchers would need to identify what material characteristics evoke which associations within a target group of potential consumers. So, to continue with the car example, researchers would eavesdrop on potential customers' conversations in dealerships and shows and register repeating adjectives they would hear. They would then create pairs of adjectival opposites, such as "traditional–modern," "timid–daring," or "feminine–masculine," and create a survey, asking participants to evaluate specific design features based on a series of those adjectival pairs, each time attaching a numeric value on a scale of 5 or 7 between the extremes. Gathering the amassed data, researchers got something like graphic equalizers for human emotions, where one could increase or decrease associations with modernity, boldness, or masculinity but playing with the shape of the side view mirror.[75]

In robotics, to identify what design features would elicit a feeling of "kindness," researchers compared a series of robots that functioned exactly the same but had different appearances. They discovered that rounded, smooth, and visually light designs were associated with "gentleness." Consequently, a metallic and only vaguely humanoid robot was deemed "kinder" than a humanoid one dressed in an angular and rough attire. One surprising discovery was that sexy ≠ caring: features that scored high on "attractiveness" scored low on "warmth."[76] In the late '80s and early '90s Japanese customers' minds, the survey showed, there was no overlap between sexual and caretaking labor; one was associated with young women and the other with middle-aged housewives, and the "signifiers" of each read as mutually exclusive.

Early attempts to identify desires and associations of potential buyers of service robots brought about a realization that roboticists had to focus on engineering human feelings as much as circuit boards and actuators, if not even more.[77] In the 1990s, fueled by the rapid developments in the field of artificial intelligence, Japanese engineers integrated principles of industrial design with the cognitive science of information processing to form a new subfield of "affective engineering" (*kansei kōgaku*).[78] The 1990s, however, felt very different from the 1980s. Retrospectively dubbed by economists a "lost decade," the post–economic bubble Japan was no longer a place of unleashed consumption. After the devastating earthquake that destroyed Kobe in January 1995 and the sarin gas attack in the Tokyo metro just two months later, the country descended into an overall sense of gloom. Robotics engineers responded by designing robots that would elicit a sense of emotional attachment and make people happier, paving the way to what analysts now call the Japanese "robotopia."[79] The Honda research into robotic locomotion that began in the 1980s eventually led to the creation of ASIMO—a walking, talking, stair-climbing, child-like robot that was first unveiled at an Aichi expo in 2005. In the years since, ASIMO has been through a technological evolution, eventually becoming able to play music and kick a soccer ball around with President Obama. In 2007, the first government of the former prime minister Shinzō Abe came up with the *Innovation 25* roadmap, which promised that by 2025, service and household robots would "revitalize" the rapidly aging society.[80] In 2022, the goalpost moved to 2050. Attempts to correct the shortcomings of the 1980s Computopia with yet another technical solution sowed the seeds of the Robotopia dreams of the present.

The robot that earned the 2019 CES (Consumer Electronics Show) Award from the Verge was an adorable blob. Is it a baby seal? An owl? A teddy bear? It does not really matter; the robot is cute and huggable. Its short arms resemble penguin wings and lack fingers, which the robot does not really need because its main purpose is to roll around the house and coo, nudging its owner for hugs. The device is LOVOT—a robot that loves you. "Our robot," says its designer, "doesn't do any work for humans and it doesn't have any [media] contents for entertainment purposes. But neither do dogs or cats. What it does is recognize you and bother you. That's the aim of our robot."[81]

Indeed, like the WASUBOT thirty-something years earlier, LOVOT does not manufacture anything. Unlike the techno-futuristic visions of mid-1980s, however, the robot does not provide any practical service either, even though that, technologically, it could have been made to at least sweep the floor like a Roomba. But the statement that LOVOT "does nothing" is simply untrue. In fact, it is a highly sophisticated machine that was created to do *a lot*—emotionally. LOVOT's appearance is designed to trigger mental/emotional associations with cute, fuzzy, and seemingly cuddly baby animals. Its coos are purposefully infantile, and when it rolls around and nudges its owners to hug it, it makes its owner feel needed, simulating a sense of active interdependency. It is warm to touch and responds to stroking and petting, soliciting the activity that releases the "hug hormone," oxytocin. It is about the size and weight of a large baby, and when cuddled, it closes its giant, lifelike eyes and seemingly falls asleep, satisfying its owner's nurturing instincts. When the owner is looking at the robot, it looks back or even comes closer. Its tiny wheels (which retract when it is picked up, so that they don't spoil the illusion of hugging an infant) are nothing like the highly sophisticated bipedal locomotion that Katō Ichirō worked so hard to build. But in terms of cognitive effect, LOVOT is a materialization of Katō's "dreams of 'My Robot.'" It can recognize up to thousand people but shows particular preference to its owner, whose unique bodily habits, like their gait, are registered by the pattern recognition software.

LOVOT is decidedly not the C-3PO look-alike bipedal tennis partner envisioned by 1980s engineers and policymakers. After the initial promise of the early aughts, it became clear that creating a robot that checks all the boxes—automaticity, versatility, locomotion, pattern recognition, language processing—was not practical. Not that it was impossible, just that it required too much human involvement to operate and was too costly to manufacture to make it a viable marketable product for home use. One by one, corporations turned to more specialized, nongeneral use models, focusing on practical tasks (such as "smart" hospital beds) while eliminating the humanoid, bipedal, and emotional components, or on remotely operated robots that are not autonomous, or—as is the case with LOVOT—giving up on everything else except emotional labor.

What the visioneers of the 1980s did get right is that emotional labor is a commodity in high demand. Present-day Japan is suffering an epidemic of loneliness, and not only of the elderly but also the younger generations whose overwhelming and all-consuming work schedules preclude the possibility of enjoying a rich, fulfilling social life. Engineers who design emotional robots—the LOVOT, the conversational Pepper, the "avatar" girlfriend by Gatebox, or the humanoid hand someone can attach to the inside of their coat to simulate the feeling of walking in a park while holding somebody's hand—are no doubt motivated by the real needs of lonely people. But the historical roots of the technologies and designs they produce reach down to corporate greed that commodified love.

The Politics of Personable Commodities

Greed was a reagent that catalyzed the transformation of robots from industrial to personal and then to personable. When we try to comprehend what motivated the sudden generous funding and promotion of humanoid robotics research in 1980s, we find that cute robots were envisioned as a commodified band-aid for the emotional wounds inflicted by the market-driven economy.

This is not to say that Katō and his like were soldiers of corporate greed. But without a market-oriented agenda, the aspirations to create a humanoid robot would be limited to a few university labs of well-established professors who no longer needed to worry about rank and funding. Instead, driven by the desire to make corporate technocratic utopia a reality in Japanese society, the government offered funding not only to university research centers but also to corporate labs that promised to develop personal robots. While robotics engineers were trying to address specific human needs, the Japanese government had a broader perspective on the role of personal robots in the market-driven economy: funding research on what made robots personable was an attempt to offset the need to invest in welfare or to develop services that would replace labor previously provided by women. For corporations, unfulfilled emotional needs meant that there was a market demand for a product that promised to fulfill these needs. It is not a coincidence that ASIMO was developed by Honda, personable pet dog Aibo by Sony, and the information robot Pepper by the telecommunication giant Softbank—the research on how to make a robot personable was a pathway to profit.

Japanese personable robots did not just *happen* to be cute; rather, robotic cuteness was the result of decades of research, funded by the Japanese government and corporations that sought to commodify emotional labor. Japanese robots' becoming personal and then personable followed changes in the kind of *laborers* robots were envisioned to replace.[82] There was no need for robots to be personable when they were meant to replace alienated factory workers, but that changed when robots came to be envisioned to replace service providers, such as a taxi driver, a hairdresser, and later a caretaker or a partner. With the realization that "service" included not only the functional tasks but also the emotional labor that accompanied it, engineers had to design robots capable of addressing customers' emotional needs. Robots first became *personal* to provide service and then *personable* to provide "service with a smile."

Personable robots, therefore, expose a particular characteristic of Japanese corporate techno-optimism—the outsized weight placed on the commodification (and fulfillment) of emotional needs. Japan is by no means exceptional among neoliberal economies in manipulating emotions for profit.[83] What stands out, however, is government investment in developing commercial products, such as cute robots, which promise to satisfy the very emotional needs that government neglected to address in its aggressive pursuit of a market-driven economy. The desire for robotic companionship is a cry for help. Burnout, exhaustion, loneliness, alienation—all create a yearning for somebody, or something, that shows that they care, can lend an ear, offer a sense of growth and fulfillment, and help enjoy life. And personable robots, if one can afford them, promise to do all of the above. The cute Japanese robots reduce the mental distress of overworked Japanese citizens, allowing the government to avoid solving pressing social problems and keeping profits flowing into corporate pockets.

NOTES

1. Air from Bach's Orchestral Suite no. 3 in D major BWV 1068. Koichi Koganezawa, Atsuo Takanishi, and Sugano Shigeki, eds., *Waseda Robotto No Ayumi*, 3rd ed. (Tokyo: Ichiro Kato Laboratory, 1991), 26–27.

2. Ozaki Shōtarō, "Mujinka Kōba to Robotto," *Ōtome-shon* 16, no. 5 (1971): 27; Nori Kageki and Masahiro Mori, "An Uncanny Mind: Masahiro Mori on the Uncanny Valley and Beyond," *IEEE Spectrum*, June 12, 2012, https://spectrum.ieee.org/an-uncanny-mind-masahiro-mori-on-the-uncanny-valley.

3. Hiromi Mizuno, "Introduction: A Kula Ring for the Flying Geese: Japan's Technology Aid and Postwar Asia," in *Engineering Asia: Technology, Colonial Development and the Cold War Order*, ed. Hiromi Mizuno, Aaron Stephen Moore, and John Paul DiMoia (New York: Bloomsbury Academic, 2018), 1–42.

4. Takashi Nishiyama, *Engineering War and Peace in Modern Japan, 1868–1964* (Baltimore: Johns Hopkins University Press, 2014).

5. Timothy S. George, *Minamata: Pollution and the Struggle for Democracy in Postwar Japan* (Cambridge, MA: Harvard University Press, 2001).

6. Andrew Gordon, *The Wages of Affluence: Labor and Management in Postwar Japan* (Cambridge, MA: Harvard University Press, 1998); Sheldon M. Garon, *The State and Labor in Modern Japan* (Berkeley: University of California Press, 1987).

7. Tessa Morris-Suzuki, *Beyond Computopia: Information, Automation and Democracy in Japan* (London: Kegan Paul, 1988), 56.

8. Morris-Suzuki, *Beyond Computopia*, 50; Masahiro Mori, "Purosesu, Ōtome-shon No Shisutemu Keikaku," *Kikai No Kenkyū* 20, no. 1 (1968): 190.

9. Yūjirō Hayashi, *Jōhōka Shakai: Haadona Shakai Kara Sofuto Shakai e* (Tokyo: Kodansha, 1969). In her 1988 book, Tessa Morris-Suzuki brilliantly analyzed the ideological and the economic premises of the *Information Society* visions. Most of the data about the *Information Society* in this article derive from her scholarship. Morris-Suzuki, *Beyond Computopia*.

10. Morris-Suzuki, *Beyond Computopia*, 10, 79.

11. Yulia Frumer, "Manufacturing Hands: Robot Fingers and Human Labour in Post-War Japan," in "Making History," special issue, *History & Technology* 38 nos. 2–3 (2022): 239–256; Mori, "Purosesu, Ōtome-shon"; Tadashi Yamashita, "Materiaruzu Handoringu No Jidōka," *Report for Manufacturing Research (Seisan Kenkyū)*, Institute of Industrial Science, University of Tokyo 17, no. 2 (1965): 31–36.

12. Yoneji Masuda, *The Information Society as Post-Industrial Society* (Tokyo: Institute for the Information Society, 1980), 146.

13. Cited in Morris-Suzuki, *Beyond Computopia*, 14, 19.

14. See, for example, the journal *Energy*, funded by Esso, or the journal *Systema*, funded by Shinko Electric.

15. Cited in Morris-Suzuki, *Beyond Computopia*, 30.

16. Tohru Moto-Oka, ed., *Fifth Generation Computer Systems: Proceedings of the International Conference on Fifth Generation Computer Systems, Tokyo Japan, October 19–22, 1981* (Amsterdam: Elsevier, 1982).

17. Masuda, *The Information Society as Post-Industrial Society*, 51, 141. Emphasis in original.

18. Sheldon M. Garon, *Molding Japanese Minds: The State in Everyday Life* (Princeton, NJ: Princeton University Press, 1997), 223.

19. Daniel Bell, *The Coming of Post-Industrial Society: A Venture in Social Forecasting* (New York: Basic Books, 1973); Alvin Toffler, *Future Shock* (London: Bodley Head, 1970); Alvin Toffler, *The Third Wave* (Toronto: Bantam Books, 1981).

20. Jeremy A. Greene, *The Doctor Who Wasn't There: Technology, History, and the Limits of Telehealth* (Chicago: University of Chicago Press, 2022), 225–227.

21. Hiromi Mizuno, *Science for the Empire: Scientific Nationalism in Modern Japan* (Stanford, CA: Stanford University Press, 2009); Simon Partner, *Assembled in Japan: Electrical Goods and the Making of the Japanese Consumer* (Berkeley: University of California Press, 1999). Citation from Morris-Suzuki, *Beyond Computopia*, 14.

22. Garon, *Molding Japanese Minds*; Yize Hu, "Resilience of Systems: Technocracy and Information Society in Japan, 1960s–1980s" (PhD dissertation, Johns Hopkins University, 2024).

23. Morris-Suzuki, *Beyond Computopia*, 47; Sheldon M. Garon, *Beyond Our Means: Why America Spends While the World Saves* (Princeton, NJ: Princeton University Press, 2012), chap. 9.

24. I am embracing the term coined by Patrick McCray in W. Patrick McCray, *The Visioneers: How a Group of Elite Scientists Pursued Space Colonies, Nanotechnologies, and a Limitless Future* (Princeton, NJ: Princeton University Press, 2012).

25. Yoshirō Hoshino, *Mai Kaa: Yoi Kuruma, Warui Kuruma Wo Miyaburu Hō* (Tokyo: Kōbunsha, 1961).

26. Akio Morita, Edwin M. Reingold, and Mitsuko Shimomura, *Made in Japan: Akio Morita and Sony* (New York: Dutton, 1986), 83.

27. Morita, Reingold, and Shimomura, *Made in Japan*, 81–82.

28. Paul Roquet, *The Immersive Enclosure: Virtual Reality in Japan* (New York: Columbia University Press, 2022), chap. 1.

29. Laine Nooney, *The Apple II Age: How the Computer Became Personal* (Chicago: University of Chicago Press, 2023).

30. Hayashi, *Jōhōka Shakai*, chap. 7.

31. Masuda, *The Information Society as Post-Industrial Society*, 148.

32. General Policy Committee of Social Policy Council, *The Information Society and Human Life: The Outlook for People's Life in the Information Society* (Tokyo: Social Policy Bureau, Economic Planning Agency of Japanese Government, 1983), 49.

33. Japanese Ministry of International Trade and Industry, *90 Nendai No Denshi Sangyō Bijon* (Tokyo: Japanese Ministry of International Trade and Industry, 1989).

34. Morris-Suzuki, *Beyond Computopia*, 174.

35. William M. Tsutsui, *Manufacturing Ideology: Scientific Management in Twentieth-Century Japan* (Princeton, NJ: Princeton University Press, 1998), chaps. 5 and 6; Masami Nomura, "The Myths of the Toyota System," *AMPO: Japan-Asia Quarterly* 25, no. 1 (1994): 18–24.

36. Morris-Suzuki, *Beyond Computopia*, 188.

37. General Policy Committee of Social Policy Council, *The Information Society and Human Life*, 51.

38. Morris-Suzuki, *Beyond Computopia*, 76–84.

39. Hayashi, *Jōhōka Shakai*, 81.

40. Mori Masahiro in reply to techno-optimism of a science fiction writer, Sakyō Komatsu. Sakyō Komatsu, Masahiro Mori, and Natsuhiko Yoshida, "Roundtable [Zadankai]," *Energy* 7, no. 4 (27), (1970): 2.

41. Ichirō Katō, Atsuo Takanishi, and Shigeki Sugano, *Mai Robotto* (Tokyo: Yomiuri Shinbunsha, 1990), 168.

42. General Policy Committee of Social Policy Council, *The Information Society and Human Life*, 56, 59.

43. International Symposium on Population Structure and Development, *Economic and Social Implications of Population Aging: Proceedings of the International Symposium on Population Structure and Development, Tokyo, 10–12 September 1987* (New York: United Nations, 1988).

44. *Dreams to My Robot—In Memory of Prof. Katō Ichirō*, documentary (Waseda University, 1994).

45. Ichirō Katō, *Sekai Hatsu no Nisoku Hokō Robotto: Baio Mekatoronikusu no Saizensen*, Tekuno saiensu (Tōkyō: Asahi Shuppansha, 1983), 62.

46. Judging from what sounds like a reply to critics, from Katō, *Sekai Hatsu*, 88.

47. Katō, *Sekai Hatsu*, 116.

48. Ichirō Katō, "Karakuri kara Mai Robotto e," *Journal of Japanese Society of Mechanical Engineering* 87, no. 792 (1984): 1245.

49. Katō, *Sekai Hatsu*, 96.

50. Katō, *Sekai Hatsu*, 73.

51. Katō, *Sekai Hatsu*, 116.

52. Ichirō Katō, "Karakuri kara Mai Robotto e," 1243.

53. Katō, *Sekai Hatsu*, 98.

54. Daniel White and Hirofumi Katsuno, "Artificial Emotional Intelligence beyond East and West," *Internet Policy Review* 11, no. 1 (2022).

55. Author's emphasis. Katō, *Sekai Hatsu*, 122; *Dreams to My Robot—In Memory of Prof. Katō Ichirō*.

56. Koganezawa, Takanishi, and Sugano Shigeki, *Waseda Robotto No Ayumi* [Development of Waseda Robot], 26; Japan Association for the International Exposition, *Tsukuba Expo'85*, 120–121.

57. Japan Association for the International Exposition, *Tsukuba Expo '85: Official Guide Book* (Tsukuba: Japan Association for the International Exposition, 1985), 22–23.

58. Susumu Tachi, Kazuo Tanie, Kiyoshi Komoriya, Yuji Hosoda, and Minoru Abe. *Study on Guide Dog (Seeing-Eye) Robot: Its Basic Plan and Some Experiments with MELDOG MARK I*, vol. 32 (Tokyo: Bulletin of Mechanical Engineering Laboratory, 1978); Susumu Tachi, Kazuo Tanie, Kiyoshi Komoriya, Yuji Hosoda, and Minoru Abe, "Guide Dog Robotics Basic Plan and Some Experiments with Meldog Mark I," *Mechanism and Machine Theory* 16, no. 1 (1981): 21–29.

59. Hirose and Ogawa, "Honda Humanoid Robots Development," 11–13; Hirose Masato, "ASIMO."

60. Katō, Takanishi, and Sugano, *Mai Robotto*, 169–171, 179–181.

61. Katō, Takanishi, and Sugano, *Mai Robotto*, 183.

62. Katō, Takanishi, and Sugano, *Mai Robotto*, 174–177.

63. "Jōhō sakugyō." Katō, Takanishi, and Sugano, *Mai Robotto*, 173.

64. Hayashi, *Jōhōka Shakai*, 81.

65. The term "emotional labor" was first coined by Arlie Hochschild in 1983, based on her research on flight attendants. In the original formulation, "emotional labor" referred to managing internal (and often quite justifiable) negative emotions for the purpose of sustaining either monetary or social economy. It was only when the term entered public discourse in the late '90s that the meaning of the phrase came to also encompass the creation of positive emotions in others. Arlie Russell Hochschild, *The Managed Heart: Commercialization of Human Feeling* (Berkeley: University of California Press, 1983).

66. Katō, Takanishi, and Sugano, *Mai Robotto*, 178.

67. Katō, Takanishi, and Sugano, *Mai Robotto*, 178–180.

68. Katō, *Sekai Hatsu no Nisoku Hokō Robotto*, 109.

69. Katō, Takanishi, and Sugano, *Mai Robotto*, 177–178.

70. Ichirō Katō, *Mujinka Kōba e no Chōsen: Kōgyōyō Robotto no Hanashi* (Tokyo: Japanese Industrial Standards Committee, 1973), 74.

71. Mitsuo Nagamachi, "Jōcho Kōgaku to Sono Ōyō," *Japan Ergonomics Society* 22, no. 6 (1986): 319.

72. Masamitsu Ōshima, "ME Kenkyū no Ayumi: Ningen Kōgaku," *Japanese Journal of Medical Electronics and Biological Engineering* 10, no. 6 (1972): 556.

73. Nagamachi, "Kansei no Hyōka," 955–958.

74. Charles Egerton Osgood, George J. Suci, and Percy H. Tannenbaum, *The Measurement of Meaning* (Urbana: University of Illinois Press, 1957). The person from whom Nagamachi learned about the SD method was a female industrial designer, Sachiko Saitō. See Sachiko Saitō, "Semantikku Diferensharu (SD) Hō," *Japan Ergonomics Society* 14, no. 6 (1978): 315–325.

75. Nagamachi, "Kansei no Hyōka," 956; Nagamachi, "Jōcho Kōgaku to Sono Ōyō," 321.

76. Matsuura Soichi, Yamamoto Kazumichi, Fujishiro Tomoko, and Iguchi Nobuhiro, "Dansu Robotto No Kaihatsu to Sono Shukanteki Hyōka," *Journal of the Robotics Society of Japan* 9, no. 2 (1991): 49–55.

77. Hiroshi Ishiguro, "Android Science: Toward a New Cross-Interdisciplinary Framework," in *Toward Social Mechanisms of Android Science: A CogSci 2005 Workshop*, Stresa, Italy, July 25–26, 2005, 1–6.

78. White and Katsuno, "Artificial Emotional Intelligence beyond East and West," 9–10.

79. Cosima Wagner, *Robotopia Nipponica—Recherchen zur Akzeptanz von Robotern in Japan* (Berlin: Tectum Verlag, 2013).

80. Jennifer Robertson, "Robo Sapiens Japanicus: Humanoid Robots and the Posthuman Family," *Critical Asian Studies* 39, no. 3 (2007): 369–398; Jennifer Robertson, "Gendering Humanoid Robots: Robo-Sexism in Japan," *Body and Society* 16, no. 2 (2010): 1–36.

81. Japan Brandvoice, "A Machine That Does Nothing? Cuddly LOVOT May Be the Most Innovative Japanese Robot Yet," *Forbes*, https://www.forbes.com/sites/japan/2019/02/06/a-machine-that-does-nothing-cuddly-lovot-may-be-the-most-innovative-japanese-robot-yet/, accessed August 28, 2022.

82. Frumer, "Manufacturing Hands."

83. Luke Stark, "Algorithmic Psychometrics and the Scalable Subject," *Social Studies of Science* 48, no. 2 (April 2018): 204–231; Luke Stark, "Affect and Emotion in DigitalSTS," in *A Field Guide for Science & Technology Studies*, ed. Janet Vertesi and David Ribes (Princeton, NJ: Princeton University Press, 2019), 117–135; Natasha Dow Schüll, *Addiction by Design: Machine Gambling in Las Vegas* (Princeton, NJ: Princeton University Press, 2012).

Afterword

From Groovy Science to Greedy Science

David Farber

Scientific culture, in all its myriad iterations, has never been monolithic. So, to reveal that an unexpectedly broad swathe of scientists and their compatriots in the Long Sixties opened the door to madcap forms of "Groovy Science" (as David Kaiser and W. Patrick McCray argued in their eponymous 2016 book), only to have that door bang closed in the 1980s with a hegemonic turn to "Greedy Science" (as is suggested in this new eponymous volume edited by Michael D. Gordin and W. Patrick McCray), is more a heuristic device than it is a totalizing claim. And yet, as the authors in *Greedy Science* demonstrate, big changes in how and why scientific research in numerous places around the world was produced and in how scientists were rewarded *did* happen in the 1980s. Not everybody in and around the scientific community joined in the rush for riches or celebrated the drive for celebrity and the monetization of scientific research. But as the authors herein write, in the 1980s, many of the formal and informal rules of the scientific enterprise changed.

In the United States, new federal legislation, court rulings, national and local planning initiatives, corporate practices, university incentives, individual financial prospects, and media opportunities reshaped the place and purpose of science and the status of scientists. At the same time, and hard to nail down, a broad cultural transformation that occurred inside and outside the United States legitimated, within segments of the scientific community, avarice, greed, personal wealth, unabashed individualism, and a general indifference or even hostility to notions of the social good. (As Prime Minister Margaret Thatcher famously stated in 1987, "There's no such thing as society. . . . People must look after themselves first.")

The chapters in this far-ranging collection, centered as they are on the notion of an emergent "greedy science" in the 1980s, bring a far-flung range of historical developments into focus. The authors of *Greedy Science* demonstrate that scientific enterprise in the United States, and in a surprising number of other places, contributed to (even as it was affected by) the neoliberal or market turn in national, regional, and local governance. In the "Eighties" era, notable elite scientists, often guided by their patrons and partners, gainsaid the idealistic notion of a noble altruistic pursuit of knowledge and cheerfully embraced, instead, a bevy of financial and personal rewards.

As the volume's editors note, this market-oriented, reward-driven orientation of the scientific enterprise is not new, at least in the United States. Sinclair Lewis's classic medical science novel, *Arrowsmith*, published in 1925 and awarded the Pulitzer Prize in 1926, ponders similar issues. The novel's research-loving protagonist is warned by one of his scientific mentors to get his head out of the clouds and accept the market-driven needs of his profession: "It's all very fine, this business of pure research: seeking the truth, unhampered by commercialism or fame-chasing. Getting to the bottom. Ignoring consequences and practical uses. But do you realize if you carry that idea far enough, a man could justify himself for doing nothing but count the cobblestones on Warehouse Avenue."[1] Still, as this volume convincingly details, in the 1980s, forces broader than individual decision-making changed the opportunity structure, social conventions, and cultural meanings that undergirded the relationship between science, scientists, and the marketplace.

This dynamic reached deep into society. In the United States, an unexpected variant showed up in how science—or, at least, "science"—was perceived and deployed by well-known and sometimes celebrated figures in poor, Black communities during "the Sixties" and then, in quite different form, during "the Eighties."

At the cusp of the 1960s and 1970s, the Black Panther Party, (in)famous for their chant, "Revolution has come/Off the Pigs," turned to science to advance their revolutionary cause. Between 1968 and 1970, the Panthers opened up thirteen urban-based public health clinics: the Peoples' Free Medical Clinics. The clinics depended on the volunteer work of doctors, nurses, and medical technicians who freely gave their time, talent, and energy to the cause.

In 1971, the Panther-run clinics began to focus on screening for sickle cell anemia—a scourge of African-descended people and a disease that had received little funding or attention from the federal government or Big Pharma. The minister of health for the Chicago Black Panther Party, Ronald "Doc" Satchel, explained, "The medical profession within this capitalist society . . . is composed generally of people working for their own benefit and advancement rather than the humane aspects of medical care." The Panthers meant to create an alternative system. In 1972, the Panthers amended their well-known "Ten Point Program," adding a new point 6: "We believe that the government must provide, free of charge, for the people, health facilities which will not only treat our illnesses, most of which have come about as a result of our oppression, but which will also develop preventive medical programs to guarantee our future survival."[2] Here, in plainest form, was an altruistic, nonpecuniary call for a people's science aimed at the welfare of "Black and oppressed people."[3]

The Panthers' quest for community-based health care and scientific research into the medical conditions that plagued Black Americans met with widespread support, at least within elements of Black America and allied, movement-oriented whites. In 1972, John Lennon and Yoko Ono cohosted an episode of the popular daytime network TV talk show, the *Mike Douglas Show*, in which they featured Black Panther leaders discussing their campaign to publicize and draw scientific research to sickle cell anemia. Robert Fullilove, a 1960s civil rights activist who became a professor of sociomedical sciences at the Columbia University School of Public Health, later wrote, "The Black Panther Party was acutely aware of health disparities. . . . The language of health disparities did not exist back then, but everyone knew it. . . . Responding to them, made the Panthers very popular."[4] The Panthers, best known for their violent rhetoric and deadly encounters with police, placed much of their emphasis in the early 1970s on community service that featured a commitment to science-based solutions to their community's needs.

By the early 1980s, the Black Panther Party was gone, destroyed both by internal divisions and brutal criminality, as well as by external governmental persecution. The Panthers' most influential leader, Huey Newton, had become a violent cocaine addict. But "the brothers on the block," the kind of tough young Black men who in the Sixties had been the main target of Black Panther recruiting efforts, did not completely repudiate the Panthers' commitment to a people's science. In their search for better lives, they embraced a people's science of a very different kind. In the 1980s, rather than attempting to bring medical science expertise to their communities or demanding that scientists altruistically turn their research to the health problems that plagued poor Black communities, some of these young men enthusiastically created their own profit-driven scientifically based enterprises. These were young men well attuned to the practices of "greedy science."

They became street chemists, proprietors of their very own (if humble) labs. In domestic kitchens or, in the case of more successful operations, purpose-built facilities, they mixed roughly equal portions of baking soda and powder cocaine (some "cooks" occasionally experimented with additional ingredients) and then poured the mixture into a big pot of boiling water. Out of that rudimentary chemical process emerged rock cocaine, better known as crack. Some kitchen labs churned out an ounce or two at a time. Other, more sophisticated operations produced many kilos. Regardless of output, these pharmaceutical operators divvied up their carefully weighed product into small plastic bags or, more often, branded plastic vials and then used a network of neighborhood sales representatives to bring crack to the underground consumer marketplace. Often, they operated without any regard for the community harm they caused. Few carefully weighed the personal risks of their illegal enterprises—the likelihood of prison time or the possibility of death at the hands of business rivals. Mostly, they simply embraced the drug business, using a rudimentary people's science to seek neighborhood fame and ill-gotten fortunes.

Crack cocaine met a lucrative market need; it provided a cheap, highly potent, fast-acting drug for people, especially poor people, who found themselves on the wrong side of Reaganomics' market imperatives. At less than five dollars a hit, crack provided an easy-seeming escape from dead-end destitution and social anomie. To meet this market, crack dealers built an illicit pharmaceutical business. While their science was basic, it rested on more than a hundred years of research and development into the properties and production of cocaine.

German scientists in the 1860s had first isolated the alkaloid that gave the leaves of the coca plant its kick.[5] Quickly thereafter, storied pharmaceutical companies, such as Merck and Parke-Davis, began refining coca leaves into pure cocaine. A broad range of scientifically minded, profit-driven entrepreneurs, pharmacists, and patent medicine purveyors legally marketed coke to people around the world. Consumers in the United States emerged as the product's biggest customers. But after nearly fifty years of wide-open cocaine sales in the United States, public health officials, progressives of various stripes, and the kind of people who decried almost all forms of intoxicants, including alcohol, ended the legal, unregulated open-market use of cocaine in the United States. Over the next half-century, this created possibilities for an illicit marketplace.

By the 1970s, a number of cocaine distributors and users were experimenting, in a rough trial-by-error process, with the drug. Some sought a cheaper coke high. In Latin American countries where coca was grown, lower-income people, while scientifically unschooled, were willing to diligently investigate the properties of their favorite high. They discovered an inexpensive, fairly accessible alternative to pricey powder cocaine. They procured unprocessed coca leaves, mashed them up, and used a solvent such as gasoline to produce a smokeable, if crude and foul-tasting, substance called *basuco*. This intermediate form of cocaine was smoked, not snorted. The smoked product went not, as with powder cocaine, to the inefficient nasal membranes but instead straight to the lungs and then to the brain, producing a faster-hitting, if short-lived, and more potent high. While few producers and users of *basuco* fully understood the science behind their product's cheap and effective high, they got the gist of it.

Seeking the same intense high without the unpleasantness and quality uncertainties associated with *basuco*, more well-to-do cocaine users experimented with smoking high-quality powder cocaine treated with ether. At least a few of the first of these experimenters were scientifically minded students at the University of California, Berkeley.[6] Freebasing, which could and did produce a dangerously flammable and tricky-to-use product, had entered the American cocaine scene. From these two tracks of cocaine experimentation came crack cocaine—a cheap, easy to make and easy to use, quality smokeable form of cocaine. People's science had produced a market champion.[7]

With the exception of some Sixties-era producers of hallucinogenic and cannabis products—Groovy Science—manufacturers and dealers of illegal drugs are not generally known as altruists. They are rarely celebrated as community leaders. What made the inner-city crack dealers of the Eighties (and well into the Nineties) different from most of their antecedents was that, like the Black Panthers, they were lauded (at least among certain of their peers) as "ghetto" heroes. Unlike the Panthers, who gained respect for their defense of the Black community and their service ideal, crack dealers were celebrated for the stacks of money they accrued and for the luxury goods they ostentatiously displayed. "Their hustle game was tight," explained an admirer. "It was all about looking fly, getting bitches, flashing the finest jewels, and non-stop flossing."[8] An entire genre of Hip Hop—Gangsta Rap—was largely devoted to praising crack dealers' material success, as well as to the murderous violence they visited upon their competitors. Biggy Smalls matter-of-factly laid out the crack dealers' guide to low(er)-risk wealth in his classic rap: "The Ten Crack Commandments" (1997). In the 1980s and thereafter, greedy science was not only legitimated in the highest reaches of the credentialed scientific community or in socially sanctioned research sites, such as universities. Greedy science was embraced even in the hard, hustling realms of the poor and the dispossessed. A little knowledge of elementary chemistry offered a lottery ticket to the trickle-down wealth neoliberal, market-driven politics promised.

In the 1980s, in disparate circles that ranged from ghetto entrepreneurs to corporate chieftains and from local economic development officers to national policymakers, the wealth-accumulating and wealth-producing possibilities of science—in both its most sophisticated and most elementary forms—gained credibility, legitimacy, and institutional support. In the United States and in many other nodes around the world during the 1980s, the long-standing if often contentious ties between the capitalist project and scientific enterprise tightened. But such relationships were not new. As historian Jefferson Cowie argued in his book, *The Great Exception*, the legitimation of greed and self-aggrandizement during the 1980s that we sum up as "the neoliberal turn" was not a particularly new development. In the United States, at least, it was a new iteration of the market-driven norms of American culture and politics.[9]

According to Cowie, the period between 1933 and the early 1970s, during which major historical actors including presidents Franklin Roosevelt and Lyndon Johnson, driven by a powerful labor union movement and myriad social change activists, turned the United States away from a winner-take-all society and toward a more egalitarian, social justice–oriented political culture, is unrepresentative of American life. In the context of this analysis, greedy science is not new. It is instead (as Sinclair Lewis suggested in 1925 in *Arrowsmith*) the normal trajectory of the scientific enterprise in America's individualistic, market-driven society. Nonetheless, this volume demonstrates, with both precision and some alarm, exactly how that repudiation of the long New Deal era and related social democratic regimes occurred in the realm of science.

As the editors of *Greedy Science* note in their introduction, economic challenges in the post-1968 world played a major role in the transformation of the political economy of the United States and most other market-based economies. Particularly in the United States and the United Kingdom, political leaders such as President Ronald Reagan and Prime Minister Margaret Thatcher argued that too many public policies created by progressive politicians were aimed—poorly, they argued, at producing greater economic equity and not enough toward generating new, globally competitive profitable enterprises. Innovators and entrepreneurs, they argued, had been handcuffed by Big Government in the form of punitive tax codes, regulatory overreach, and a lack of targeted support. As a result, economic growth had faltered and the breakthrough technologies that generated new industries and whole economic sectors had suffered. Market-driven policies and the individual financial rewards those policies guaranteed, neoliberal champions argued, would unshackle the innovators, including scientists, who had the talent and ambition to make the world anew.

Near the end of his presidency, speaking before some of the nation's most celebrated scientists, Ronald Reagan claimed that his policies had worked as he had expected and that they had accelerated scientific innovation and scientists' opportunities. At the 1988 Presentation Ceremony for the National Medals of Science and Technology, Reagan explained that his economic policies had been a boon for scientists and for innovation. After joking about his age—noting that he remembered the horse-and-buggy days—he praised the assembled scientists, calling them "the builders, the dreamers, the heroes"; he extolled them for the progress they had brought to the world. Then, he moved the spotlight off them and onto his economic policies, stating that their success was made possible, at least in part, by his administration's market-driven approach to innovation. Those policies, he insisted, had played a significant role in unshackling their genius and fueling their innovations: "I'm convinced that perhaps the most important action we've taken has involved knocking down the barriers to progress that government itself had erected . . . hundreds of thousands of new businesses have been formed, many of them linked to specific new technologies. All of this represents the application of knowledge to human needs on a massive scale—not by government, but by committed individuals, acting in freedom."[10] Reagan then briefly paused, assuming that his homage to free market science would earn him a round of applause. At least in that room, at that moment, the scientists and their families in attendance did not respond; they sat on their hands.

Still, Reagan had a point. Scientists in the 1980s, whether greedy or not, were incentivized by infusions of private venture capital. Around the world, policy regimes, driven by the links between scientific innovation and wealth production, provided new opportunities for scientists and their corporate and university backers. A culture that championed heroic individualists gave elite scientists new avenues for attaining fame and fortune. Through a variety of innovative, commercialized breakthroughs, the neoliberal-driven scientific enterprise of the 1980s era played a major role in invigorating the global economy, even as it contributed to new levels of wealth and income inequality. And while the low-rent kitchen chemists who brought crack cocaine to the marketplace are not on anyone's list for any kind of Nobel Prize, they too recognized that greedy science offered them a path to riches and celebrity, even at the risk of imprisonment or even death. In the 1980s, the rules of the game changed. And science and scientists of all kinds wrestled with the possibilities. Some of them resisted. Some kept their heads down. And some happily participated in creating (or re-creating) a fiercely inequitable realm in which their gifts and their talents were richly rewarded.

NOTES

1. Sinclair Lewis, *Arrowsmith* (New York: Harcourt Brace, 1925), 121.

2. Both quotes are found in Angola 3 News, "Medical Self Defense and the Black Panther Party: An Interview with Alondra Nelson," *Truthout*, January 17, 2012.

3. For the best history of the Panthers' public health turn, see Alondra Nelson, *Body and Soul: The Black Panther Party and the Fight against Medical Discrimination* (Minneapolis: University of Minnesota Press, 2011).

4. "The Black Panther Party Stands for Health," Columbia Mailman School of Public Health, February 23, 2016, https://www.publichealth.columbia.edu/news/black-panther-party-stands-health.

5. In this section, I am drawing directly on my book, David Farber, *Crack* (Cambridge, UK: Cambridge University Press, 2019), chap. 1.

6. Farber, *Crack*, 52–53.

7. Farber, *Crack*, chap. 2.

8. Farber, *Crack*, 144.

9. Jefferson Cowie, *The Great Exception: The New Deal and the Limits of American Politics* (Princeton, NJ: Princeton University Press, 2016).

10. President Ronald Reagan, "Remarks at the Presentation Ceremony for the National Medals of Science and Technology," July 15, 1988, https://www.reaganlibrary.gov/archives/speech/remarks-presentation-ceremony-national-medals-science-and-technology-1. Video recordings of the speech are available online.

Contributors

Jon Agar
Professor of Science and Technology Studies
University College London

Angela N. H. Creager
Thomas M. Siebel Professor of the History of Science
Princeton University

David Farber
Roy A. Roberts Distinguished Professor of History
University of Kansas

Yulia Frumer
Bo Jung and Soon Young Kim Associate Professor of East Asian
    Science
Johns Hopkins University

Cathy Gere
Professor of History of Science and Medicine
University of California, San Diego

Michael D. Gordin
Rosengarten Professor of Modern and Contemporary History
Dean of the College
Princeton University

W. Patrick McCray
Professor of History
University of California, Santa Barbara

Cyrus C. M. Mody
Chair in the History of Science, Technology, and Innovation
Maastricht University

Laine Nooney
Assistant Professor of Media and Information Industries
New York University

Myrna Perez
Associate Professor of Classics and Religious Studies; Women's,
Gender, and Sexuality Studies
Ohio University

Robin Wolfe Scheffler
Associate Professor in the Science, Technology, and Society Program
Massachusetts Institute of Technology

Asif Siddiqi
Professor of History
Fordham University

Matthew Stanley
Professor at the Gallatin School of Individualized Study
New York University

Hallam Stevens
Professor of Interdisciplinary Studies in the College of Arts, Society,
and Education
James Cook University

Margaret A. Weitekamp
Curator, Social and Cultural History of Spaceflight
Department Chair, Space History Department
National Air and Space Museum, Smithsonian Institution

Peter Westwick
Adjunct Professor (Research) of History
University of Southern California
Director, Aerospace History Project
Huntington-USC Institute on California and the West

Index

Page numbers in *italics* refer to figures.

1980s era: capitalism in, 2, 7, 64; definition and description of, 6–9; greed in, 1–2, 6–9; historical study of, 7, 9, 10. *See also* capitalism; greed; greedy science; neoliberalism

academic entrepreneurship. *See* entrepreneurship
advertisements: Japanese corporate, 328; pharmaceutical, 149; space industry, 124, 126, 292; UK scientists in, 105–6, *106*; for VisiCalc, 80–81, 83
aerospace industry. *See* space industry
Afanas'ev, Sergei, 280–82, 284, 287, 291
Agar, Jon, 11, 97
AIDS: combination therapy for, 178–79; denials of, 182–83, 184; inaction not addressing, 175–76; pharmaceutical industry and, 12, 168–72, 175–85, 187; political activism on, 64, 168–72, 175–85, 187; South African activism on, 170, 179–85; surrogate endpoints as markers of, 169, 176–77, 178, 184
AIDS Coalition to Unleash Power (ACT UP), 168–72, 175–79, 181, 184
Akiyama, Toyohiro, 141
Alvarez, Luis and Walter/Alvarez hypothesis, 298–99, 300
Alvey program, 101, 104, 110
Alzheimer's drugs, 184–85, 189n43
Ambrosino, Michael, 54
American Institute of Aeronautics and Astronautics (AIAA), 312–15

American Museum of Natural History (AMNH), 45, 49, 50, 54, 56
Ames, Bruce / Ames test, 152–53, 155–56, 163–64nn38–39
Ang Mo Kio Industrial Estate, 229–30
apocalyptic narratives: on asteroid impacts, 13–14, 297–318; on Soviet military and weapons, 269, 317
Apple Computer: entrepreneurship of, 92n53; initial public offering by, 249; in Singapore, 230; in Texas, 253; VisiCalc and, 11, 78, 80, 82–84, 91n25
Arrowsmith (Lewis), 350, 355
artificial intelligence (AI): Japanese investment in, 326, 333, 339; oil industry investments in, 20, 29–32. *See also* Japanese humanoid robots
ASIMO, 340, 342
asteroid impacts lobbying, 297–318; 1980s as context for, 298, 306, 311; asteroid sighting and, 312; books for, 307–10, 312; catastrophism focus in, 307–10; competition for attention on, 311, 314–15; government contract goals in, 315–18; greed and, 298, 304, 312, 317; mass extinction events and, 299–300, 303, 309–10; nuclear power dangers referenced in, 305, 306, 308–9, 310; nuclear weapons and, 300–304, 305, 310, 313, 315–18; overview of, 13–14, 297–98; precise danger warnings in, 304–10; recommended actions in, 309–10, 313; risk assessments in, 305–7, 310, 313; total destruction warnings in, 298–304

Baker, James and Janet, 31

Barber, Edwin, 51–53

Barnett, Allen, 25–26

Batam development, 231–35, *232*

Bayh-Dole Act (1980), 8, 169, 197

Beck, Ulrich, 306, 307, 310

Benson, Johan, 312, 314

Bintan development, 231, *232*, 234–35

biocapital, 168–69, 170, 184, 188n39

bioconstitutionalism, 170, 179–83, 188n39

Biogen: Alzheimer's drugs by, 184, 185; Cambridge location for, 195, 197, 201–6, 214n30; oil industry investments in, 26, 39–40n63

Biopolis, 206, 217–18, 219

biotechnology, 147–61; biocapital of, 168; Cambridge region; chemical research and regulation in, 148, 149–54; deregulation of, 148, 160–61; entrepreneurship in, 202; environmental regulation for carcinogens and, 148, 149–54, 159; government funding for, 198–99, 201, 212n11, 213n22; greed and, 147–48, 167–68, 194–95, 197–99, 210; Human Genome Project in, 7, 148–49, 157; human testing in, 157–59; Japanese, 325–26; neoliberalism influencing, 149; oil industry investment in, 8, 20, 26–29, 39–40n63, 170, 199; overview of, 12, 147–49; patents for, 8, 27–28, 108–9, 148, 159, 166n76, 180, 197–98; pharmaceutical industry and, 168, 209; privatization in, 12, 147–48, 161n4, 163n25, 166n82; Texan, 260; tumor virology research in, 148, 154–57; UK, 108–9; valuation of companies in, 195, 197–98, 209; venture capital for, 147, 195, 197, 199–202, 206. *See also* Cambridge biotechnology community; pharmaceutical industry; Singapore

Black Panther health care networks, 47, 64, 351–52, 354

Bloch, Erich, 5

Böer, Karl-Wolfgang, 25, 38n45

Boesky, Ivan, 1, 33

Boyer, Herbert, 8, 200

brand: Cambridge locational, 209, 210, 216n68; of Gould, 11, 47–48, 55, 61, 66–67; Japanese

corporate, 326; Personal Software name change for, 88; Singaporean, 221; space industry, 124, 125, 132

breast cancer, 158–59, 160

Bricklin, Dan, 75–79, 82, 88, 90n20

Bush, George H. W. and administration, 255, 257, 258

Bush, George W., and administration, 217

Cambridge biotechnology community, 193–210; Biogen in, 195, 197, 201–6, 214n30; greed and, 194–95, 197–99, 210; Hoechst-MGH partnership in, 195, 199–201, 213n19; map of, *196*; overview of, 12, 194–95; patents for, 180; real estate and, 195, 206–10, 214–15n48, 215n51, 215n59, 215n61; regulatory restrictions on, 194, 203, 205, 207, 214n30, 214n44; scientist recruitment and, 198–99, 202–3, 210, 213n16; tacit knowledge transfers in, 194–95, 201, 204, 207, 208; University Park in, 195, 206–10, 214–15n48, 215n51, 215n59, 215n61; valuation of companies in, 195, 197–98, 209

cancer and carcinogens, 148–61; chemical research and regulation on, 148, 149–54; educational information on, 159, *160*; environmental regulation for, 148, 149–54, 159; government funding of research on, 212n11; heredity and, 148–49, 157–59, *160*; human testing and, 157–59, 173; personal responsibility for, 149, 159–61; tumor virology research on, 148, 154–57

capital: neoliberal movement of, 3, 63; social, 6; start-up, 79; UK entrepreneurship and, 107. *See also* biocapital; fame; neoliberalism; venture capital; wealth

capitalism: 1980s era of, 2, 7, 64; casino, 168, 177; greedy science and, 354–55; pharmaceutical industry incentives of, 169; Singaporean, 218, 219–21, 223, 230, 237, 239–40; Soviet, 2, 239; VisiCalc and, 87–89. *See also* neoliberalism

Carter, Jimmy, and administration, 8, 154, 257

celebrity. *See* fame

Celltech, 108–9

Chakrabarty, Ananda Mohan, 8, 27–28, 180, 197

Chapman, Clark, 300, 304–13, 315

China, People's Republic of, 2, 219, 235–37

climate change: asteroid impacts creating, 299, 302; global warming and, 311; nuclear weapons causing, 302; oil industry stance on, 34, 256–57; Thatcher on, 98, 111–12, 113–17, 118

Clinton, Bill, and administration, 7, 58, 258

Clinton, Hillary, 58

Coca-Cola, 133

cocaine, 14, 352–54, 356

colorectal cancers, 157–58

computer industry: Cambridge region, 193; entrepreneurship in, 11, 72–89; Japanese, 325–27, 329–30, 333; Texan, 252–54, 256, 259; UK Alvey program in, 101, 104, 110. *See also* Japanese humanoid robots; Silicon Valley; VisiCalc

consumer psychology, 338–39

Cosmic Catastrophes (Chapman and Morrison), 307–10, 312

Cosmos: A Personal Voyage, 9

crack cocaine, 14, 352–53, 354, 356

Crazy Rich Asians (movie), 217, 219

Creager, Angela N. H., 12, 147

creationism, 47, 56, 61–66, 71n61, 261

customer-contractor principle, 99, 109

DART (Double Asteroid Redirection Test), 297, 317–18

Dawkins, Richard, 57, 65

Dawson, Harry S. "Terry," 314–15

deregulation: biotechnology, 148, 160–61; neoliberalism and, 3–4, 8, 149, 172, 173–74, 184; pharmaceutical, 12, 64, 168–69, 170, 172, 173–74, 175–78, 184–85

Diamond v. Chakrabarty, 8, 27–28, 180, 197

Doll, Richard, 159, 166n82

drugs: cocaine as, 14, 352–54, 356. *See also* biotechnology; pharmaceutical industry

educational initiatives, 123, 124, 130–34, 142, 159, *160*

Energiia, 282, 287, 289, 291, 295n38

enterprise software, 76

entrepreneurship: biotechnology, 202; computer industry, 11, 72–89; neoliberalism and, 8; oil industry, 10–11, 19–35, 250; overview of, 10–11; pharmaceutical, 169; Texas science community, 250–52, 259; UK scientific, 98, 100, 104, 107–10, 118. *See also* computer industry; Gould, Stephen Jay; oil industry

environmental issues: asteroid impacts creating, 299–300, 302, 305; carcinogens as, 148, 149–54, 159; Japanese industrialism creating, 325; nuclear weapons creating, 302, 303, 305. *See* climate change

ethical issues, 147–48, 173, 174–75. *See also* moral economy

evolution, Gould on. *See* Gould, Stephen Jay

Falwell, Jerry, 63

fame: 1980s changes in, 8–9, 356; Gould's, 11, 46–48, 51–67, 303; Sagan's, 6, 9, 302–3; social capital and, 6

Farber, David, 14, 349

Federal Advisory Committee Act (FACA), 257–58

Feigenbaum, Ed, 31, 42n101

feminism, 60–61, 113

Flawn, Peter, 250–51, 255, 264n25

Food, Drug, and Cosmetics Act, Kefauver-Harris Drug Amendments, 173–74

Food and Drug Administration (FDA): AIDS drugs under, 169, 175–76, 178, 184; Alzheimer's drugs under, 184–85; cancer research and regulations under, 149, 158; deregulation and dismantling of, 173–74; thalidomide under, 173, 174

food industry, 28–29, 119n19

Forest City, 207–8, 215n51, 215n59

Fossey, Dian, 6

Frankston, Bob, 77–79, 86, 88, 90n20, 91n28

free markets, 3, 7–8, 74, 356. *See also* capitalism; neoliberalism

Friedman, Milton, 3, 174, 175, 185

Frumer, Yulia, 14, 322

Fylstra, Dan, 77–80, 88, 90n20, 91n28

GEC, 104, 120n21

Gehrels, Tom, 300

Gekko, Gordon (character), 1, 6, 33, 106

gender issues: Gould on, 61, 66; Japanese labor issues and, 330, 331–32, 334–35, 339; leftist scientists and, 60–61; Perutz on, 113

Genentech, 8, 26–28, 147, 167–68, 197

genetic mutations, study of. *See* biotechnology; cancer and carcinogens

Gere, Cathy, 12, 167

Gilbert, Walter, 202–3, 205

globalization: AIDS activism on, 184; biotechnology regional hubs and, 210; intellectual property and, 179, 180, 182; neoliberal promotion of, 7, 149; scientist migration and, 9; zones for science and technology driving, 238–39. *See also* neoliberalism

Glushko, Valentin, 284, 287–89, 291

Goodman, Howard M., 200, 213n20

Gorbachev, Mikhail, 141, 271–72, 277, 280–81, 286, 289, 291

Gordin, Michael D., 1

Gould, Stephen Jay, 45–67; background of, 45–46, 48, 54–55; books of, 46, 51–54, 56–57, 58–60; brand of, 11, 47–48, 55, 61, 66–67; celebrity of, 11, 46–48, 51–67, 303; correspondence of, 57–58, 70n49; creationism and, 47, 56, 61–66, 71n61; greed as lens to assess, 66–67; health of, 54, 56, 70n36; leftist politics via writing of, 11, 46–54, 58–61, 65–67; magazine articles on, 56, 57; *Natural History* column of, 46–47, 49–51, 54–55, 58–59, 61, 66–67; *NOVA* special on, 46, 54–56; origin story of, 45–46, 54–55, 56, 68n4; overview of, 11, 45–48; social justice stance of, 47, 58–67

government: asteroid impacts contracts by, 315–18; biotechnology funding by, 198–99, 201, 212n11, 213n22; Japanese industry and, 326, 333, 342–43; oil industry research with, 23, 30; science funding by, 5, 8–9, 101–2, 108, 111, 120n40; Texas science contracts by, 248, 258–59. *See also* environmental issues; military; politics; privatization; Singapore; space industry

Greater Boston Area biotech community. *See* Cambridge biotechnology community

greed: 1980s era of, 1–2, 6–9; definition and description of, 2, 4, 6, 103–4, 129; neoliberalism versus, 3–5; science and technology effects of, 5. *See also* 1980s era; greedy science; neoliberalism

greedy science: 1980s era of, 6–9; capitalism and, 354–55; funding of, 5, 8–9, 101–2; groovy science versus, 10, 258, 349, 354; historical persistence of, 350, 354–55; knowledge production and, 4–5; moral economy and, 6, 212n12; overview of, 2–6, 10–14, 349–56; science-technology confluence in, 5; social capital and, 6. *See also* 1980s era; capitalism; entrepreneurship; government military; ethical issues; military; fame; neoliberalism; privatization; regional scientific communities; speculations; wealth

Green, Cecil, 256

Groovy Science/groovy science, 10, 258, 349, 354

Guise, George, 110–12

Hackerman, Norman, 251

Harrar, Linda, 54–55

Hart, Peter, 32

Hawking, Stephen, 6, 103

Hayashi Yūjirō, 325, 327, 329, 336–37

Hayek, Friedrich von, 3, 99

Hayes-Michie, Jean, 30

Helin, Eleanor, 300

Hoechst AG, 195, 199–201, 213n19

Holmes, Elizabeth, 168

Houston Area Research Center (HARC), 24, 252, 254, 256, 259, 260

Human Genome Project, 7, 148–49, 157

IBM, 84, 88, 252, 329

Icahn, Carl, 33

India, 9, 140, 170, 181, 237

Indonesia, 219, 223, 231–35, *232*

Information Society, 324, 325–27, 329–32, 334

intellectual property, 8, 179, 180, 182. *See also* patents

Interkosmos program, 124, 139–43

Japanese humanoid robots, 322–43; appearance of, 337, 339, 340–41, 342; ascendancy of research on, 323–24, 332–36; automation or industrial robots versus, 323, 325–26, 329, 330–32, 333–34, 336; consumer psychology and, 338–39; demographics and, 332; economic context for rise of, 324–25; emotions and, 337–43, 346n65; gender issues and, 330, 331–32, 334–35, 339; greed and, 324, 325, 333, 341–42; household and caregiving responsibilities and, 324, 331–32, 334–37, 339–40; Information Society and, 324, 325–27, 329–32, 334; information work by, 336–37; labor issues and, 324, 325–26, 330–32, 336–37, 342–43; limitations of techno-utopia leading to, 329–32; neoliberalism and, 324, 326–27, 342–43; overview of, 14, 322–24; personable, 337–43; "personal" as commodity underlying, 328–29, 343; politics and, 342–43; service labor and, 324, 331, 334–37, 342–43

Jarvis, Gregory, 123–24, 139
Jennings, Peter R., 77, 79–80, 88
junk bonds, 4, 86–87
Jurong Industrial Estate (JIE), 221–23
Jurong Town Corporation (JTC), 217–40; Batam and Bintan developments by, 231–35, *232*; Biopolis by, 206, 217–18, 219; business developments by, 227–31; offshore developments by, 231–37, *232*; origins of, 221–24; overview of, 217–21; Singapore Science Park by, 219, 224–27, 242n51; subsidiaries and affiliates of, 230–31, 233, 235, 238, *238*, 239; Suzhou development by, 235–37; technology park model by, 220, 231, 237–38, 240; zones for science and technology by, 238–40. *See also* Singapore

Kapor, Mitch, 88, 92–93n69
Katō Ichirō, 322, 332–37, 341, 342
Kefauver-Harris Drug Amendments, 173–74
Kennedy, John F., and administration, 125, 248–49
Kilby, Jack, 24–25, 38n39
"killer app" hypothesis, 73, 89, 89n4
King, Stephen, 57–58
Kramer, Larry, 175

LaHaye, Tim, 62–63
Lamont-Havers, Ronald, 201

Lathrop, Jay, 24
leveraged buyouts, 4, 86–87
Lewis, Sinclair, 350, 355
Lotus 1-2-3, 87, 88, 92–93n69
LOVOT, 340–41

Mandela, Nelson, 179, 180, 183
Massachusetts General Hospital (MGH), 195, 200–201, 213n19, 213n22
Masuda Yoneji, 326, 327, 329
Mbeki, Thabo, 182–83
McAuliffe, Christa, 6, 123–24
McCray, W. Patrick, 1
McDonnell Aircraft Corporation/McDonnell Douglas, 125–26, *127*, 134–37, *136*, *138*
McLean v. Arkansas (1981), 62
Medical Research Council (MRC), 108–9
Medicines and Related Substances Bill (South Africa, 1997), 180–82, 184
mergers and acquisitions, 4, 33–34, 86–87, 88
Michie, Donald, 29–30
Microelectronics and Computer Consortium (MCC), 252–54, 256, 260, 263n24
military: asteroid impacts lobbying and, 300, 301, 304, 313; science and technology funding by, 5, 102; Soviet, 13, 269–92, 301; space industry and, 129, 275, 279–89, 291, 304. *See also* weapons industry
Milken, Michael, 33, 86–87
missile defense system. *See* Strategic Defense Initiative
MIT, University Park, 195, 206–10, 214–15n48, 215n51, 215n59, 215n61
Mitchell, George, 24, 251–52, 254, 256, 261–62
Mody, Cyrus C., 10–11, 19
Monsanto, 29, 40n63
moral economy, 6, 212n12. *See also* ethical issues
Morrison, David, 302, 304–15
MTV (Music Television), 141–42, 292
mutations, genetic, study of. *See* biotechnology; cancer and carcinogens

Nagamachi, Matsuo, 338
National Academy of Sciences (NAS), 28, 150, 256–58, 260

National Aeronautics and Space Administration (NASA), 123–43; advertising and marketing with, 124, 126, 128, 132–34, 137, *138*; asteroid impacts research and, 297, 299–302, 304–5, 309, 312, 314–16; budget hearings for, 314–15; nuclear weapons and, 301–2; oil industry influence on, 20, 249; payload specialists for, 133–39, *136*, *138*, 141; private industry collaboration with, 11, 123–39, *127*, *136*, *138*, 142–43; Soviet Union and, 124, 139–42; space shuttle program under, 11, 123–30, *127*, 133–39, 145n19, 145n22, 145n25; Texas location for, 249; Young Astronaut program of, 124, 130–34, 142

National Research and Development Corporation (NRDC), 108–9

National Science Foundation (NSF), 5, 25

Natural History, 46–47, 49–51, 54–55, 58–59, 61, 66–67

neoliberalism: ascendancy of, 7–8, 63, 355–56; biotechnology influenced by, 149; definition and description of, 3–4, 63, 119n7, 149; deregulation and, 3–4, 8, 149, 172, 173–74, 184; free markets and, 3, 7–8, 74, 356; greed versus, 3–5; Japanese industry and, 324, 326–27, 342–43; oil industry and, 30, 34; pharmaceutical industry influenced by, 169, 171, 172, 173–74, 177–78, 184; politics and, 3, 7–8, 63–64, 100, 147, 258; privatization under, 4, 8, 11–12; Texas science community and, 258–59, 261; VisiCalc and, 74. *See also* capitalism; deregulation; privatization; Reaganism; Thatcherism

Nixon, Richard, and administration, 148

Nooney, Laine, 11, 72

NOVA, 46, 54–56, 62

nuclear power: asteroid impact risk versus dangers of, 305, 306, 308–9, 310; fusion research for, 23, 33–34, 38n27, 105; oil industry and, 23, 33–34, 38n27; radiation and, 151, 163n25; Texas science community and, 254; UK, 114, 116

nuclear weapons: asteroid impacts and, 300–304, 305, 310, 313, 315–18; nuclear winter from, 302–4, 305, 310, 317; radiation and, 148, 150; UK, 114. *See* Strategic Defensive Initiative

O'Donnell, Peter, 253–54, 255, 256–58, 261–62

oil industry, 19–35; artificial intelligence investments, 20, 29–32; biotechnology investments, 8, 20, 26–29, 39–40n63, 170, 199; drilling, exploration, and production by, 21–22, 30; entrepreneurship support, 10–11, 19–35, 250; greed in, 33–35, 250; hostile takeovers in, 33–34; neoliberalism and, 30, 34; nuclear power and, 23, 33–34, 38n27; oil shocks in, 3, 22, 34, 80, 250; overview of, 10–11, 19–20; philanthropy of, 20–21, 36n7; privatization in, 30; short-term versus long-term investments by, 32–34; Singaporean, 223, 241n9, 242n24; solar industry and, 20, 23–26, 33–34; Texas science community and, 247–48, 249–52, 256–57, 259, 260–61; venture capital ties to, 21, 27, 31

Organization of the Petroleum Exporting Countries (OPEC), 22, 250

Ovshinsky, Stanford, 25, 38n39

p53 (tumor suppressor gene), 156–57

Paris Is Burning (documentary), 186

Patent and Trademark Act (1980). *See* Bayh-Dole Act (1980)

patents: biotechnology, 8, 27–28, 108–9, 148, 159, 166n76, 180, 197–98; pharmaceutical, 170, 180–82; solar industry, 24, 25–26; space industry, 134; UK, 108–9

Pepsi-Cola, 133, 141–42, 292

Perez, Myrna, 11, 45

Perot, Ross, 248, 252, 255, 258

personal computing industry. *See* computer industry; VisiCalc

personal responsibility: for cancer risks, 149, 159–61; for Japanese social welfare, 326–27; for pharmaceutical intake, 170, 172, 174, 184–85; UK focus on, 98, 110–13, 114–17, 118, 121n54, 350

Personal Software (later VisiCorp), 77–80, 83, 85, 87–89

Perutz, Max, 110–13

pharmaceutical industry, 167–87; 1980s context for, 167–68, 169–70, 185–87; advertising by, 149; AIDS drugs by, 12, 168–72, 175–85, 187; Alzheimer's drugs by, 184–85, 189n43;

biocapital of, 168–69, 170, 184; bioconstitutionalism and, 170, 179–83; biotechnology and, 168, 209; catastrophes in, 172–73; cocaine by, 353; combination therapy findings of, 178–79; corruption of research in, 178, 184; deregulation or regulatory reduction for, 12, 64, 168–69, 170, 172, 173–74, 175–78, 184–85; ethics and informed consent in, 173, 174–75; greed and, 167–68, 169–70, 177, 187; inaction by, 175–76; neoliberalism influencing, 169, 171, 172, 173–74, 177–78, 184; patents for, 170, 180–82; personal responsibility of patients versus, 170, 172, 174, 184–85; political activism to influence, 64, 168–72, 175–85, 187; prices charged by, 180–82; private organization influence on, 12, 167–87; regulatory tightening on, 172–75, 178, 185; South African activism on, 170, 179–85; surrogate endpoints in, 169, 176–77, 178, 184; thalidomide by, 172–73, 174, 187n10; UK, 102. *See also* biotechnology; Singapore

philanthropy, oil industry, 20–21, 36n7

Pickens, T. Boone, 33

Pilot, Agnes, 56, 70n49

politics: AIDS activism and, 64, 168–72, 175–85, 187; biotechnology issues and, 150; Christian Right, 47, 61–66, 71n61; Gould's writing and, 11, 46–54, 58–67; Japanese humanoid robots and, 342–43; neoliberalism and, 3, 7–8, 63–64, 100, 147, 258; oil industry and, 22, 247, 261; space industry influenced by, 125, 140–41, 143; Texas science community and, 247, 255–56, 257–58, 261–62. *See also* asteroid impacts lobbying; neoliberalism; Reaganism; Thatcherism; *specific politicians*

Porter, W. Arthur "Skip," 24, 251

Press, Frank, 256–57

privatization: in biotechnology, 12, 147–48, 161n4, 163n25, 166n82; neoliberalism and, 4, 8, 11–12; in oil industry, 30; overview of, 11–12; in pharmaceutical industry, 12, 167–87; in Singapore, 230–31; in UK, 11, 30, 100–102, 111–13, 118, 141. *See also* space industry; United Kingdom (UK) science

Proxmire, William, 303–4

Quayle, Dan, 314

racial issues: Black Panther health care networks and, 47, 64, 351–52, 354; cocaine dealers and, 352–54; Gould on, 46, 48, 56, 59–61, 66–67; leftist scientists and, 61; pharmaceutical research and, 173, 174, 175

radiation, 148, 150–52, 163n25

Reagan, Ronald, and administration: 1980s ascendancy of, 7–8; AIDS drugs under, 177; asteroid impact lobbying under, 298, 301, 310; biotechnology and environmental regulation under, 154, 201; Gould's transformation under, 11; greedy science under, 355–56; neoliberalism (Reaganism) of, 3, 7–8, 177, 355–56; NSF appointment by, 5; space industry under, 125, 128, 130, 131, 283–84, 314–15; Thatcher on, 115. *See also* Strategic Defense Initiative

Reaganism, 3, 7–8, 177, 355–56. *See also* neoliberalism

real estate: Cambridge, 195, 206–10, 214–15n48, 215n51, 215n59, 215n61; Indonesian, 234; Singapore, 206, 219–21, 222–31, 237–38

real estate investment trusts (REITs), 210

regional scientific communities: overview of, 12–13, 193–94. *See* Cambridge biotechnology community; Singapore; Texas science community

religious issues: creationism as, 47, 56, 61–66, 71n61, 261; personal responsibility and, 115, 121n54

responsibility: intergenerational, 116; social, 21, 115–16. *See also* personal responsibility

retinoblastoma, 157

robots. *See* Japanese humanoid robots

Rous sarcoma virus (RSV), 154–55

royalties, 78–79, 88, 91n31, 250

Sagan, Carl, 6, 9, 302–3, 307, 310, 316–17

Sagdeev, Roal'd, 277, 278, 279, 281, 289

Schafer, Ray, 202–3

Scheffler, Robin Wolfe, 12, 193

Schulman, Sarah, 175–76

science and technology. *See* greedy science

SDI. *See* Strategic Defense Initiative

semantic differential, 338–39

Sharman, Helen, 141

Sharp, Philip, 203–4

Shoemaker, Gene and Carolyn, 299–302, 305

Siddiqi, Asif, 13, 269

Silicon Valley: 1980s context for, 9; neoliberalism and, 258; regional scientific communities' comparison to, 12, 193, 220–21, 240, 247, 249, 251–52, 260

simian virus 40 (SV40), 154, 156, 165n58

Singapore, 217–40; Biopolis in, 206, 217–18, 219; business developments in, 227–31; capitalism and state capitalism in, 218, 219–21, 223, 230, 237, 239–40; industrialization and, 218–24, 229–30, 232–33; Nobel Prize winners in, 240n1; offshore developments by, 231–37, 232; oil industry in, 223, 241n9, 242n24; origins of JTC in, 221–24; overview of, 13, 217–21; privatization in, 230–31; real estate and landlordism in, 206, 219–21, 222–31, 237–38; Singapore Science Park in, 219, 224–27, 242n51; technology park model by, 220, 231, 237–38, 240; wealth and economic goals in, 217–18, 219–20, 227–28, 240; zones for science and technology in, 238–40. See also Jurong Town Corporation

Slovic, Paul, 306–7

Smirnov, Leonid, 280–81, 283, 287, 288, 291

social capital, 6. See also fame; wealth

social justice, 47, 58–67, 355

social responsibility, 21, 115–16

Software Arts, 79, 88

solar energy, 20, 23–26, 33–34

somatic mutation theory, 150–51, 162n22

South Africa, AIDS activism in, 170, 179–85

Soviet Union: asteroid impacts response by, 301, 304, 310, 313; capitalism in, 2, 239; dissolution of, 7, 271; greed in, 271–74, 279, 284, 291; military-industrial leadership change in, 280–82, 287, 288, 291; military of, 13, 269–92, 301; physicists in, 271, 273–74, 276–79, 288–89, 290–91; space industry in, 124, 139–43, 271, 273–76, 279–89, 291–92, 295–96n38, 296n50; Strategic Defense Initiative response by, 13, 270–92

SpaceCamp (movie), 132–33

space industry, 123–43; advertising and marketing of, 124, 126, 128, 132–34, 137, *138*, 141–42, 292; asteroid impact defense and, 297, 299–302, 304–5, 309, 312–16; commercialization and, 124, 125–30, *127*, 141–43, 292, 315; greed and, 124, 129–30, 143; military and, 129, 275, 279–89, 291, 304; overview of, 11, 123–24, 142–43; payload specialists for, 133–39, *136*, *138*, 141; private industry collaboration with, 11, 123–39, *127*, *136*, *138*, 141–43; satellites of, 126, 275, 285; Soviet, 124, 139–43, 271, 273–76, 279–89, 291–92, 295–96n38, 296n50; space shuttle program in, 11, 123–30, *127*, 133–39, 145n19, 145n22, 145n25, 282–83, 291; Texan, 248–49, 259, 260; UK, 102, 141; Young Astronaut program in, 124, 130–34, 142

speculations: on asteroid impacts, 13–14, 297–318; on Japanese humanoid robots, 14, 322–43; on Soviet Strategic Defense Initiative response, 13, 270–92; VisiCalc calculations for, 81, 85–86, 89

spreadsheets, 72–73, 76, 85–87, 89. See also Lotus 1-2-3; VisiCalc

Staley, Peter, 176

Stanley, Matthew, 13, 297

Star Wars. See Strategic Defense Initiative

state capitalism, Singaporean, 219–21, 223, 230, 237, 239–40

Stevens, Hallam, 12–13, 217

Strategic Defense Initiative (SDI, "Star Wars"): context and justification for, 269–70; government contracts for, 259, 265n42; opposition to, 265n43, 274–76; physicists on, 271, 273–79, 288–89, 290–91; Soviet response to, 13, 270–92; space industry and, 271, 273–76, 279–89, 291–92, 295–96n38, 296n50; Texas science community and, 259; unknown/unlikely efficacy of, 270, 272–75, 277–78, 280, 285, 290

Students for a Democratic Society (SDS), 48

Sumitomo, 323, 335

Sunder Rajan, Kaushik, 168, 170, 182

Superconducting Super Collider (SSC), 246, 254–56, 257, 259, 260
Suzhou development, 235–37

Tagliaferri, Edward, 312–13, 315
Teller, Edward, 276, 316
Temasek Holdings, 230, 239, 243n72, 245n110
Ternes, Alan, 49–51
Texas Engineering Experiment Station (TEES), 24, 251
Texas Instruments (TI), 24, 33, 248, 249, 252, 260
Texas science community, 246–62; aerospace industry and, 248–49, 259, 260; disadvantages of, 260–61; electronics industry and, 248, 252–54, 256, 260; entrepreneurship in, 250–52, 259; government contracts for, 248, 258–59; Microelectronics and Computer Consortium in, 252–54, 256, 260, 263n24; models for, 249, 252; money for, 249–50, 253–54, 256–57, 258–59, 260; motivations to develop, 249–50; National Academy of Sciences and, 256–58, 260; neoliberalism and, 258–59, 261; oil industry and, 247–48, 249–52, 256–57, 259, 260–61; overview of, 13, 63–64, 246–47, 259–62; politics and, 247, 255–56, 257–58, 261–62; prior industries setting stage for, 247–49; Superconducting Super Collider in, 246, 254–56, 257, 259, 260
Thailand, 219, 235
thalidomide, 172–73, 174, 187n10
Thatcher, Margaret, and administration: 1980s ascendancy of, 6–8; background of, 98–100; on climate change, 98, 111–12, 113–17, 118; on greed, 97–98, *103*, 103–4, 117–18; greedy science under, 11, 63, 97–118, 355; neoliberalism of, 3, 7–8, 100; oil industry under, 30; on personal responsibility, 98, 110–13, 114–17, 118, 121n54, 350; privatization under, 11, 30, 100–102, 111–13, 118; science policy under, 106–13; on society, 3, 65, 97, 350; Soviet relations with, 271. *See also* Thatcherism
Thatcherism: background of, 98–100; definition and description of, 3, 100–101; greed

and, 97–98, *103*, 117–18; neoliberalism and, 3, 7–8, 100; oil industry and, 30; personal responsibility in, 98, 115, 118; privatization under, 100–101, 118; science policy influenced by, 106–13. *See also* neoliberalism
time-sharing systems, 75, 76–77, 78, 79, 83, 91n35
Toxic Substances Control Act (TSCA, 1976), 148, 153
Trade-Related Aspects of Intellectual Property Rights (TRIPS) agreement, 180, 182
Treatment Action Campaign (TAC), 181–83, 184, 185
Treatment Action Group (TAG), 178
tumor virology research, 148, 154–57
Tuskegee Study of Untreated Syphilis in the Male Negro, 173, 174, 175

United Kingdom (UK) science, 97–118; 1980s context for, 101–3; biotechnology in, 108–9; climate change stance in, 98, 111–12, 113–17, 118; entrepreneurship and, 98, 100, 104, 107–10, 118; greed and, *103*, 103–6, *106*, 117–18; greed-science intersection for, 104–6, *106*; oil industry and, 30; overview of, 11, 63, 97–98, 117–18; patents in, 108–9; personal responsibility in, 98, 110–13, 114–17, 118; privatization in, 11, 30, 100–102, 111–13, 118, 141; space industry in, 102, 141; Thatcherism and, 97–101, 117–18; Thatcherite policy for, 106–13
UNIVAC, 90n14
University of California, San Francisco (UCSF), 200
University Park, 195, 206–10, 214–15n48, 215n51, 215n59, 215n61
Ustinov, Dmitrii, 280–81, 283, 285–86, 287, 291

Velikhov, Evgenii Pavlovich/Velikhov commission, 277–80, 282, 284, 288–89, 290–91
Velikovsky, Immanuel, 307, 308
venture capital: biotechnology, 147, 195, 197, 199–202, 206; greedy science and, 356; oil industry, 21, 27, 31; pharmaceutical, 168; UK, 107

VisiCalc, 72–89; 1980s as context for, 74; advertisements for, 80–81, 83; competitors to, 84, 88, 92–93n54, 93n69; decline and demise of, 88–89, 92n54; finances of, 78–79, 85, 88–89, 91n31, 92n56; greediness and, 74, 81, 86–89; imagining and making of, 75–78; "killer app" hypothesis on, 73, 89, 89n4; lawsuit over, 87–88; market influence of, 73–74, 85, 87, 89, 89n4; media coverage of, 73, 84, 92n54; name of, 78, 91n28; neoliberalism and, 74; overview of, 11, 72–74; packaging of, 80, 91n39; price of, 79; publishing and marketing of, 78–82, 83, 87–88, 91n39; as spreadsheet, 72–73, 76, 85–87, 89; uses and functionality of, 82–87

WABOT, 333
WABOT-2, 322, 335. *See also* WASUBOT

Walker, Charles D. "Charlie," 134–39, *136, 138*
WASUBOT, 322–23, 335, 340
Watson, James, 198
wealth: 1980s changes in, 8–9, 356; Gould's, 67; greed and greedy science defined by, 2, 98, 103; Singaporean, 217–18, 219–20, 227–28, 240; social capital and, 6; Thatcherism and, 100, 107, 118
weapons industry: asteroid impacts response in, 315–18; Soviet response to SDI in, 13, 270–92. *See also* military; nuclear weapons
Weinberg, Alvin, 308–9
Weinberg, Steven, 251, 264n34
Weitekamp, Margaret A., 11, 123
Westwick, Peter, 13, 246

Young Astronaut program, 124, 130–34, 142

www.ingramcontent.com/pod-product-compliance
Lightning Source LLC
Chambersburg PA
CBHW020910210326
41598CB00018B/1824